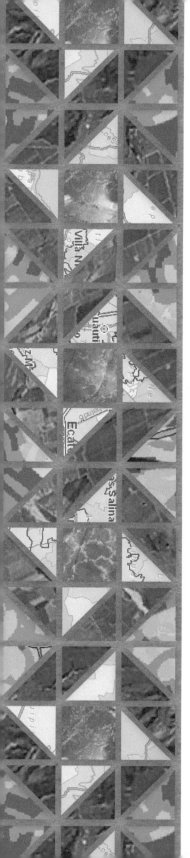

Land Administration for Sustainable Development

Ian Williamson

Stig Enemark

Jude Wallace

Abbas Rajabifard

ESRI Press Academic
REDLANDS, CALIFORNIA

ESRI Press, 380 New York Street, Redlands, California 92373-8100

Copyright 2010 ESRI

All rights reserved. First edition 2010

14 13 12 11 10   1 2 3 4 5 6 7 8 9 10

Printed in the United States of America

Library of Congress Cataloging-in-Publication Data
Land administration for sustainable development / Ian Williamson ... [et al.]. — 1st ed.
   p. cm.
   Includes bibliographical references and index.
   ISBN 978-1-58948-041-4 (pbk. : alk. paper)
   1. Land use—Management. 2. Sustainable development. I. Williamson, I. P.
HD111.L195 2009
333.73'16—dc22   2009030118

Ask for ESRI Press titles at your local bookstore or order by calling 1-800-447-9778. You can also shop online at www.esri.com/esripress. Outside the United States, contact your local ESRI distributor.

ESRI Press titles are distributed to the trade by the following:

In North America:

Ingram Publisher Services
Toll-free telephone: 1-800-648-3104
Toll-free fax: 1-800-838-1149
E-mail: customerservice@ingrampublisherservices.com

Cover design and production **Fred Estrada**
Interior design and production **Monica McGregor**
Image editing **Jay Loteria and Monica McGregor**
Editing **Carolyn Schatz**
Copyediting and proofreading **Julia Nelson**
Permissions **Brian Harris**
Printing coordination **Cliff Crabbe and Lilia Arias**

Quilt mosaic cover images courtesy of Landsat.org, Global Observatory for Ecosystem Services, Michigan State University; Ramon Perez; Earth Satellite Corporation, from ESRI Data and Maps 2008; Alachua County, Florida

Contents

List of tables ———————————————————————————————— VI

Foreword ————————————————————————————————— VIII

Preface ——————————————————————————————————— IX

Acknowledgments ——————————————————————————————— XII

About the authors —————————————————————————————— XIII

List of abbreviations ——————————————————————————— XIV

Part 1 Introducing land administration ————————————————— 2

Chapter 1 Setting the scene————————————————————————— 4

1.1 Integrated land administration ——————————————————— 5

1.2 Why build a land administration system? ——————————————— 15

1.3 The changing nature of land administration systems ————————— 26

1.4 Land reform ——————————————————————————————— 29

1.5 Good governance ————————————————————————————— 30

1.6 Ten principles of land administration ——————————————— 33

Chapter 2 People and land administration ————————————————— 36

2.1 People and land——————————————————————————————— 37

2.2 Historical evolution ———————————————————————————— 58

Part 2 A new theory ———————————————————————————————— 68

Chapter 3 The discipline of land administration ——————————————— 70

3.1 Evolution of land administration as a discipline ——————————— 71

3.2 Land administration and sustainable development —————————— 84

3.3 Incorporation of restrictions and responsibilities in LAS ————————— 88

Chapter 4 Land administration processes ———————————————————— 94

4.1 Importance of land administration processes ————————————— 95

4.2 Core land administration processes————————————————————— 97

4.3 Examples of tenure processes ——————————————————————— 99

4.4 Reforming LAS by improving process management———————————— 112

Chapter 5 Modern land administration theory ————————————————— 114

5.1 Designing LAS to manage land and resources ————————————— 115

5.2 The cadastre as an engine of LAS ————————————————————— 127

Part 3 Building modern systems — 134

Chapter 6 Building land markets — 136

6.1 A land administration view of land markets — 137

6.2 Building infrastructure to support formal markets — 150

6.3 Land valuation and taxation — 163

Chapter 7 Managing the use of land — 170

7.1 Land use — 171

7.2 Planning control systems — 176

7.3 Urban land-use planning and regulations — 179

7.4 Rural planning and sectoral land-use regulations — 185

7.5 Land consolidation and readjustment — 189

7.6 Integrated land-use management — 192

7.7 Land development — 194

Chapter 8 Marine administration — 204

8.1 The need to improve marine administration — 205

8.2 Challenges in building marine administration systems — 206

8.3 Existing marine administration — 207

8.4 The marine cadastre concept — 210

8.5 Marine registers — 214

8.6 Developing a marine SDI — 216

8.7 Using the land management paradigm to meet marine needs — 222

Chapter 9 SDIs and technology — 224

9.1 Why do land administration systems need an SDI? — 225

9.2 Introducing the SDI — 229

9.3 Integrating information about the natural and built environments — 233

9.4 Making ICT choices — 238

9.5 Land administration and cadastral data modeling — 256

9.6 Maintaining momentum — 261

Chapter 10 Worldwide land administration activities — 262

10.1 Land administration projects — 263

10.2 Recent land administration and cadastral activities — 269

10.3 The Worldwide Cadastral Template Project — 271

10.4 Improving capacity to make global comparisons — 290

Part 4 Implementation — 292

Chapter 11 Capacity building and institutional development — 294

11.1 The modern capacity-building concept — 296
11.2 Capacity development — 297
11.3 Capacity-building issues in land administration — 301
11.4 Institutional capacity in land management — 303
11.5 Education and training in land administration — 307

Chapter 12 The land administration toolbox — 314

12.1 Using land administration tools — 315
12.2 General tools — 320
12.3 Professional tools — 333
12.4 Emerging tools — 387

Chapter 13 Project management and evaluation — 402

13.1 Project context — 403
13.2 Designing and building land administration systems — 404
13.3 Evaluating and monitoring land administration systems — 429

Part 5 The future of land administration — 434

Chapter 14 Future trends — 436

14.1 The land administration journey — 437
14.2 LAS supporting sustainable development — 439
14.3 LAS to support spatially enabled society — 440
14.4 LAS issues in the next decade — 443
14.5 The challenges ahead — 445

Glossary — 448
Reference list — 458
Index — 472

Tables

CHAPTER 1

    Table 1.1 Traditional benefits of LAS —————————————————— 17

    Table 1.2 Ten principles of land administration —————————————— 34

CHAPTER 2

    Table 2.1 Array of concepts of land —————————————————— 40

    Table 2.2 Cadastral components ——————————————————— 65

    Table 2.3 General relationships between land registries and cadastres ————— 66

CHAPTER 4

    Table 4.1 Processes for second-class surveys in systematic land titling ————101

    Table 4.2 Simple land transfer process ————————————————— 102

    Table 4.3 Simple mortgage process —————————————————— 103

    Table 4.4 Formal and informal transfers of land ————————————— 104

    Table 4.5 Boundary determination processes —————————————— 109

    Table 4.6 Simple cadastral surveying process —————————————— 110

CHAPTER 6

    Table 6.1 Simplified characteristics of evolutionary stages of land markets ——— 152

    Table 6.2 Evolution of infrastructure and tools in LAS —————————— 155

CHAPTER 8

    Table 8.1 Range of activities in the marine environment —————————— 214

    Table 8.2 Marine resource titling standards ——————————————— 217

CHAPTER 9

    Table 9.1 Integration issues ————————————————————— 236

CHAPTER 10

Table 10.1 Types of worldwide land administration and related projects ———————— 264

Table 10.2 Principles and associated indicators ——————————————————— 275

Table 10.3 Matrix of registration system vs. registration method ————————— 276

Table 10.4 Matrix of registration system vs. establishment approach ——————— 277

Table 10.5 Land parcels and population and number of strata titles per million ——— 278

Table 10.6 Averages of parcel registration data ————————————————— 283

Table 10.7 Number of full-time equivalent professional land surveyors per million population ——— 287

CHAPTER 11

Table 11.1 The new capacity-building approach ————————————————— 299

Table 11.2 Capacity building in land administration ———————————————— 301

Table 11.3 Comprehensive approach to institutional development ———————— 304

CHAPTER 12

Table 12.1 The land administration toolbox ——————————————————— 318

Table 12.2 Tenure tools ——————————————————————————— 333

Table 12.3 Differences among registration systems ——————————————— 341

Table 12.4 Comparison of systematic and sporadic titling ———————————— 351

Table 12.5 Building tenure tools—————————————————————————— 384

Table 12.6 Characteristics of pro-poor vs. market tools ———————————— 388

Table 12.7 Gender land tools —————————————————————————— 397

CHAPTER 13

Table 13.1 A LogFrame analysis for titling projects ——————————————— 417

Table 13.2 The LogFrame matrix ———————————————————————— 420

Table 13.3 Summary of evaluation framework for LAS —————————————— 431

Foreword

When people think about geography, they generally think about land. So, it is not a great leap to see the connection between geographic and land information systems, and how they work together to achieve effective land administration. Land administration systems (LAS) in turn drive the way toward sustainable patterns of land use across the globe.

Land Administration for Sustainable Development details this journey. The book, by renowned experts in the field Ian Williamson and Stig Enemark, and their coauthors Jude Wallace and Abbas Rajabifard, chronicles how land administration systems have evolved from linking cadastral land records to demonstrating their inherent power for sharing spatial information that can change the world. Advances in spatially based technologies have helped land administration systems to bring about development that is just, equitable, and ultimately sustainable.

The book reflects the philosophy of Hernando de Soto, an author and Peruvian economist who extols the value of empowering the poor through property ownership. Poverty reduction, gender equality, and social justice are important themes of the book as it shows how securing land tenure and managing the use of land can transform society.

Land is not just the earth that people walk on. It is fundamentally the way people think about place. Thus, land administration is not just about land—it is about people. *Land Administration for Sustainable Development* explores why it is imperative for society to build the capacity to manage land for the public good. It presents ten principles of land administration, along with a toolbox of best practices for realizing the land management paradigm of land tenure, land value, land use, and land development. Finally, it points the way toward meeting the challenges that land administration systems face to ensure the vision of economic development, social justice, environmental protection, and good governance.

This is a book for people who want to learn about the theory and processes of land administration as they relate to land markets and to the world we live in. For as much as land is a consumable good, it is also a spiritual place, a natural resource, and an environmental wonder. I hope you will enjoy this book, which provides the career wisdom of four scholars who have devoted themselves to sharing their knowledge to make the world a better place.

Jack Dangermond
President, ESRI

Preface

Imagine a country without any basic administration of land. Imagine that tenure to land and property cannot be secured and that mortgage loans cannot be established as a basis for property improvement and business development. Imagine that the use and development of land is not controlled through overall planning policies and regulations. And imagine a slum area of 250 hectares (about 1 square mile) with more than 1 million inhabitants lacking the most basic occupation rights and without basic water and sanitary services.

Land administration systems (LAS) are designed to address these problems by providing a basic infrastructure for implementing land-related policies and land management strategies with the aim of ensuring social equity, economic growth, and environmental protection. A system may involve an advanced conceptual framework supported by sophisticated information and communications technology (ICT) models as in many developed countries, or it may rely on very fragmented and basically analog approaches that are found in less developed countries.

Until the past couple of years, the developed world often took land administration for granted and paid little attention to it. But the recent global economic collapse has sharply focused world attention on mortgage policies and processes and their related complex commodities, as well as on the need for adequate and timely land information. Simply put, information about land and land-market processes that can be derived from effective LAS plays a critical role in all economies.

The preceding examples are just some of the issues that motivated us to write this book. This book is intended for a wide audience. Nonexperts and those unfamiliar with LAS may find it useful to enhance their basic understanding of landownership, land markets, and the environmental and social issues concerned with land. Politicians and senior government officials may find it useful as they tackle problems of economic development, environmental and resource management, poverty alleviation, social equity, and managing indigenous rights, particularly from a sustainable development perspective. Land administrators and others working in land-related professional fields may benefit from the theory and toolbox approach to assist in improving or reforming LAS. Finally, the academic community—instructors and students at the university and college level—may find it a useful book that explores both theory and practice by looking at the administration of land holistically, as well as exploring the institutional, policy, and technical aspects of designing, building, and managing LAS.

For more than three decades, Ian Williamson, Stig Enemark, and Jude Wallace have been fascinated by land issues. The vision for the book came from Ian and Stig, who initially wanted to document their lives' work in the land-related field. Both have a strong cadastral background with Ian having strength in institutions, particularly in the English-speaking world, and Stig bringing knowledge of European systems with a focus on land management. They recognized the need for a strong legal perspective, which was provided by Jude, who has spent a lifetime working as a land policy lawyer. All recognized the need for solid technical support, with the expertise provided by Abbas Rajabifard, who has many years of experience in spatial data infrastructure (SDI) and geographic information systems (GIS). However, the end result is a book written collaboratively with all authors taking responsibility for the entire text.

The collective vision was to write a practical book with a strong and universal theoretical foundation that explores the systems that administer the ways people relate to land. This cannot be done successfully without a major focus on building the capacity of people and institutions. Building and maintaining these capacities are at the heart of modern land administration.

An overall theme of the book is therefore about developing land administration capacity to manage change. For many countries, meeting the challenges of poverty alleviation, economic development, environmental sustainability, and management of rapidly growing cities are immediate concerns. For more developed countries, the pressing issues are updating and integration of agencies within relatively successful LAS and putting land information to work for emergency management, environmental protection, economic decision making, and so on.

The objective was to write a book that was equally of use to both less developed and developed countries. This global context necessitated a holistic view of land administration as a central component of the land management paradigm. The book offers this paradigm as the theoretical basis for delivering such a holistic approach to LAS in support of sustainable development. While the book recognizes that all countries or jurisdictions are unique and have their own needs, it highlights ten principles of land administration that are applicable to all.

Land administration is not a new discipline. It has evolved out of the cadastre and land registration fields with their specific focus on security of land rights. While the land management paradigm is the central theme of the book, embracing the four land administration functions (land tenure, land value, land use, and land development), the role of the cadastre as the engine of LAS is underscored throughout.

We hoped to write a book that could be easily read and understood by nonexperts in the field, politicians, and senior government officials, as well as being of interest to students, land administrators, and land-related professionals. We acknowledge that "a picture is worth a thousand words" and include many photographs, pictures, diagrams, and figures throughout.

The book develops several themes that make it stand apart from other books on the subject. The most important involves the adoption of a toolbox of best practices for designing LAS with general, professional, and emerging tools that are tailored to specific country needs. Also, there is a focus on using common land administration processes as a key to understanding and improving systems. The book further explores the relationship between land administration and land markets, the central economic driver for most countries. The book concludes by emphasizing the importance of land administration to the spatial enablement of society, where government uses place as the key means of organizing information related to activities ranging from health, transportation, and the environment to immigration, taxation, and defense, and when location and spatial information are available to citizens and businesses to support these activities.

Ian Williamson
Stig Enemark
Jude Wallace
Abbas Rajabifard

Acknowledgments

The authors gratefully acknowledge the assistance, advice, and support provided by many friends and colleagues across the globe in gathering the information for this book and in its publication. As professionals, researchers, and academics, we have benefited from the international experiences and knowledge of colleagues in academic institutions, professional organizations, governments, and the private sector globally through discussions, joint research, and professional projects.

We also acknowledge the work of many graduate students and researchers internationally who have contributed to this book, particularly from the Center for Spatial Data Infrastructures and Land Administration in the Department of Geomatics at the University of Melbourne, Australia, and the Department of Development and Planning at Aalborg University, Denmark.

We are grateful for the support and work of colleagues in the International Federation of Surveyors; from United Nations agencies involved in land systems, especially UN–HABITAT, the Food and Agriculture Organization, and the World Bank; and the United Nations-supported Permanent Committee on GIS Infrastructure for Asia and the Pacific (PCGIAP). Their experience is well known among professionals, though many people remain unaware of the depth of their contributions to global knowledge and best practices. The international work of the people in these organizations forms a working background for our efforts to explain how land administration systems (LAS) might deliver sustainable development.

The staff at ESRI Press have been very supportive and understanding during the inevitable challenges in writing this book. We wish to express our gratitude and thanks for their commitment and professionalism.

The discipline of land administration theory and practice does not provide single answers to the many complex land management questions facing our world today. Our opinions will be debated, and we acknowledge this book is just one more step in the land administration journey. We look forward to improving our ideas and encourage all to participate in helping to build better systems for managing and administering land.

Ian Williamson
Stig Enemark
Jude Wallace
Abbas Rajabifard

About the authors

Collectively, the authors have more than a century of experience in government, the private sector, and universities in designing, building, and managing land administration systems (LAS) through research, international cooperation, and consultancies in both developed and less developed countries.

Ian Williamson is a professional land surveyor and chartered engineer. He is professor of surveying and land information at the Center for Spatial Data Infrastructures and Land Administration in the Department of Geomatics at the University of Melbourne, Australia. Williamson is well known for his experience in cadastre, land administration, and spatial data infrastructures (SDIs) with his strengths being policy, technical, and institutional issues as well as practical strategies to build these systems.

Stig Enemark is a professional land surveyor. He is currently president of the International Federation of Surveyors (FIG). He is professor of land management and problem-based learning in the Department of Development and Planning at Aalborg University, Denmark. Enemark has wide experience in land administration with his strengths being the relationship of land administration to land management and related educational and capacity building issues.

Jude Wallace is a land-policy lawyer. She is a senior fellow at the Center for Spatial Data Infrastructures and Land Administration in the Department of Geomatics at the University of Melbourne, Australia. Wallace has spent a lifetime exploring the legal and policy aspects of land in a global context. Her expertise ranges from improving the most modern land administration systems to developing pro-poor land strategies.

Abbas Rajabifard is a professional land surveyor and chartered engineer. He is currently president of the Global Spatial Data Infrastructure Association. Rajabifard is an associate professor and director of the Center for Spatial Data infrastructures and Land Administration in the Department of Geomatics at the University of Melbourne, Australia. His experience includes technology and SDIs that support LAS.

Abbreviations

ADB Asia Development Bank

API application programming interface

APR Asia and the Pacific Region

ATKIS Authoritative Topographic and Cartographic Information System (German)

AusAID Australian Agency for International Development

BOND British Overseas NGOs for Development

BOOT build, operate, own, and transfer

BPN Badan Pertanahan Nasional (National Land Agency) (Indonesia)

BPU basic property unit

CASLE Commonwealth Association for Surveying and Land Economy

CCDM core cadastral domain model

CIDA Canadian International Development Aid

CLGE Council of European Geodetic Surveyors

CODI Committee on Development Information

CPD continuing professional development

DBMS database management system

DCDB digital cadastral database

DFID Department of International Development (England)

EDM electronic distance measurement

EFT electronic funds transfer

e-LAS e-land administration system

EU European Union

EUROGI European Umbrella Organization for Geographic Information

FAO Food and Agriculture Organization of the United Nations

FIG International Federation of Surveyors

GIS geographic information system(s)

GLTN Global Land Tool Network

GML Geography Markup Language

GNAF Geocoded National Address File (Australia)

GPS Global Positioning System

GRET Groupe de Recherche et d'Echanges Technologiques (Research and Technological Exchange Group) (France)

GSDI Global Spatial Data Infrastructure

GTZ Deutsche Gesellschaft fur Technische Zusammenarbeit (German Technical Cooperation)

HRD human resource development

IAG International Association of Geodesy

ICA International Cartographic Association

ICT information and communications technology

ID identification

IDPs internally displaced persons

IHO International Hydrographic Organization

IIED International Institute for Environment and Development

IP Internet Protocol

IPRs intellectual property rights

ISO International Organization for Standardization

ISPRS International Society for Photogrammetry and Remote Sensing

IT information technology

ITC International Institute for Geoinformation Science and Earth Observation (the Netherlands)

ITQ international transferable quotas

KPI key performance indicator

LAPs land administration projects

LAS land administration systems

LFA Logical Framework Analysis

LIN land information network

LIS land information system

LURC land-use right certificate (Vietnam)

MBI market-based instruments

MDGs Millennium Development Goals

MOLA Meeting of Officials on Land Administration

NGO nongovernmental organization

NLTB Native Land Trust Board (Fiji)

NRC National Research Council (United States)

NSDI national spatial data infrastructure

NSW New South Wales (Australia)

OECD Organization for Economic Cooperation and Development

OGC Open Geospatial Consortium Inc.

OICRF International Office of Cadastre and Land Records (the Netherlands)

P2P peer-to-peer

PCGIAP Permanent Committee on GIS Infrastructure for Asia and the Pacific

PCIDEA Permanent Committee for Spatial Data Infrastructure for the Americas

PIN personal identification number

PMO Project Management Office

PPAT Pejabat Pembaut Akte Tanah (land deed official) (Indonesia)

PPPs public-private partnerships

PSMA Ltd. Public Sector Mapping Agency (Australia)

QA quality assurance

RORs rights, obligations, and restrictions

RRRs rights, restrictions, and responsibilities

SAPs Structural Adjustment Programs

SDI spatial data infrastructure

SEG spatially enabled government

SGD Surveyor General's Department (Swaziland)

SIDA Swedish International Development Cooperation Agency

SOA service-oriented architecture

SOAP Simple Object Access Protocol

STDM social tenure domain model

SWOT strengths, weaknesses, opportunities, and threats

TCP/IP Transmission Control Protocol/Internet Protocol

TLTP Thailand Land Titling Project

UDDI universal description, discovery, and integration

UK United Kingdom

UML Unified Modeling Language

UN United Nations

UNCLOS United Nations Convention on the Law of the Sea

UNDP United Nations Development Programme

UNECA United Nations Economic Commission for Africa

UNECE United Nations Economic Commission for Europe

UNEP United Nations Environment Programme

UNESCAP United Nations Economic and Social Commission for Asia and the Pacific

UN–HABITAT United Nations Agency for Human Settlements

UNRCC United Nations Regional Cartographic Conferences

UNRCC-AP United Nations Regional Cartographic Conference for Asia and the Pacific

USAID United States Agency for International Development

WCS Web Coverage Service

WFS Web Feature Service

WMS Web Map Service

WPLA Working Party on Land Administration (UNECE)

WSDL Web Service Description Language

Land Administration for Sustainable Development

Part 1

Introducing land administration

Part 1 of this book introduces the concept and principles of land administration in addition to providing an overview of the structure and objectives of the book. It explains how the concept of land administration has evolved and continues to evolve as part of a wider land management paradigm. The ingredients of land administration systems (LAS) and the reasons for building and reforming LAS are explored. The differences between land administration and land reform are emphasized, as is the central role of good governance in building and operating successful LAS. Ten principles of land administration that are equally applicable to developed and less developed systems are presented in chapter 1.

A key to understanding the role of LAS in society is understanding the evolving relationship of people to land and how these relationships in different jurisdictions and countries have dictated how specific LAS evolve, as described in chapter 2. A historical perspective of land administration is introduced along with its key components to help set the scene for the rest of the book. The different perceptions of land and how they affect the resulting administration of land are discussed. Lastly, the cadastral concept is introduced and its central role in LAS explained, particularly the cadastre's relationship to land registries and its evolving multipurpose role.

Chapter 1
Setting the scene

1.1 Integrated land administration

1.2 Why build a land administration system?

1.3 The changing nature of land administration systems

1.4 Land reform

1.5 Good governance

1.6 Ten principles of land administration

1

1.1 Integrated land administration

A NEW FRAMEWORK

A land administration system provides a country with the infrastructure to implement land-related policies and land management strategies. "Land," in modern administration, includes resources and buildings as well as the marine environment—essentially, the land itself and all things on it, attached to it, or under the surface.

Each country has its own system, but this book is primarily about how to organize successful systems and improve existing ones. This exploration of land administration systems (LAS) provides an integrated framework to aid decision makers in making choices about improvement of systems. The book is based on the organized systems used throughout modern Western economies where the latest technologies are available, but it is also applicable to developing countries struggling to build even rudimentary systems. The improvement of integrated land administration involves four basic ingredients in the design of any national approach:

- The **land management paradigm**, with its four core administration functions
- The **common processes** found in every system
- A **toolbox approach**, offering tools and implementation options
- A role for land administration in supporting **sustainable development**

The **land management paradigm** can be used by any organization, especially national governments, to design, construct, and monitor LAS. The core idea is to move beyond mapping, cadastral surveying, and land registration to use land administration as a means of achieving sustainable development. These familiar processes need to be approached holistically and strategically integrated to deliver, or assist delivery of, the four main functions of land management: land tenure, land value, land use, and land development. If the organizations and institutions responsible for administering these processes are multipurpose, flexible, and robust, they can assist the larger tasks of managing land, as well as dealing with global land and resource issues. The land management paradigm encourages developed countries to aim for improved governance, e-democracy, and knowledge management and developing countries to implement food and land security, while improving governance, and, in many cases, building effective land markets.

While the theoretical framework offered by the land management paradigm is universal, implementation may vary depending on local, regional, and national circumstances. In this book, the enigma of open-ended opportunities for implementation is solved by applying an engineering approach that relates design of LAS to management of local practices and processes. **Common processes** are found in all countries and include dividing up land, allocating it for identifiable and secure uses, distributing land parcels, tracking changes, and so forth. Variations in how countries undertake these processes underscore the remarkable versatility of LAS.

But among all the variations, market-based approaches predominate, both in theory and in practice. This popularity arises from the relative success of markets in managing the common processes of land administration while, at the same time, improving governance, transparency, and economic wealth for the countries where land administration is successful. Market-based approaches provide best-practice models for improvement of many national LAS where governments seek economic improvement. The tools used in market-based systems are therefore frequently related to general economic development. This relationship is, however, far from self-evident. Market-based approaches are creatures of their history and culture. Applying them to other situations requires foresight, planning, and negotiation.

This leads to the third ingredient of good LAS design: the **toolbox approach**. The land administration toolbox for any country contains a variety of tools and options to implement them. The tools and how they are implemented reflect the capacity and history of the country. The selection of tools discussed in this book reflects the historical focus of land administration theory and practice in cadastral and registration activities. It includes general tools such as land policies, land markets, and legal infrastructure; professional tools related to tenure, registration systems, and boundaries; and emerging tools such as pro-poor land management and gender equity.

There are, of course, many other tools. Valuation, planning, and development tools raise separate and distinct issues. Many countries include land-use planning and valuation activities in formal LAS. Other countries rely on separate institutions and professions to perform these functions and define LAS more narrowly. For this reason, the book does not discuss the professional tools used to perform functions of valuation, use, and development although these topics are introduced. For all LAS, however, these functions need to be undertaken in the context of the land management paradigm and integrated with the tenure function. The design of a tool by an agency engaged in any of the four primary functions needs to reflect its integration with the others. The cadastre remains a most important tool, because it is capable of supporting all functions in the land management paradigm (noting that the cadastre is more correctly a number of tools within one conceptual framework). Indeed, any LAS designed to support sustainable development will make the cadastre its most important tool.

The list of tools and their design will change over time, as will the suitability of any particular tool for use in national LAS. The appropriate options to deliver LAS will also change. To successfully use the toolbox approach, the LAS designer must understand the local situation, diagnose steps for improvement, and select the appropriate tools and options. Usually, the steps can be clarified by international best practices explained in well-documented case studies, United Nations and World Bank reports and publications, and a wide variety of books and reports.

One of the major problems with LAS design, even in countries with successful systems, is the isolation of various components and agencies. This is generally known as the problem of "silos." Another problem is reliance on single-tool solutions to remedy complex situations. The toolbox approach addresses both these problems. It requires that each tool be considered in the context of all the others and that it be tested against the overall land management paradigm. It relies on using methods and options appropriate to a particular situation, compared with a "one size fits all" suite of policy and technical options.

Figure 1.1 Even a traditional village environment such as that in Mozambique can benefit from effective land administration.

The options now available for implementing the tools at hand vary widely and will continue to evolve. The essential theme of this book is to inform the design of LAS by starting with the broad context of the land management paradigm, observing the common processes that are being used, and then choosing the appropriate tools to manage these processes according to a well-grounded understanding of what is appropriate for local circumstances in the light of international best practices.

In practice, from a land administration design perspective, LAS problems are universally shared. Whether or not a country uses private property as the foundation of its land rights, land security and land management are overriding imperatives for the new role of land administration in supporting **sustainable development**. Whether a country is economically successful or resource hungry, betterment and improvement of existing systems are essential. Thus, an overarching theme is developing land administration capacity to manage change. For many countries, such as Kenya, Vietnam, and Mozambique, alleviating poverty, furthering economic development and environmental sustainability, and managing rapidly growing cities pose pressing challenges. The protection of traditional ways of life is also an overarching policy (figure 1.1). For more developed countries, the immediate concerns involve updating and integrating agencies in existing, relatively successful LAS and putting land information to work to support emergency management, environmental protection, and economic decision making. Iran (figure 1.2), for example, struggles to manage urban sprawl, while Chile (figure 1.3) needs LAS to aid delivery of sustainable agriculture.

Figure 1.2 Tehran, Iran, needs land administration to deal with the challenges posed by urban sprawl.

The theoretical concept of a land administration role in delivery of sustainable development relies on using the land management paradigm to guide the selection of tools for managing common processes. Within this framework, a wide range of options and opportunities is available to LAS designers and land-use policy makers. One tool, however, is fundamental: the cadastre, or more simply, the land parcel map. The history and influence of the cadastre, particularly after World War II, demonstrates that modern cadastres have a much more significant role than their original designers envisaged. Within the constant that land administration should be used to deliver sustainable development, the cadastre has extended purposes. Two functionalities of the modern cadastre underpin this philosophy: Cadastres provide the authoritative description of how people relate to specific land and property, and they provide the basic and authoritative spatial information in digital land information systems (LIS).

Even with the help of a clear theoretical framework, an explanation of how cadastres should be used within LAS to support sustainable development is far from easy. Cadastres take on many shapes and sizes. Some countries, for example, the United States, do not yet use a national cadastre, though most assiduously collect parcel information in some form or another. Other countries do not have the resources to build high-end cadastres, and need a well-designed, incremental approach. To deal with varietal situations, this book categorizes cadastres as three general types, depending on their history and function: the European or German approach, the Torrens title approach, and the French/Latin approach (see chapter 5, "Modern land administration theory"). The focus here is on the European, map-based cadastre with integrated land registration functions. The utility of this tool in land management is seen both in its successful use by its European inventors and in the contrast of lack of land management capacity in countries that use other approaches.

The analysis of land markets in this book shows how LAS organized markets to build economies in developed countries and to accelerate wealth creation by systematically converting land into an open-ended range of commodities. Internationally, market advancement will remain the driver for LAS change. But it should go beyond that. Sustainable development is much more urgent—economic wealth is only one part of the equation. Unless countries adopt LAS informed by the land management paradigm, they cannot manage their future effectively. Our argument is that planned responses to the availability of land and resources will help manage the social, economic, and environmental consequences of human behavior. Only then will nations be able to deal with the water, salinity, warming and cooling, and land and resource access issues facing the globe. Even more important is improvement of the global and national capacity to handle population growth and movement, burgeoning urban slums, and the alleviation of poverty.

Thus, this theory of land administration assumes that resources applied to building a cadastre can pervasively improve an entire LAS, and eventually public and private administration in general, while simultaneously improving land-based services to government, businesses, and the public. Whether the question is how to set up LAS or how to adapt an existing system, designers need to take into account the dynamism in land use, people's attitudes, institutions, and technology—and its potential. An ability to foresee what will happen in the future is helpful for managing this dynamism. The final chapter delves into how spatially enabled governments and societies inform a new vision of land administration. The spectacular growth in spatial technologies affords governments the ability to use this expanded information to focus on sustainable development. This hopeful scenario is offered to challenge those engaged in land administration and related activities, and to provide a clear direction for furthering excellence in LAS.

The theoretical framework for LAS will always be open-ended. Because the framework is under construction, rather than a precise recipe, guidance is offered in the form of **ten land administration principles** (see section 1.6). These principles show how each part of LAS should be designed and integrated. They ensure that people dealing with land-related questions can identify the best tools and options for local LAS. The themes are generic and apply regardless of capacity, economic models, or government arrangements. These statements help define both a generic modern LAS and a system suitable for local circumstances.

Figure 1.3 Land administration has a new role in supporting mixed rural land uses that ensure sustainable agriculture in places like Chile.

Primarily, the book is a "how to" guide, building on sixty years of development of an academic discipline in land administration that grew out of land surveying for cadastral purposes to incorporate multidisciplined approaches to land issues. The discipline now engages planners, valuers, political scientists, sociologists, human geographers, anthropologists, lawyers, land and resource economists, and many others. The expansion of the discipline came from the realization that holistic approaches to land management are essential to secure tenure, improve peace and order in a community, and deliver sustainable development. Achievement of these goals is, in practice, far from easy. Experience suggests that improving LAS design and operations can contribute to their success.

STRUCTURE OF THE BOOK

The book has five parts:

- ◆ **Part 1** Introducing land administration
- ◆ **Part 2** A new theory
- ◆ **Part 3** Building modern systems
- ◆ **Part 4** Implementation
- ◆ **Part 5** The future of land administration

PART 1 INTRODUCING LAND ADMINISTRATION

Chapter 1 explains the approach of the book and its themes. The central activities in land administration are designing, building, managing, and monitoring systems. This chapter explores the difference between land administration and land reform. LAS are seen as fundamental to delivery of global sustainable development. The reasons for building LAS are explained. Ten principles of LAS design distill recent developments in land administration theory and practice into a short but comprehensive description of modern LAS, capable of being used by countries at all stages of development.

Chapter 2 describes how groups of people think about land and the different approaches they take to land administration. These sociological aspects influence how people build systems to organize their unique approaches. These land administration responses to human experience, especially those influenced by colonialism, are described so that the modern concept of a multipurpose cadastre can be seen in its historical context.

PART 2 A NEW THEORY

Chapter 3 explains the relationship between land administration and sustainable development. This broad approach shows how national interests are no longer the only input: International imperatives for sustainable development are making greater impact on national systems, though implementation is highly variable. Within the wide range of approaches, some tools are commonly used, and the cadastre remains fundamental. Even the earliest systems used basic tools of maps and lists. Land administration still relies on maps and records of land usage (as distinct from planning and zoning) and landownership. Modern LAS rely on well-built, technically designed cadastres, which are unique for every system. The result is that the development of land administration as a distinct discipline changes over time depending on both local and international pressures and influences. The evolution of land administration as a discipline is discussed.

Chapter 4 deals with the basic functions of LAS. While historical analyses are useful, the better approach to understanding a particular LAS involves analysis of its core processes. Tenure processes are illustrative of general approaches used over recent decades to achieve security and sustainability. Basic land administration processes include the transfer of land (through transactions to buy, sell, lease, and mortgage as well as through social changes) and land titling. The land administration functions supporting land tenures and their related processes are the core of the book.

Chapter 5 identifies modern land administration theory. The most important feature is placing land administration within the land management paradigm, so that the processes and institutions in any LAS are focused on delivering sustainable development as their ultimate goal, not on delivering outcomes defined by a silo agency, such as a land registry or cadastral and mapping office. The broad design of LAS allows seamless inclusion of marine areas and other resources. The key tool, the cadastre, is given the formative role in building this approach.

PART 3 BUILDING MODERN SYSTEMS

Chapter 6 focuses on using LAS to build land markets. It approaches the formalization of market activities in five stages. An important but neglected component, the cognitive capacity of the beneficiaries of the formal land market, is explained. Land valuation and taxation systems are briefly described within the overarching task of designing complete and effective LAS.

Chapter 7 discusses managing the use of land. The concept of land use is introduced together with planning control systems. Urban and rural land-use planning and regulations are reviewed in the context of the land management paradigm. The roles of land consolidation and readjustment and integrated land-use management are described. Finally, land development is discussed as part of the paradigm.

Chapter 8 introduces marine administration by recognizing that administration of land and resources does not stop at the water's edge. It explores the extension of administration into coastal zones, seabeds, and marine areas. The concepts of the marine cadastre, marine SDIs, and marine registers are introduced and discussed.

Chapter 9 provides an introduction to how an SDI can be integrated into overall LAS, together with associated spatial technologies. Universal questions about land are linked into the new technological horizon in which spatial information, including information about land and resources, is a national asset, provided it is well managed. The concept of an SDI and the technical architecture supporting it are part of the modern land administration world.

Chapter 10 provides a global perspective of the variety of land administration activities worldwide and of the emerging analytical and comparative literature.

PART 4 IMPLEMENTATION

Chapter 11 highlights the importance of capacity building as a key component of building LAS. It covers the human dimensions of social, government, and individual capacity to devise and run land administration processes capable of meeting land management goals. The need to develop competencies is given prominence as the key to sustainable administration systems. The modern capacity building concept is explored along with capacity development in the context of land administration. Institutional capacity in land management is discussed together with the need for education and research in land administration.

Chapter 12 introduces the toolbox approach that is the core of the book. The early parts of the book are designed to help decision makers understand how tools are developed and what tools might be useful for a local land administration system. Given that LAS in any country or jurisdiction represent a unique response to local customs and traditions, laws, and institutional and governance arrangements, the "one size fits all" approach is unreliable. On the other hand, established and proven policies and strategies, along with the toolbox approach, are proposed to guide development and reform of LAS. What tenures should be available? How should boundaries be identified? What technology should be used? How should land information be collected and accessed? The list of questions is open-ended, but each country has particular concerns that require specific solutions. This chapter presents basic information about the various tools and implementation options and how they can be integrated into a robust and adaptable national system.

Chapter 13 discusses project management and evaluation with respect to land administration. The project-based approach draws the tools together and allows policy makers and system designers to identify the policies, tools, and systems needed amid the choices already identified. The project cycle; the importance of a LAS vision and objectives; the need to understand existing LAS, the components in LAS, and land administration projects (LAP); the use of best practices and case studies; and, most importantly, the need to engage the community and stakeholders are covered.

PART 5 THE FUTURE OF LAND ADMINISTRATION

Chapter 14 looks at future trends in land administration. It reviews the land administration journey with a particular focus on the role that land administration can play in sustainable development and in supporting a spatially enabled society. It recognizes the inherent dynamism of land administration and the importance of planning its future directions. Globalism, population growth, and government accountability are universally driving change. The challenges ahead, including the impact of new technologies, especially spatial technologies, are discussed. These technologies are likely to extend the capacity to deliver sustainable development objectives if local systems are capable of absorbing them. The trends identified by experts need to be built into planning processes to ensure that LAS remain capable of accommodating new situations and providing effective ways to deal with changing scenarios.

1.2 Why build a land administration system?

INCORPORATION OF INFORMAL LAS INTO FORMAL SYSTEMS

The basic reason that societies manage land is to satisfy human needs. Having a secure home, or even a secure place to sleep or work, satisfies fundamental necessities of life, just as guaranteeing a harvest to the sower of grain delivers food security. Consequently, land is managed by all settled societies, whether they explicitly acknowledge it or not. The systems used can be formal or informal, and either will work well if circumstances permit. From the perspective of land administration theory, the variety of informal systems defies attempts to categorize them. These systems do not institutionalize most of the tools in the toolbox. They use very different options to deliver the tools they use, and they produce results that are unique to the situation. Informal systems are the most common. Even developed nations have informal systems used among slum dwellers, traditional peoples, and other groups. Incorporation of these informal systems into a regional or national LAS framework is an overarching and crosscutting theme in the discipline. Many informal systems are under threat, mostly from population increases, but also as a result of environmental changes, war and dislocation, encroachment on resources, and general transition from traditional to less traditional social, economic, and political orders. LAS design needs to be sensitive to these threats and patterns of change among informal

Figure 1.4 An informal
settlement in Vietnam is
an example of the types of
challenges posed by LAS that
develops informally.

administration systems such as those embodied by the informal settlements found in Vietnam (figure 1.4). Each tool needs to be designed with the operation of informal systems in mind.

TRADITIONAL BENEFITS OF LAS

While informal systems constantly emerge and change, the global trend is to manage land through formal systems. The reasons for formalizing land administration are complex and have changed radically over the past century. Most countries still seek the traditional benefits of LAS (table 1.1). These traditional reasons for supporting LAS have wide support in the literature (GTZ 1998; DFID 2003; ILC 2004; UNECE 2005c).

GREATER BENEFITS OF MODERN LAS

While the traditional benefits remain the predominant incentives for a country's investment in LAS, even more compelling reasons flow from global environmental issues and population increases. Also, while the traditional benefits inform the mission statements of the agencies running LAS in developed countries, a modern LAS approach requires these agencies to operate beyond their immediate silos, deliver larger economic benefits, enhance the capacity of land information, and support regional, not just jurisdictional, environmental management. Thus, the broader benefits identified as follows are relevant to all nations.

TABLE 1.1 – TRADITIONAL BENEFITS OF LAS

Support for governance and rule of law	The formalization of processes used for land management engages the public and business, and, in turn, this engagement leads to their support for the institutions of government.
Alleviation of poverty	A primary means of alleviating poverty lies in recognizing the homes and workplaces of the poor and their agricultural land as assets worthy of protection.
Security of tenure	This is the method of protecting people's associations with land. It is the fundamental benefit of formal land administration. Ensuring security throughout the range of tenures used in a country helps provide social stability and incentives for reasonable land use. Conversion of some of the rights into property is the core process of commoditization of land needed for effective markets.
Support for formal land markets	Security and regularity in land arrangements are essential for successful, organized land markets. LAS manage the transparent processes that assist land exchange and build capital out of land.
Security for credit	International financing norms and banking practices require secure ownership of land and robust credit tenures (that is, tenures which support security interests in land) that can only exist in formal LAS.
Support for land and property taxation	Land taxation takes many forms, including tax on passive land holding, on land-based activities, and on transactions. However, all taxation systems, including personal and company taxation, benefit from national LAS.
Protection of state lands	The coherence of national LAS is dependent on its coverage of all land. Thus, management of public land is assisted by LAS.
Management of land disputes	Stability in access to land requires defined boundaries, titles, and interests. If LAS provide simple, effective processes for achieving these outcomes, land disputes are reduced. The systems also need additional dispute management processes to cover breakdown caused by administrative failure, corruption, fraud, forgery, or transaction flaws.
Improvement of land planning	Land planning is the key to land management, whether the planning is institutionalized within government or achieved by some other means. Impacts of modern rural and urban land uses affect adjoining land and beyond. These impacts need to be understood and managed by effective land planning assisted by LAS.

Continued on next page

Continued from previous page

TABLE 1.1 – TRADITIONAL BENEFITS OF LAS	
Development of infrastructure	Construction of power grids, gas supply lines, sewerage systems, roads, and the many other infrastructure elements that contribute to successful land use require LAS to balance private rights with these large-scale infrastructure projects, whether provided by public or private agencies.
Management of resources and environment	Integration of land and resource uses is a difficult aspect of LAS design. Land and resource titles require complicated and mutually compatible administrative and legal structures to ensure sustainability in the short and long term.
Management of information and statistical data	Each agency needs to appreciate the importance that the information generated through its processes holds for the public, businesses, and government in general. More importantly, everyone needs to understand the fundamental importance of integrated land information for sustainable development.

MANAGING HOW PEOPLE THINK ABOUT LAND

Attempts to transport modern tools of cadastres and registration systems from Western democracies to other countries have resulted in both successes and failures. Analyses of these experiences raise the issue of how LAS interact with their intended beneficiaries. Especially since 2000, analysis of LAPs and other endeavors to improve LAS have identified a primary, but often neglected, function of LAS: management of the cognitive framework used by a society to understand land and to give significance and meaning to land-related activities. A cognitive awareness of land is unique to every nation, and often to local areas and specific groups within nations. It influences the relationships among land uses, institutions, administrations, and people. Realizing the importance of the cognitive aspects of land led to improved international understanding of how to build a land administration system to fit the context of its intended beneficiaries. A growing analytical literature dealing with the transportability of market-based systems and their associated technical tools (Bromley 2006; Lavigne Delville 2002a) highlights fundamentally different normative realities and the problems of blending them into LAS design to achieve a sustainable result. Demand-driven service models, capacity building, transparency, accountability, conformity with local ideas of land, and incorporation of spiritual and social meanings of land are some of the changes in LAS design flowing from better understanding of the cognitive aspects of land.

DELIVERING SUSTAINABLE DEVELOPMENT

The three dimensions of sustainable development—economic, environmental, and social—which form the "triple bottom line," are at the heart of several decades of reform and have made a global impact on land administration. Increasingly, the bottom line now includes a fourth dimension of good governance. While land administrators can and should play a role in contributing to sustainability objectives (UN-FIG Bathurst Declaration 1999; Williamson, Enemark, and Wallace, eds. 2006), the ability to link the systems to sustainability has been poor and presents many challenges. As a result, a continuing theme for modern LAS is the exploration of the strategies and technologies to deliver sustainable development objectives, particularly through delivery of information in a form that can be used for sustainability accounting—the emerging systems for monitoring and evaluating achievement of sustainability objectives and initiatives.

BUILDING ECONOMIES, NOT JUST LAND MARKETS

Countries with highly successful economies use formal systems containing all the tools in the land administration toolbox. These wealthy and successful economies thrive on regular, predictable, and institutionalized access to land. They provide reliable and trusted institutions to manage land and to deliver security of tenure, equity in land distribution, sensible and attractive development, and fair land taxation (see chapter 6, "Building land markets"). Productivity in the agricultural sectors is much higher. Credit is widely available at comparatively low rates. Personal wealth in the form of real estate assets grows. Business investment in land increases. Countries seeking similar economic advantages tend to modify their local systems to emulate those in successful countries and generally adopt options tried and tested by those countries to institutionalize their own land administration tools.

Much of the literature on land administration and cadastres takes the objective of LAS supporting efficient and effective land markets for granted. But what is a land market in a modern economy? Since LAS was first developed, land commodities and trading patterns have undergone substantial change: Commodities are now complex, international in design, and run by corporations rather than individuals. Markets continually evolve, primarily in response to economic vitality and sustainable development objectives. Developments in information and communications technology also drive land markets. Modern land markets involve a complex and dynamic range of activities, processes, and opportunities, and are impacted by a new range of restrictions and responsibilities imposed on land and land-based activities. Are current LAS capable of supporting modern markets that trade in complex commodities, such as water rights, mortgage-backed securities, utility infrastructure, land information, and the vertical villages in high-rise condominium developments?

ACHIEVING SOCIAL GOALS

There is surely no need to argue that effective land administration improves the lives of people who enjoy its fruits. A comparison of the living experiences in developed countries versus the standard of living of people in undeveloped countries is enough. However, whether we can transfer these social and political effects through land administration tools is a real question. More and more, research is showing that while delivery of security of tenure is the overarching goal, other social goals flow out of protecting people's relationships to land. LAS replace personal protection of land with formal systems, allowing people to leave their homes and crops—their property—to seek markets for their labor and produce. Children who would otherwise mind the home can attend school (Burns 2006). Nutrition and food security are improved, especially for the rural poor, but also for the urban poor through small garden plots. Some newly emerging research on containment of land disputation will likely add to these positive results.

The most significant social goal for LAS is gender equity. Increasing the access of women to land is a goal consistently sought by land projects. Delivery is another question. The pursuit of gender equity has significantly improved the knowledge of status quo opportunities for women in terms of ownership and has generated innovative ideas about increasing women's access (Giovarelli 2006). In the developing world, more than half of all women work in agriculture, but most own no land (figure 1.5). There is, therefore, much work to be done.

MANAGING CRISES

World population is estimated to be 10 billion by 2030, up from 2 billion in 1950, and 6.5 billion in 2000. The population of cities in developing countries will double from 2 billion to 4 billion in the next thirty years. To prevent people from living in slums, developing nations must every week between now and 2036 create the equivalent of a city capable of housing 1 million people (UN–HABITAT 2006a). Water is even more problematic than land. One person in five has no access to potable water. North America's largest aquifer, the Ogallala, is being depleted at a rate of 12 billion cubic meters a year. Between 1991 and 1996, the water table beneath the North China Plain fell by an average of 1.5 meters a year. The Aral Sea in Central Asia, once the world's fourth largest inland sea and one of its most fertile regions, is now a toxic desert. Land disputation infects the social fabric of many nations.

This is a small part of a litany of hard issues faced by national governments and international development agencies. Every day, similar observations cross the newswires. Earthquakes, tsunamis, cyclones, hurricanes and other disasters, and human conflict and war add to the challenges. No

Figure 1.5 Achieving gender equity in land administration is a fundamental issue in places like Malawi.

matter where we start our analysis, the world clearly needs better land and resource management through effective administration. And our responses must be much more carefully designed.

BUILDING MODERN CITIES

A cityscape of even fifteen years ago is nothing like the modern, crowded high-rise megacities like Hong Kong (figure 1.6) that have spread throughout the world. The most successful economies of the world clearly benefit from a land management capacity delivered by well-developed LAS. Successful provision of utilities, organized land allocation, robust property rights, and high levels of land taxation are features of cities in developed economies. These qualities help generate the wealth needed to build urban infrastructure capable of delivering reasonable urban environments with high human and business densities.

By contrast, cities that respond haphazardly to mass rural population movements experience many problems. UN–HABITAT, the UN agency for human settlements (www.unhabitat.org), predicts that in many countries, especially in Africa, more people will eventually live in these unmanaged cities, many without adequate water or sanitation, than in managed cities, unless substantial counteraction is taken. Unchecked, demand leads to an inability to provide services or to facilitate and coordinate ordered growth. Jakarta, Indonesia; Lagos, Nigeria; Manila, the Philippines; Kabul, Afghanistan; Tehran, Iran; Mexico City, Mexico, and many other burgeoning urban areas are veritable case histories of cities faced with severe management challenges.

These unmanaged megacities are in desperate need of administrative infrastructure. All would benefit from a large-scale cadastral map, even of the most basic kind, and a path toward a land administration system that is capable of implementing the land management paradigm. Bangkok, Thailand's experience in using such a map illustrates the utility of a systematic approach (Bishop et al. 2000).

DELIVERING LAND INFORMATION FOR GOVERNANCE AND SUSTAINABILITY

Information about land is a major asset of government and is essential for informed policy making in the public and private sectors. The information, in itself, is valuable, even if not sold. In fact, the economic worth of land information is probably greater if it is freely available. The questions of who collects the information and how it is made available are vital to LAS operations. Many countries, including Indonesia, Malaysia, Laos, and China, regard maps and plans as quasi military information and impose substantial restrictions on their availability. Another group, including the United States and New Zealand, makes land and spatial information, including digital maps, generally available at little or no cost to stimulate the economy. And still another group, including Australia and European countries, generally pursues a cost recovery path and relies on the primary audience for land information to pay an estimated price reflecting the cost of maintenance and sometimes data collection. Other common limitations on access to land information in market systems include privacy policies and laws, licensing arrangements, pricing systems (in regard to whether the cost is capital outlay or a tax-deductible and routine business expenditure), and difficulties of access.

Whatever policy decisions about restrictions to access are taken, land and spatial information is a national asset capable of being used to improve the opportunities of citizens and businesses, especially when the processes are in digital form. The availability of information, especially through the creation of an SDI, plays a vital role in a nation's use of land and spatial information. The transparency of land registry operations, given that they document private ownership of land, is important to a nation's public credibility and ability to monitor subsequent changes in landownership and secondary transactions. The development of e-government also makes land information more important.

Accessibility of land information can transform the way governments and private sectors do business in modern economies. In the future, technology-driven, spatially enabled LAS will service a larger range of functions by matching people and activities to places and locations, basically through the spatial identification of a land parcel in a cadastral map. Location or place will

Figure 1.6 Busy high-rise megacities like Hong Kong require a robust LAS.

relate to many more land administration activities and associated data, such as management of restrictions and responsibilities, new forms of tenure, and complex commodity trading. Modern LAS need to be designed in a way that recognizes the potential of land information and capitalizes on its increasing value (see chapter 14, "Future trends").

ENCOURAGING THE USE OF NEW TECHNOLOGY

The next generation of LAS will benefit from advances in spatial and information and communications technology. While a great deal of land administration practice will still concern policy, and institutional and legal issues, technology will stimulate development of entirely new concepts and approaches. Trends in access to land information provided by LAS, particularly through the Internet, the impact of geographic information systems, and the development of appropriate cadastral data models, are now being absorbed by the mainstream.

The next generation of LAS will depend on SDIs to facilitate integration of built and natural environmental databases—a precondition for analyzing sustainable development issues. Currently, integration is difficult: Built (mainly cadastral) and natural (mainly topographic) datasets were developed for different reasons using specific data models and are often managed by independent organizations.

The engagement of the private sector in land administration, especially through new technical products, will also increase.

REDUCING THE DIVIDE BETWEEN RICH AND POOR NATIONS

The contrast between rich and poor nations is readily apparent from a land administration perspective (De Soto 2000). Poor countries need more, not less, comparative land management capacity. While titling land can retrieve the lost capital of the poor, integration of the land administration functions in organized LAS is essential to accommodate planning and other issues experienced by poorer nations. Failure to build a robust infrastructure for land management will also have severe consequences for rapidly developing economies like India's and China's. Escalation of their need for better land management will compound their inability to deliver it because they have not taken the time to plan and build land management infrastructure.

DELIVERING THE MILLENNIUM DEVELOPMENT GOALS

Since 2000, delivery of security of tenure has been driven by the Millennium Development Goals (MDGs) adopted by 189 UN member countries and numerous international organizations as a focus for foreign aid. The goals are

1. Eradicate extreme poverty and hunger
2. Achieve universal primary education
3. Promote gender equality and empower women
4. Reduce child mortality
5. Improve maternal health
6. Combat HIV/AIDS, malaria, and other diseases
7. Ensure environmental sustainability
8. Develop a global partnership for development

The MDGs, especially goal 7, require social and environmental outputs, not merely economic outputs, and require LAS for delivery (Enemark 2006a). Implementation of global and national land policy at this level requires much more people-based, social information, in addition to information about processes relating directly to land. Newer kinds of information build the capacity of land policy makers and administrators to take local conditions into account, while being aware of intercountry comparisons and world best practices. Women's de jure and de facto access to land, inheritance systems and the capacity of formal LAS to reflect them, the relationship between land and resource tenures, the nature of land disputes, and the performance of

Figure 1.7 In the Philippines, access to basic services can come informally.

related markets in money, agricultural products, and agrarian labor are now additional starting points for information collection, process management, and LAS design.

PROVIDING A FRAMEWORK FOR DELIVERY OF BASIC SERVICES

Western countries are able to provide utilities and services to their homes and businesses in predictable and orderly ways. This capacity arises because they organize access to land. However, millions of people live in places where organized access to land and provision of basic services is not possible, and informal systems such as those in the Philippines are used instead (figure 1.7).

Access to clean water and sanitation is especially problematic in crowded urban slums. The delivery of these basic facilities requires a concerted approach to organizing access to water and sanitation facilities, which is only possible if land itself is organized. The development of new approaches for finance and governance of access to clean drinking water and basic sanitation anticipates recognition of water and sanitation as basic human rights (Tipping, Adom, and Tibaijuka 2005) and envisions concerted global approaches to satisfying these rights. These goals cannot be satisfied outside the national LAS framework.

1.3 The changing nature of land administration systems

Modern land administration, its theories, and tools need to be understood by a diverse audience, including policy makers, administrators, students, and professionals. Their choices about designing, building, and managing LAS and about determining when a system is working effectively will be crucial to national development. These tasks are complicated because the world of land administration involves constant change, reflecting the changes in social, political, and economic systems that influence the way governments and other organizations do things. Moreover, three other influences make LAS especially dynamic. The systems are simultaneously

- at the center of sustainable development issues;
- the place where new technologies challenge existing service delivery and institutional operations;
- often involve a clash between national and international trends.

Given these pressures, the success of LAS requires its designers to identify and address institutional, legal, technical, and knowledge transfer issues, while understanding how land is used within communities. An engineering focus for designing, building, and managing LAS is needed to manage this broad array of issues. Project management; the role of pilot projects; the evaluation and monitoring of LAS; the role of government, private, academic, and nongovernmental organization (NGO) sectors, and public engagement are all important. Moreover, a major commitment to capacity building and institutional development—the overriding components of sustainable LAS—is crucial. The engineering focus therefore expands to incorporate multidisciplinary approaches, especially to take account of the relationships among LAS, the people and businesses they serve, and the governments that build or oversee the systems within the regional and international framework.

Like any evolving discipline, land administration generates discussions, debates, and points of view about how things might be done. These debates generate theory and research that build the discipline, and improve responses by governments to their most pressing and complex land issues. In general, land administration debates revolve around three kinds of issues:

1. *When can LAS tools be successfully transported?* The first kind of issue is generated by land markets and attempts by governments and LAPs to transport familiar land administration tools, particularly systems for land titling, cadastres, and property-based land rights. These tools support the healthy economies of the

thirty-five or so developed countries that have effective formal and free land markets. These familiar market-based tools took hundreds of years to build. They are clever and sophisticated and extremely expensive to install and manage. These tools are deeply embedded in the government of their source countries. Transporting them successfully to other countries, even to those where land markets are planned, involves adapting them to the "best fit" in different contexts. Especially since 1990, improved understanding of how the tools work, and the part that people play in supporting them, has inspired robust and inventive approaches in countries seeking to use a land market approach to improve land management. The case histories of conversion of the centralized land organization in postcommunist countries to market approaches, and the titling programs of successful Asian economies, especially in Thailand and Malaysia, illustrate what can be done.

2. *How can LAS help solve poverty?* The second kind of issue involves upgrading security of tenure, food security, and sustainable livelihoods where land market approaches are not possible or are problematic—for example, in newly occupied peri-urban slums, indigenous and traditionally held land, or postconflict countries. Common contexts involve highly centralized governments, countries experiencing limited governance capacity and endemic mass poverty, and postconflict situations. Responses to these issues of poverty and capacity by the fraternity of aid workers, economists, engineers, sociologists, lawyers, and many others are helping to identify new tools, technologies, and land management approaches to improve land access and organization. Generally, these new ideas encourage flexible and localized approaches to tenures, planning, and provision of basic amenities, especially water and waste systems, for millions of people.

3. *What is land administration?* The third kind of issue involves what constitutes land administration. The most commonly accepted definition of land administration is set out in the United Nations Economic Commission for Europe (UNECE) Land Administration Guidelines (1996): "Land administration: the processes of recording and disseminating information about ownership, value, and use of land when implementing land management policies." Even in a traditional sense, the coverage is broad. Jon Lindsay (2002) saw land administration as management of a system of land rights, including a broad range of subjects:

   ◆ Procedures by which land rights are allocated or recognized

   ◆ The definition and delimitation of boundaries between parcels

- The recording of information about land rights, <u>rights holders</u>, and parcels

- Procedures governing transactions in land, including sales, mortgages, leases, and dispositions

- The resolution of uncertainty or adjudication of disputes concerning land rights and boundaries

- Institutions and processes for the planning, controlling, and monitoring of land use

- Land valuation and taxation procedures

Together, these subjects describe a widely agreed framework for approaching land administration. However, LAS that are capable of producing information and performing functions to deliver sustainable development have an even broader scope. LAS that operate at this higher policy level must include even more subjects, particularly

- Procedures for public engagement

- Support for the cognitive function of LAS by integrating systems with the way their intended beneficiaries think about land

- Management of restrictions on land

- Technologies for land management and information

- Support for trading in complex and secondary commodities

- Support for the management of utilities and provision of services (electricity, drainage, sewerage, communications)

- Monitoring and evaluation processes

- Sustainability accounting

This wider coverage goes beyond a government focus, though government remains the agency responsible for designing, monitoring, and reforming the overall system. As yet, no country has built a land administration system that fully addresses the needs of sustainable development. This broader program for LAS also identifies one of the major issues faced by countries seeking better land management—human resources. Even highly developed nations lack sufficient people with the professional and technical capacities to support their systems.

All participants in these debates, and indeed many other debates about land and resources, assume that constant improvement in land management capacity is necessary, and that an organized approach can help. The overall land policy choice is a question for each nation and its people. Whatever it is, the land administration response should be to drive the processes and functions of the system toward delivery of sustainable development. This book therefore encompasses all the basic approaches to land policy: traditional, centralist, diversified, and market-based. It is written particularly for countries seeking improvement paths based on a land market approach. In international experience, this is the most common policy choice for advancement. In other words, the land policy direction of nation states generally involves more, not less, of a land market approach, with the intent being to use land to generate national and individual wealth, alleviate poverty, ensure food and land security, and assist equitable land distribution. The market-based approach used here recognizes that many people, including groups in countries with highly successful land markets, do not need or want individual titles, though they certainly require secure access to land. It also recognises that modern land management requires highly developed and successfully implemented restrictions on private ownership.

1.4 Land reform

Land administration projects are different from land-reform projects, though in many practical situations, the distinction is blurred. Many land administration activities are undertaken as part of projects aimed at improving national or regional administration of government and social justice. The growth of international development aid gave land and its administration great significance (Bruce et al. 2006). The contrast between countries capable of organizing land and those where land and food security are tenuous led to concerted attempts to improve LAS design. The predominant reasons articulated for stabilizing and improving administration of land are economic, but, more and more, humanitarian reasons are included. The poor need water and food security and housing. Countries need to manage movement of the rural poor to cities. The estimated 2.7 billion people living on incomes below or around the international poverty line of $2.60 per day remains an overpowering challenge to governments to better organize land and its uses. In other words, the drivers for modern LAS in developing nations emphasize the contrast between living conditions for those with predictable land arrangements and for those without. Here, LAS design strives to deliver predictability, security, and the accoutrements of sanitation, water and housing, using whatever tools, formal or informal, are appropriate.

Land projects of another kind are also undertaken. Land reform programs aimed at redistributing or reconfiguring land are very common (Lindsay 2002). Land reform, consolidation, restitution, and redistribution are complex processes that inevitably involve politics. They presuppose capacity for both land policy making and land administration of some kind. These processes complicate policy implementation by their relationship with the exercise of power and political activities, especially because of their potential to raise levels of land disputation. The discipline of land administration does not provide an analysis of when and to whom to redistribute land and resources. Rather, it defines the administrative institutions and processes suitable to implement these political decisions. Thus, land administration is not land reform, but it is an important precondition to successful reforms.

Perhaps the most monumental efforts in land redistribution and reform followed the failure of command economies in Central and Eastern Europe, leading to applications by ten countries in 1997 for membership in the European Union (EU)—Estonia, Latvia, Lithuania, Poland, the Czech Republic, Slovakia, Hungary, Slovenia, Romania, and Bulgaria.

Substantial rebuilding of their LAS was needed to reflect the EU standards of functioning market economies, including management of competitive pressures and market forces within the processes of returning state and collectively owned land to private ownership. While each country took a divergent implementation path, with varying degrees of success, they all needed to establish LAS to achieve the objectives of the *Acquis Communautaire* (the "rules" of the EU) (Bogaerts, Williamson, and Fendel 2002; Bruce et al. 2006). Successful implementation of political decisions of how consolidation was to be performed and in whose favor consolidation worked depended on legal and administrative support (Dale and Baldwin 1998, 2000). Land administration in accession countries was recognized as a key component in strategies to achieve the protection of human rights, the Common Agriculture Policy, and an effective free market. The success and longevity of these political processes required carefully designed LAS to minimize disputes and reinforce change. The levels of success were mixed, but the efforts demonstrated that the key features of LAS that facilitate political change are transparency, accessibility, and reliability.

1.5 Good governance

GOOD GOVERNANCE IN LAND ADMINISTRATION

Governance is the process of governing. Land administration, therefore, is essentially about good governance. The UNECE land administration principles (2005c) are built on the assumption that "sustainable development is dependent on the State having overall responsibility for managing information about the ownership, value, and use of land." The land management paradigm extends this connection by demanding an even wider approach to governance in land administration, in which the government builds infrastructures for management of land in addition to management of information. Thus, the paradigm builds governance directly into land administration.

Governance refers to the manner in which power is exercised by governments in managing a country's social, economic, and spatial resources. It simply means the processes of decision making and the processes by which decisions are implemented. This indicates that government is just one of the actors in governance. The concept of governance includes formal as well as informal actors involved in decision making and implementation of decisions made, and the formal and informal structures that have been set in place to arrive at and implement the decision.

Good governance is a qualitative term or an ideal that may be difficult to achieve. The term includes a number of characteristics—i.e., as identified in the UN–HABITAT Global Campaign on Urban Governance (2002). The characteristics or norms are as follows:

- ◆ **Sustainability:** Social, economic, and environmental needs must be balanced while being responsive to the present and future needs of society.
- ◆ **Subsidiarity:** Allocation of authority at the closest appropriate level must be consistent with efficient and cost-effective services.
- ◆ **Equity of access:** Women and men must participate as equals in all decision-making, priority-setting, and resource allocation processes.
- ◆ **Efficiency:** Public services and local economic development must be financially sound and cost-effective.
- ◆ **Transparency and accountability:** Decisions taken and their enforcement must follow rules and regulations. Information must be freely available and directly accessible.
- ◆ **Civic engagement and citizenship:** Citizens must be empowered to participate effectively in decision-making processes.
- ◆ **Security:** All stakeholders must strive for prevention of crime and disasters. Security also implies freedom from persecution and forced evictions and provision of land tenure security.

Once the adjective "good" is added, a normative debate begins. Different people, organizations, and government authorities will define "good governance" according to their own experience and interests. For example, it may be argued that issues such as rule of law, responsiveness, participation, and consensus orientation should be added to the preceding list. The term good governance can also be viewed in several contexts such as corporate, institutional, national, and local governance.

Of these, the standards of transparency, equity, accountability, subsidiarity, and also participation are especially important to sustainable LAS. These standards, in turn, have an impact on the most basic of human needs: the production of food. As the Food and Agriculture Organization (FAO) says:

> *"Adequate institutional arrangements are required to determine rights and access to rural resources, such as land, water, trees, and wildlife, as a prerequisite to agricultural development and food security. Many countries specifically require advice on such institutional arrangements for property rights, on how to ensure more equitable access by women and men to natural resources, on functioning land markets and land administration to take account of mortgage-secured credit for investment, and (on) good governance of land and natural resources." (2007)*

These general considerations link land administration to governance, so that land stabilization is seen as essential to successful nationhood and civic capacity. The FAO projects and themes on governance illustrate the connection (2006). In its study on Good Governance in Land Tenure and Administration, FAO remarks:

> *"The message to land administrators is that they cannot pursue technical excellence in isolation. Their skills and techniques should serve the interests of society as a whole. … Land administrators act as guardians of the rights to land and the people who hold those rights. In doing so, they act to stabilize public order and provide the preconditions of a thriving economy." (2007)*

The major international agencies demonstrate that successful land administration requires accountable government. Sustainable systems require that the institutions that interact with the citizens who are its intended beneficiaries do so in ways that build their confidence, particularly by negating disputes and managing points of tension relating to landownership, use, and availability. The major engagement should involve policy formation and implementation to ensure that the system reflects the cognitive capacity of the beneficiaries and their beliefs about land. A national capacity to create laws through legislation and subordinate legislation is also necessary for sustainable LAS. For nations on the development track, rule by law, rather than rule by the elite or ad hoc responses to circumstances, is essential. These conditions apply even if the nation's administration horizon includes land held in social tenures that rely on informal systems of land management.

For successful governance, institutions need to be stable, transparent, and free of corruption. Weak governance in land administration leads to massive overregulation and production of conflicting and gap-ridden bodies of laws, standards, and documents. There is little cohesion and mutual reinforcement of legal and economic norms. Sadly, LAS in developing countries more often exhibits corruption in the collection of fees; multiple rent-seeking and unnecessary processes; delivery of multiple and ineffective titles to parcels; arbitrary allocation of land; and negligible capacity for planning or controlling building standards. Repeated problems in developing countries include legitimation of mass land theft; failure to police uncontrolled evictions; inability to manage interaction among competing tenure holders, especially between landowners and users and resource takers; and inability to manage state assets. Weak governance will never be able to manage the transition of the world's populations from rural to urban areas.

To be sure, good governance is central to delivery of appropriate, effective, and efficient land administration in both developing and developed countries.

1.6 Ten principles of land administration

Despite the uniqueness of local systems, the range of cognitive frameworks about land, and the difficulties in transferring institutions, design of robust and successful LAS is possible. The **ten principles of land administration** in table 1.2 set boundaries for designers, builders, and managers of LAS to help them make decisions about their local system. Overall, the principles are written with the goal of making establishment and reform of LAS easier. The principles implement the modern philosophy in land administration — to develop and manage assets and resources within the land management paradigm to deliver sustainable development. They are universally applicable. Countries at the early stages of development will not be able to use the full array of technical options or professional skills, but they can improve land management through appropriately designed LAS.

The principles reflect a holistic approach to LAS and focus on sustainable development as the overriding policy for any national system, irrespective of whether a country implements

TABLE 1.2 – TEN PRINCIPLES OF LAND ADMINISTRATION

1. LAS	**LAS provide the infrastructure for implementation of land polices and land management strategies in support of sustainable development.** The infrastructure includes institutional arrangements, a legal framework, processes, standards, land information, management and dissemination systems, and technologies required to support allocation, land markets, valuation, control of use, and development of interests in land.
2. Land management paradigm	**The land management paradigm provides a conceptual framework for understanding and innovation in land administration systems.** The paradigm is the set of principles and practices that define land management as a discipline. The principles and practices relate to the four functions of LAS—namely, land tenure, land valuation, land use, and land development, and their interactions. These four functions underpin the operation of efficient land markets and effective land use management. "Land" encompasses the natural and built environments, including land and water resources.
3 People and institutions	**LAS are all about engagement of people within the unique social and institutional fabric of each country.** This encompasses good governance, capacity building, institutional development, social interaction, and a focus on users, not providers. LAS should be reengineered to better serve the needs of users, such as citizens, governments, and businesses. Engagement with society, and the ways people think about land, are at its core. This should be achieved through good governance in decision making and implementation. This requires building the necessary capacity of individuals, organizations, and wider society to perform functions effectively, efficiently, and sustainably.
4 Rights, restrictions, and responsibilities	**LAS form the basis for conceptualizing rights, restrictions, and responsibilities (RRRs) related to policies, places, and people.** Rights are normally concerned with ownership and tenure whereas restrictions usually control use and activities on land. Responsibilities relate more to a social, ethical commitment or attitude toward environmental sustainability and good husbandry. RRRs must be designed to suit the individual needs of each country or jurisdiction and must be balanced among different levels of government, from local to national.
5. Cadastre	**The cadastre is at the core of LAS that provide spatial integrity and unique identification of every land parcel.** Cadastres are large-scale representations of how the community breaks up its land into usable pieces, usually called parcels. Most cadastres provide security of tenure by recording land rights in a land registry. The spatial integrity within the cadastre is usually provided by a cadastral map that is updated by cadastral surveys. Unique parcel identification provides the link between the cadastral map and the land registry and serves as the basis of LAS and the land information it generates, especially when it is digital and geocoded. The cadastre should ideally include all land in a jurisdiction: public, private, communal, and open space.
6. LAS are dynamic	**LAS dynamism has four dimensions:** The first involves changes to reflect the continual evolution of people-to-land relationships. This evolution can be caused by economic, social, and environmental forces. The second dimension is evolving ICT and globalization, and their effect on the design and operation of LAS. The third dimension is the dynamic nature of the information within LAS, such as changes in ownership, valuation, land use, and the land parcel through subdivision. The fourth dimension involves changes in the use of land information.

Continued on facing page

Continued from previous page

TABLE 1.2 – TEN PRINCIPLES OF LAND ADMINISTRATION

7. Processes	**LAS include a set of processes that manage change.** The key processes concern land transfer, mutation, creation and distribution of interests, valuation, and land development. The processes, including their actors and obligations, explain how LAS operate as a basis for comparison and improvement. While individual institutions, laws, technologies, or separate activities within LAS, such as property in land, a land registry, specific piece of legislation, or technology for cadastral surveying, are important in their own right, the processes are central to overall understanding of how LAS operate.
8. Technology	**Technology offers opportunities for improved efficiency of LAS and spatial enablement in terms of land issues.** The potential of technology is far ahead of the capacity of institutions to respond. Technology offers improvements in the collection, storage, management, and dissemination of land information. At the same time, developments in ICT offer the potential for spatial enablement in terms of land issues by using location or place as the key organizer for human activity.
9. Spatial data infrastructure	**Efficient and effective LAS that support sustainable development require an SDI to operate.** The SDI is the enabling platform that links people to information. It supports the integration of natural (primarily topographic) and built (primarily land parcel or cadastral) environmental data as a prerequisite for sustainable development. The SDI also permits the aggregation of land information from the local to the national level.
10. Measures for success	**A successful land administration system is measured by its ability to manage and administer land efficiently, effectively, and at low cost.** The success of a land administration system is not determined by the complexity of legal frameworks or the sophistication of technological solutions. Success lies in adopting appropriate laws, institutions, processes, and technologies designed for the specific needs of the country or jurisdiction.

property institutions, communal land arrangements, or socializes land. They highlight the importance of information and participation of people in the process. They set the framework in which the historical development of familiar ingredients, such as cadastres and land registries, can be meshed with recent innovations, particularly incorporation of social tenures, new complex commodities appearing in highly organized land markets, and the technical potential of spatial information.

Chapter 2
People and land administration

2.1 People and land

2.2 Historical evolution

2

2.1 People and land

EVERYONE IS THE SAME BUT DIFFERENT

People must relate to land in some way. These relationships tend to get more and more organized as they evolve. Land administration is the study of how people organize land. It includes the way people think about land, the institutions and agencies people build, and the processes these institutions and agencies manage. While the variations are considerable, organizational and administrational principles have a remarkable consistency across the globe. Use of maps, creation of concepts, and practical approaches to identifying land are virtually universal. In countries with a better capacity to organize, land administration is highly developed, professionalized, and institutionalized. The history of these well-organized systems is virtually the story of the development of land administration as a coherent, unique discipline.

Evolution of land administration revolves around management of land parcels—that is, the small units of land used by people in their daily lives. The cadastre, a concept central to modern land administration, is the principal instrument used to manage parcels and, at its core, involves a land registry. In one form or another, a cadastre is essential. Indeed, the power of cadastres to improve land management and contribute to good governance is even greater in modern land administration.

CONCEPTS OF LAND

Land has both physical (buildings and resources) and cognitive (theory and concept) aspects. LAS are vital for management of both, though management of the physical assets predominates in both theory and practice. This is because the influential Western systems successfully integrate LAS with the concepts and ideas about land understood in their communities, so that explicit analysis of local cognitive aspects of land is no longer necessary. However, congruence between the physical and cognitive aspects remains an essential undercurrent even in Western systems to ensure LAS can perform their most significant task—management of how people think about land.

Consider a recent innovation in LAS in many countries of popularizing commodities that consist of cubes of airspace. Here, the empty cube underpins strata or condominium titles. Obviously, the physical boundaries of walls, floors, and ceilings are immediately visible. They define the physical parameters of the commodity. But the essential feature of the commodity is the bundle of rights, restrictions, and responsibilities attached to the airspace. If the building is demolished, this collection of attachments to airspace remains the commodity. It is realizable because of the records and the cognitive appreciation of their meaning shared among owners and everyone else. All commoditized rights in land are abstract in this sense. They exist in people's minds as ideas verified by the record base, the land, and people's behavior.

Around the world, cognitive approaches to land are remarkably variable, reflecting the different ways people think. Their organization of thinking through normative systems and processes is developed amid their unique social responses to the local landscape. Table 2.1 shows an array of real-life, people-to-land concepts. Most societies, and indeed individuals within a society, use a "multiple choice" approach. They mix and match concepts of land to suit their changing lifestyles or needs at the moment. What is fanciful according to one set of norms is real and actualized by another. The variety of concepts of land is unlimited and ever changing (figure 2.1).

The challenges for LAS designers involve understanding the array of concepts of land used in a particular society, selecting those that work most successfully according to identified land

Figure 2.1 The concept of land, as in this Alaskan landscape—a spiritual place, natural resource, environmental wonder, physical space, consumable good—is in the eye of the beholder.

policies, and institutionalizing these concepts. Difficulties arise because commodities are not the only abstractions involved in land. Cultural and spiritual meanings of land are also vital. The modern trend in LAS design is to respect the cultural origins, as well as the colonial experiences, of nations to ensure that the administrative processes match the ways people think and their plans for the future (Bromley 2006). Given their market orientation, the prototype tools used in modern LAS (particularly registers of rights and cadastres) focus on property rights. LAS in non-Western countries, and those that serve groups that do not rely on land markets, need to reflect different cognitive aspects, particularly the spiritual, ancestral, and social meanings of land. The prototype tools need to be universally adapted to incorporate the remarkable variety of ways people think about and act in relation to land.

MANAGING THE EVOLUTION OF CONCEPTS OF LAND

Synchronizing a land administrative system with its cognitive impact is one of the most difficult and underexplored aspects of system design. To improve civil governance through land administration, the system needs to reinforce the cognitive understanding shared by members of the group about land. The congruity between LAS and the ways people think about land is therefore a key component of successful systems. While internalization of LAS by their beneficiaries can be relatively assumed in successful market-based systems, nonmarket situations expose enormous discontinuities. In developing and transitional countries, this mutual understanding needs to be carefully established. In this way, the "people" components of LAS design are fundamental to its success and sustainability.

TABLE 2.1 – ARRAY OF CONCEPTS OF LAND

Land as terra firma	The ground on which we live
	Natural resources—everything living, except people, including wild animals and plants
	Broad meaning—nature and its manifestations, including air, water bodies, soil, and subsoil
Land as physical space	The surface and area upon which life takes place
	Fixed in quantity
	Cannot be destroyed or increased
	Includes entire surface of Earth: oceans, mountains, valleys, and plains
	Includes cubic space: airspace, subsurface space, and associated minerals and gases
	Units of land—in regions or spatial entities, ranging from single parcels to suburbs, countries, up to and including the entire planet
Land as a deity (spiritual)	The source of all life and sustainer of all life
	By extension, the fountain of fertility and the final resting place of every person, therefore the abode of ancestral spirits
	A deity that possesses itself and owns everybody and everything, and exercises certain controls over people who use it
Land as a community	The natural ecological community for which individuals have special rights and responsibilities
	The group of individuals living in a particular area with common interests associated with their individual and collective good
	Concepts of "home," and "fatherland" and "motherland"
	Land as a location or situation
	Location with respect to land markets, geographic features, other resources, and given names for identification
	Significance of place in determining value and use on the basis of location, accessibility, strategic importance, and so on
Land as a property institution	An institution articulating private rights to own land as a basis for trading, established and sanctioned by a society
	Property held by the state on behalf of the people in centralized economies
Land as a factor of production	As a factor in economics, along with labor, capital, and management as factors of production
	As a "nature given" source of food, fiber, building materials, minerals, energy resources, and other raw materials used by society
Land as capital	In classical economics, land is a durable "free gift of nature" and capital is expendable past savings, the stored-up production of people
	Sometimes, land is regarded as capital itself because of the ability to raise capital funds using land as collateral

Continued on facing page

Continued from previous page

TABLE 2.1 – ARRAY OF CONCEPTS OF LAND	
Land as a consumption good	A consumer good produced by human enterprise Parks and recreation sites, developed building lots, a factor of production
Land as a commodity	A formative commodity in simple land markets "Unbundled" land—the new concept of extending commercial opportunities in land, unlimited by spatial parameters, multiplying interests out of land as separate tradable commodities—for example, water, minerals, and complex commodities A system of wealth acceleration and economic growth
Land as a human right	Exhortative claims for rights in land are fundamental political tools. The formative claim is the Universal Declaration of Human Rights, Article 17: "Everyone has the right to own property ...." (UN 1948) National constitutions frequently transform the exhortations into legal rights
Land as nature	Natural environments, features associated with the workings of nature without human effort Access to sunlight, rainfall, wind, climatic conditions, soils, topography, and so on Comparative qualities and quantities of natural resources such as mineral deposits, forests, water, fish, sunlight, rainfall
Land as a resource	A means of support or provision The sum total of the natural and man-made resources over which possession of land gives control A means of support, source of wealth, power, status, and revenue Includes human improvements attached to land
Land as environment	A place requiring management to preserve its capacity to sustain life, carrying restrictions and responsibilities

Land tenure is the generic concept used in land administration theory to explain how people approach and think about land. Though tenures fall into various types or general classifications, each one is unique. What is treated as land in a particular tenure varies among nations, and among communities within those nations. Land for a community might be just its surface (Indonesia), with buildings held in separate ownership (Russia), a cube of airspace as in condominiums in many cities, grazing opportunities of the colonized Masai of east Africa, products of forests, or the European, and now global, norm of everything above and below the surface, with alternative ownership for various deposits of minerals and petroleum (figure 2.2). Land tenure can encompass just about any arrangement of land that humans are capable of

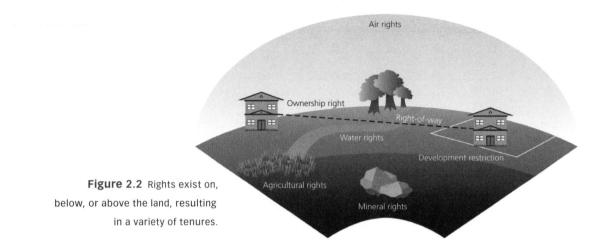

Figure 2.2 Rights exist on, below, or above the land, resulting in a variety of tenures.

creating. Only when a highly respected, confidently run administrative system manages these tenures can they be understood and managed by the people using them. Countries relying on social tenures develop elaborate systems of entitlement and management, frequently passing these on to future generations through ceremonies and ancestral song lines. Countries enjoying vibrant land markets use a theory of property to create commodities in spheres of space and related sets of opportunities. Property theory is also applied to commoditize interests in resources, separately from land. Mining, forestry, petroleum, fishing, and other interests are commoditized, and typically administered in systems separate from LAS.

In Western systems, individual property rights underpin a great deal of LAS design theory. The cultural concepts of private ownership, and the tenures they entail, are assumed in the technical solutions that focus on individually owned parcels. However, these Western approaches, and the economic analysis of land that supports them, do not exhaust the capacity of LAS design, or use of the tools selected for implementation. Modern LAS are sufficiently flexible to incorporate land held in social, informal, and transitional tenures. How this is done depends on the local experiences and responses to immediate challenges. Virtually each successful democracy with a thriving land market manages a wide array of tenure types, including social tenures, such as Maori titles in New Zealand, Aboriginal land rights in Australia, Inuit rights in Canada, Indian rights in the United States, and Lap rights in Finland and Sweden. They are also capable of including non-parcel-related land information—for example, restrictions on noise emissions according to time and decibel levels, such as in Australian environmental protection systems.

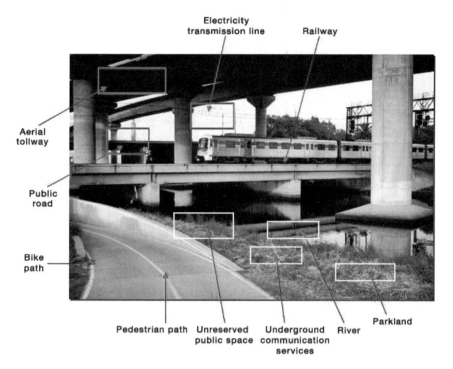

Figure 2.3 Modern urban infrastructure in Melbourne, Australia, is complex, containing many layers of titles for different uses.

Improvements in the capacity to incorporate non-parcel-based tenure systems in LAS are one of the achievements of sustainable development policy and new technical tools that have become available. The inclusion of all people-to-land relationships within national LAS is the theme behind new betterment paths, incremental improvements in security of tenures, and inclusion of indigenous land managed under multiple, competing normative systems. New cadastral tools, such as the Social Tenure Cadastral Domain Model (Augustinus, Lemmen, and Van Oosterom 2006), and land identification opportunities by Global Positioning Systems (GPS) and spatially enabled systems assist these broader, non-parcel-based approaches. The Global Land Tools Network (GLTN) of UN–HABITAT was formed in 2005 to coordinate these activities.

Modern cities not only change the way we live, they change our concept of land. Examples include the aerial walkways in downtown Minneapolis–St. Paul, Minnesota, and the underground pedestrian tunnels in Toronto, Ontario. Demands for space in major cities generate complex multiple uses in messy horizontal configurations. Figure 2.3 contains many visible layers of horizontal titles variously laid out to differentiate the railway, river, bike and pedestrian paths, road, parkland, unreserved public land, and a major aerial freeway. What is below the ground adds an additional order of complexity. Retrieval of space over highways and freeways or along river banks for high-density multiple uses and complex installation of services below ground are common in urban areas. These processes stimulate changes to the concept of land, development of new vocabularies of property rights, and changes in LAS. Thus, the most successful LAS integrate flexible approaches to registration and cadastres to create an administrative infrastructure for management of large-scale assets, such as utility supply lines (the Netherlands). Overlaying the pipes and wire information with cadastral parcel and building outlines in digital viewing systems allows for flexible approaches to land information management.

The predominant feature of modern cities is the high-rise building. The search for technical solutions to the digital representation of the third dimension—height—in a 3D cadastre (Stoter 2004) is part of the challenge of building modern LAS capable of reflecting these new ways of looking at, thinking about, and using land. Even without the convenience of the third dimension in a national cadastre, LAS are used to deliver development opportunities associated with high-density land use such as those in modern Bangkok, Thailand (figure 2.4). The utility of these developments is substantially enhanced if LAS can provide tenure security along with enhanced development opportunities.

EARLY LAND ADMINISTRATION TOOLS

Land administration initiatives must reflect the remarkable variety of approaches that people take to land. The need for administration starts with a degree of stability in the people-to-land relationship, associated with a basic form of territoriality.

> *"Territoriality is the primary expression of social power. Its changing function helps us to understand the historical relationship between society and space. … Perhaps, throughout history, one of the strongest drivers for territoriality and associated expansionist claims is the desire for commercial growth." (Grant 1997)*

In the early stages of human settlement, territorial sovereignty allowed land to become the undisputed primary source of wealth and power. Organization was essential. The utility and durability of land-use maps made them popular organization tools and ensured their place in human history and land administration. The universality of using maps to show how a community arranged its land comes from the capacity of pictures to tell a thousand words and the neutrality of their "language." As social and commercial organizations became more complex, records became more formal. They eventually provided some security of ownership since they were legally constructed, publicly acknowledged, and widely respected. By contrast, societies

Figure 2.4 High-density cities such as Bangkok, Thailand, require multiple land uses.

that did not develop the capacity to make durable records relied on the oral transfer of land information and ceremonial allocation. Among informally organized peoples, occupation of land had to take place in the presence of the chief and elders (Larsson 1996). These systems were just as complicated as their more formal cousins.

Examples remain today of very early maps recorded on the walls of caves, but portable maps also evolved, the first carved on small stones, then recorded on parchment and paper. The history of these maps tells us a great deal about how people related to land over time. The relationships between people and land, maps of these arrangements, and LAS in general are all interrelated. Change to one aspect induces changes to the others.

Recording of land arrangements to protect ownership, and to tax land holdings and produce, has a long history. Documentation of ownership and taxation of land use remain basic functions of land administration. Ancient records show that a practice of recognizing individual or family ownership of land is as old as taxing landownership and use. The earliest records of

Figure 2.5 Land surveyors are shown on the Tomb of Menna in ancient Egypt, circa 1500 BCE.

landownership date back to the Royal Registry of ancient Egypt (figure 2.5), which was created in about 3000 BCE. In China in 700 CE, the taxation system was based on crop yields and land survey records. The Romans carried out a survey in 300 CE to create a register of the lands they controlled and to use the records as a basis for tax collection (Larsson 1996; Steudler 2004, 7–10).

Registers of land holdings were also used to organize feudal systems of tenure. The European feudal system was extended to England by the conquest of the Normans in 1066. Power in the feudal system was vested in the institutional and legal structures that were put in place by the combined interests of landholders and the sovereign (Davies and Fouracre 1995). The system required dues to flow from the serfs or laborers, through lords, to the king. All land was owned directly or indirectly by the king, who granted use of these lands to his subjects (and their heirs) in return for their rendering of military or other services.

The Domesday Book was created in 1086 to record assets according to a landowner's name, tenure, area, and particulars for assessment of the land for the purposes of extracting feudal dues. The result was one of the earliest attempts to create a national inventory for fiscal purposes and to record the territory of the kingdom. There were no maps in the register, suggesting the beginnings of the English reliance on metes and bounds descriptions to describe and identify boundaries, rather than the European approach of cadastral surveying and mapping.

While land taxation remained an imperative for many countries, the European approach that relied on a primary tool, the cadastre, became the predominant model in the history of land administration.

DEVELOPMENT OF THE EUROPEAN-STYLE CADASTRE

Land administration evolved over time in response to changes in people-to-land relationships. Western land administration trends, shown in figure 2.6, followed changes in society, reflecting increasingly complicated attitudes toward land, first as personal security, then as property and wealth, and ultimately as a scarce community resource for environmental protection and sustainable development.

Western LAS developed in four general stages through which the cadastral tools matured into the modern multipurpose cadastre. The symbiotic relationship between people-to-land relationships in Europe and the broad design of LAS through these four stages is shown figuratively in figure 2.7. The four stages of LAS and cadastral development are

◆ The cadastre as a fiscal tool

◆ The cadastre as a land market tool

◆ The cadastre as a planning tool

◆ The cadastre as a land management tool—the multipurpose cadastre

At each stage, additional functions were added to the cadastre, until it transformed from a mere land administration tool into a fundamental layer of spatial information for sustainable

Ting, L., and I. P. Williamson. 1999. Cadastral Trends: A Synthesis. *The Australian Surveyor*, 4(1) 46–54, used with permission.

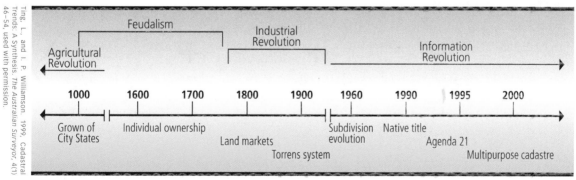

NOT TO SCALE

Figure 2.6 European LAS has evolved from the days of feudalism through the Industrial Revolution to the "information revolution."

development. The stages are worth examining in detail to explain the relevance of European history and to illustrate how the stages might be condensed to permit introduction of a multipurpose cadastre to a non-European country.

The cadastre as a fiscal tool: The first European approaches to mapping were driven by fiscal imperatives. The Swedish Land Survey of the early seventeenth century relied on maps (Larsson 1991). In the eighteenth century, mapping was used to support taxation in parts of northern Italy and the Austro-Hungarian Empire. The Teresian Cadastre (named after Empress Maria Teresa) provided durable and extensive land information as the basis for taxing the nobility. Mapping became more common after 1807 when Napoleon Bonaparte established the foundation of the European-style cadastre. He ordered the creation of maps and cadastral records of 100 million parcels in the Napoleonic Empire. Differences between the older maps and these records lie in the use of scientific measurements, systematic marking of individual parcels of land, and the diagrammatic representation of the results of these processes. During the Napoleonic era, particular entities were given the task of registering transfers and deeds of ownership. The records showed the physical location of parcels of land as accurately as techniques at the time would allow, as well as landownership across France, arranged by parcel numbers, area, land use, and land values per owner.

The efficiency of the taxation system ensured its spread throughout Europe so that state treasuries could rely on revenues generated from taxing particular uses of land. Various means of calculating the value of land according to products, production capacity, and soil types were

Ting, L., and I. P. Williamson. 1999. Cadastral Trends: A Synthesis. *The Australian Surveyor*, 4(1) 46–54, used with permission.

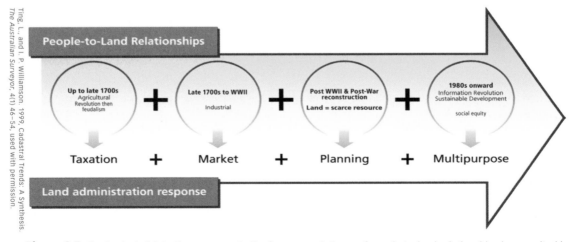

Figure 2.7 The land administration response to the four general stages of people-to-land relationships has resulted in the need for a multipurpose cadastre to be used in LAS.

used. These fiscally driven combinations of registry records or text information and scientifically prepared maps or cadastres laid the foundation for modern cadastral systems (figure 2.8). The combination in various forms remains fundamental to LAS.

The cadastre as a land market tool: The growing reliance on private landownership in Europe changed the role of the cadastre. The relationship between recording information relating to land and the institution of private property is expressed in the formal processes of land registration that identify private interests in association with a cadastre. The legal functions of the cadastre eventually became more important than its fiscal functions, though both functions require accuracy and reliability in the record base. Countries with a well-established fiscal cadastre, such as France, developed separate deeds registration programs. The German approach took cadastral records further and developed land registration rather than deeds registration. This approach to registration delivered registration of land or titles, not deeds, and used the concept of a "Grundbuch" (land book) in which each page recorded ownership of a parcel on the principle of a folio (Steudler 2004, 10). The folio was given a unique number and (ideally) contained all the information about the parcel.

England's common-law approach was different yet again. The English method of describing land continued to rely on text or word descriptions of metes and bounds for more than a

The cadastral system in Denmark

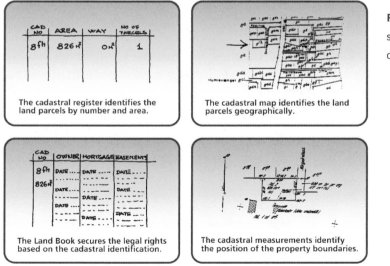

The cadastral register identifies the land parcels by number and area.

The cadastral map identifies the land parcels geographically.

The Land Book secures the legal rights based on the cadastral identification.

The cadastral measurements identify the position of the property boundaries.

Figure 2.8 Denmark's cadastral system illustrates the use of cadastral records and maps.

thousand years, in contrast with what happened elsewhere in Europe where maps, and formal cadastres, were developed based on boundary surveys.

While England lacked the taxation model that drove creation of well-organized cadastres across the Channel, it developed similar property theory and administrative systems to facilitate the growth of land markets. The Industrial Revolution came at a time of agricultural change as well as industrial invention (Ting et al. 1998). Significant land management changes led to improved productivity as the enclosure movement of the 1700s spread across Europe and England to create larger and more productive plots. (Enclosure was the controversial process of taking common lands used for traditional purposes, such as communal farming, grazing, hunting, and access to timber and other resources, and fencing the lands to be placed in private ownership. The enclosure movement incorporated long narrow strips of commonly farmed land into more productive parcels farmed by individual landowners—figure 2.9). In England, for example, about 7 million acres of land was enclosed between 1760 and 1845; this land was made more productive by mixed agriculture, including crop rotation and alternating arable and pasture uses (Toynbee 1884). Land was important in itself but became even more so as a source of capital that could facilitate mobility and investment. The land administration and property law systems, which were designed to preserve attachment of the aristocracy to land in perpetuity, became too cumbersome and unwieldy. A variety of methods were used to overcome these limitations and redistribute land by reinterpreting existing instruments, including the collapsing feudal tenures; introducing more flexible interests in the Statute of Uses; and overriding strict settlements via the Settled Land Act of 1882.

Deeds of ownership evolved to prove ownership, so that the owner, as opposed to others, could remain on the land, and also to prove title and capacity to deal with the land in market-based transactions. Deeds registry records, in addition to the deeds, proved ownership and established the confidence necessary among strangers to facilitate marketing of land, through sale, lease, and mortgage. Overall, the systems of deed titling and recording were still cumbersome when compared with the more streamlined land title systems that developed in Europe. The English colonies, including the colonial United States, adopted similar deeds conveyancing and registration systems to support their land markets.

An incremental improvement in systems design arrived from the unlikely source of South Australia, an infant British colony settled in 1836, which decided to eliminate lawyers from the conveyancing process by introducing a German-style rigorous and simple land registration system (Raff 2003). This was called the Torrens system after its major parliamentary proponent, Robert Torrens, who, after three tries, achieved passage of the formative legislation in 1858.

The Torrens system suited nineteenth-century paradigms in a young country like Australia with its large tracts of unsurveyed and untitled land. Its simplicity of having a government guarantee as well as showing the description of the parcel, the registered proprietor, and any encumbrances (i.e., mortgages) on one piece of paper (see "Registration system tools" in section 12.3 and figures 12.8 and 12.9 for detail) encouraged its adoption in other countries where the needs and history were quite different. Essentially, it spread through the Commonwealth of Nations but not in America, though some twenty states instituted small-scale versions (ten of which are still in use, in Hawaii, Massachusetts, Minnesota, New York, Colorado, Georgia, North Carolina, Ohio, Virginia, and Washington). Its introduction constituted a profound legal change in response to social and political needs that generated even greater changes in land markets and land administration, including surveying methods. The Torrens system was revolutionary in its ability to deliver certainty, along with cheaper and speedier land registration. It replaced the preexisting deeds conveyance method, where lawyers and notaries traced the title through all documents relating to historical transactions back as far as necessary to determine whether a good title was passed through each time. The Torrens system created a land register and a very rudimentary cadastre, usually based on a charting map where parcels were approximately plotted, that used the most recent entries to unequivocally describe not only all parcels of land, but also the people who held important interests. The Torrens-based LAS, together with the government guarantee of accuracy of ownership information, greatly assured the desires of a colonial society for rapid settlement of a vast land with vigorous land markets. Its applicability in developing countries is, however, problematic, largely because its successful operation presupposes capacity to deliver good governance.

The cadastre as a planning tool: The post–World War II reconstruction period and the subsequent population boom stimulated better spatial planning, particularly in urban areas. Land administration laws and systems increasingly needed to manage broad subdivisions. The growth of urban satellite cities with high-density housing, and increasing pressure on infrastructure by the sheer numbers of the urban population, necessitated better urban planning. Regulation of land use in a community involves more than the recognition of spillover effects on contiguous land; other objectives include provision of public amenities that are unlikely to be privately produced and increased efficiency by guiding development and redevelopment of land for desirable purposes (Courtney 1983). The cadastre, as the record of land parcels and registry of ownership, became a useful tool (when teamed with large-scale maps) for city planning and the delivery of vital services such as electricity, water, sewerage, and so on. Thus, a focus on planning was added to the preexisting applications of the cadastre as a fiscal and land market tool.

The cadastre as a land management tool: The 1980s saw a different twist in concerns about land scarcity. The focus turned to wider issues of environmental degradation and sustainable development, as well as social equity in land distribution. These issues brought new considerations into the economic paradigm, moving it from a short-term focus to a broader framework. Planning issues widened to include more community interests and deepened to address the need for comprehensive information about the impact of land uses on neighboring environments. The demand for more complex information was assisted by technical developments in GIS and satellite monitoring. In LAS, the multipurpose cadastre arrived (McLaughlin 1975). For example, the solution to problems faced by low-value agricultural lands in New South Wales, Australia, included sustainable land use, comprehensive integrated datasets to allow for better decision making, simplified cost-effective operation of the cadastre, and clearly defined, easily relocatable parcel boundaries supported by an appropriate low-cost cadastral survey system (Harcombe and Williamson 1998).

Similar multipurpose approaches, using different tools, appeared throughout Western European countries. The development of the Danish system described in figure 2.9 shows a typical development of the European-style cadastre and its historical relationship with the enclosure movement.

INTERNATIONAL ADOPTION OF THE MULTIPURPOSE CADASTRE

The changes in people-to-land relationships, especially the commoditization of land, gave much greater significance to the role of the cadastre in land administration theory, especially because the tool became synonymous with best practice. Generic definitions of a cadastre were therefore needed. In 1980, the U.S. National Research Council (NRC) published a study, "The need for a multipurpose cadastre," which integrated mapping and cadastral survey functions, using the geodetic reference framework throughout the record base, illustrated in figure 2.10. This realization of the importance of a well-defined and effective cadastral system capable of underpinning administration of government in multiple areas, especially in land tenure and value records, began a new era in the discipline of land administration. Now, the issues revolve around how to define and build multipurpose cadastres, rather than on why they should be built.

The NRC study and development in the Maritime Provinces in Canada (McLaughlin 1975) also broke new ground by establishing the role of a vision in land administration theory. Though a distant reality in 1980, the vision of the multipurpose cadastre directed and harmonized efforts to modernize well-established, and even rigid, approaches to surveying and institutional

EVOLUTION OF THE DANISH CADASTRE

Established in 1844, the Danish cadastre was designed to assist collection of land taxes from agricultural holdings based on a valuation of the quality of the soil.

As a result of the enclosure movement in the late 1700s, the former feudalistic society was changed into a market-based society with private landownership. The necessary maps were surveyed by plane table at a scale of 1:4,000. The resulting property framework from the enclosure movement formed the basis for the cadastral maps established in the early 1800s. Each map normally includes a village area and the surrounding cultivated areas. As a result, the maps are "island maps" and not based on any local or national grid.

The parcels within each village area were numbered and recorded in the cadastral register showing the parcel areas, parcel numbers, and the valuation based on the quality of the soil. The present cadastral framework is still based on these historic village areas.

From the beginning, the cadastre consisted of two parts: the cadastral register and the cadastral maps. Both of these components have been updated continually ever since. The land registry system was established in 1845 at the local district courts for recording and protecting legal rights of ownership, mortgage, and easements, based on the cadastral identification.

In the late 1800s, the Danish cadastre changed from a fiscal cadastre primarily used as a basis for land valuation and taxation to a legal cadastre supporting a growing land market. This evolution

Continued on next page

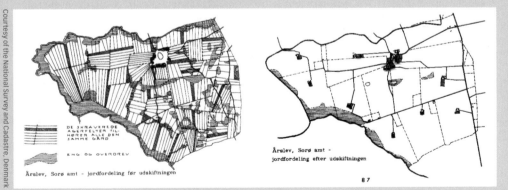

Courtesy of the National Survey and Cadastre, Denmark

Figure 2.9 Development of the Danish cadastre is based on the enclosure movement of the 1700s. The map to the left shows common farming in the village area. The map to the right shows the new agricultural holdings that came about as a result of the enclosure movement. These structures can still be found in today's topographic maps.

EVOLUTION OF THE DANISH CADASTRE *Continued from previous page*

was completed in the first years of the 1900s when taxation was based on market value. Simultaneously, in the 1920s, a new land book system of registration was established. This system of title registration was based on cadastral identification, and a close interaction between the two systems was established.

During the first half of the 1900s, land was increasingly seen in Denmark as a commodity, with a focus on agricultural production and the Industrial Revolution. Land-use regulations based on cadastral information were introduced to simultaneously improve agricultural productivity and sustain the social living conditions in rural areas. In fact, the old yielding valuation unit was used to control development in the rural areas until the late 1960s.

The 1960s introduced a close interaction between the cadastral process (e.g., subdivision) and land-use regulations. Property formation or change of property boundaries required documentation showing the approval of the future land use according to relevant planning regulations and land-use laws. Land was increasingly seen as a scarce community resource, and zoning and planning regulations were introduced to control land development. Environmental concerns appeared in the late 1970s and became a major issue in Denmark. Today, comprehensive planning and environmental protection are seen as the main tools to secure sustainable development.

New land administration infrastructure based on modern IT opportunities evolved to support these processes of sustainable land management. The cadastral register and cadastral maps are now computerized and form a basic layer for management of all land rights, restrictions, and responsibilities (RRRs). The development of the digital cadastral database is presented in figure 12.21.

arrangements. The vision was so idealistic that it proved to be practically impossible to implement (Cowen and Craig 2003). The NRC vision reflected the situation in its home base, the United States, in 1980, rather than Europe, because it downplayed the land registration functions that underpin cadastral organizations in most Continental countries.

In due course, the modern, generic European model of the multipurpose cadastre was articulated by the International Federation of Surveyors (FIG) in 1995 and similarly focused on land information as the main deliverable or outcome:

> "A parcel-based and up-to-date **land information system** containing a record of
> interests in land (e.g., rights, restrictions, and responsibilities). It usually includes a
> geometric description of land parcels linked to other records describing the nature of

the interests, the ownership or control of those interests, and often the value of the
parcel and its improvements. It may be established for fiscal purposes (e.g., valuation
and equitable taxation), legal purposes (conveyancing), to assist in the management of
land and land use (e.g., for planning and other administrative purposes), and enables
sustainable development and environmental protection." (Emphasis added)

Describing what a cadastre looks like (figure 2.11) in a way that covers its many local versions is much harder than actually building one. The institutional arrangements of a country are highly influential in the design of its local cadastre. In both Australia and Europe, cadastral systems are now closely linked with land-valuation systems. Generally, European cadastral systems originally supported land valuation for taxation purposes, with links to registration systems coming later. In Australia, the reverse was usually the case. Despite the different historical paths, however, the end results closely relate land registration and land valuation and are very similar (Williamson 1985).

In North America, while multipurpose concepts continued to evolve in Canada, the parcel-based cadastre diminished in importance in the United States, despite the NRC's 1980 study. Institutional arrangements, in particular, the control of the land parcel data layer by more than 3,232 counties, lack of federal capacity and interest; and divergence among state approaches were the underlying factors. The federal agencies that relied on parcel-based information, especially the U.S. Bureau of Statistics and Department of Homeland Security, needed to use alternative spatial databases, at huge cost to taxpayers (Cowen and Craig 2003). Similar duplication issues existed in Australia, though there, national approaches have recently

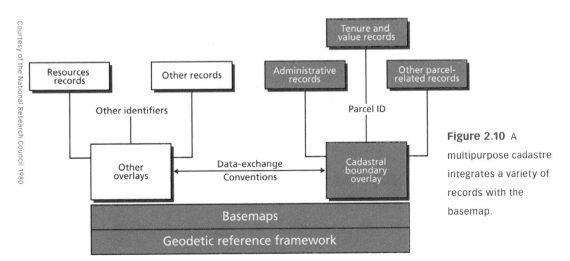

Courtesy of the National Research Council 1980

Figure 2.10 A multipurpose cadastre integrates a variety of records with the basemap.

successfully integrated parcel information and addresses through a geocoded national address file (GNAF). The need for a national approach to cadastral issues in the United States is still recognized and was the subject of a two-year NRC study titled "Land parcel databases: A national vision" (National Research Council 2007).

These historical developments clearly demonstrate the success of multipurpose cadastres as a fundamental land management tool whose acceptance (or an equivalent large-scale parcel map of some kind—see figures 5.3, 12.19, and 12.21) is now almost universal, except in countries where private land registries are still used. Italy, Spain, Greece, Portugal, and most Latin American countries still do not use national cadastres. In some developing countries where surveying and other technical skills are lacking, cadastral construction relies on other options, including aerial photos and satellite images, or sometimes, hand-drawn, isolated parcel drawings combined into composite area maps.

Inevitably, different countries are at different stages on the evolutionary cadastral continuum, reflecting national social, institutional, legal, and economic circumstances. However, common

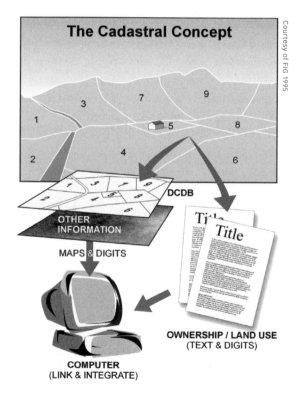

Figure 2.11 The cadastral concept as adopted by the International Federation of Surveyors makes land information the primary deliverable in the modern European model.

Courtesy of FIG 1995

principles or essential elements of a modern cadastre are identified and underpin the design all cadastral systems, whether they originate in a deeds or title registration system or as stand-alone tools. In summary, common cadastral elements include

- ◆ A complete cadastre or cadastral map showing all land parcels in a jurisdiction, irrespective of ownership
- ◆ A register or series of registers listing information about the land parcels
- ◆ A unique identifier for each parcel that links the parcel to the record(s) in the register(s)
- ◆ Dynamism (both in the maps and registers) and capacity for continuous updating
- ◆ High reliability of information in both the maps and registers, preferably supported by some legal sanction or government guarantee
- ◆ Public access to the cadastre
- ◆ Inclusion of the large-scale cadastral mapping system into a wider mapping system for a state or country, using the same control network
- ◆ Support for the spatial integrity of the cadastral mapping system by a cadastral survey system that ensures an unambiguous definition of the parcel both on the map and on the ground
- ◆ Access to and visibility of land information through information and communications technology (ICT) tools

A set of cadastral principles was proposed for Spain, Portugal, and Latin American countries in the Declaration on Cadastre in Latin America, presented at the Permanent Committee on Cadastres in Latin America in 2006 for acceptance by each member nation (EU 2006). Construction of cadastres for these countries is driven by unique factors, including management of agricultural activity and provision of infrastructure; problems in construction are also being confronted (Erba 2004). This declaration announced that the cadastre is the responsibility of government and cannot be privately owned, following the European model of building the cadastre as government infrastructure. In the Mediterranean countries and Latin America, land administration processes are predominantly undertaken by specialized professionals and their small businesses outside government, according to commercial imperatives that impede construction of the expensive infrastructure of national LAS.

The most important modern influence on the design and utility of cadastres is their capacity to support land management for sustainable development. Because they represent the ways people actually use land and present this information in large scale, they form the core layer of SDI. When cadastres contain geocoded data and are held in digital form, the information they contain becomes useful for agencies other than the cadastral and registration agencies that maintain it. The information provides reliable and authoritative data about the identity, ownership, and uses of land in a country and becomes truly multipurpose (see chapter 9, "SDIs and technology"). The potential uses of cadastres of this kind go far beyond government administration.

While most cadastral systems can be measured against these principles, elements, and emerging trends, the best way to understand any particular system from the perspective of reforming or improving it is to examine its operations and processes (see chapter 4, "Land administration processes"). This is because the design of any national cadastre necessarily reflects its local history and capacity. Two aspects of history are important to LAS design: the original legal tradition of a country and its colonial experience.

2.2 Historical evolution

TRADITIONS AND SOURCES

While an effective cadastre is regarded as essential to modern LAS (Bogaerts, Williamson, and Fendel 2002; FIG 1996), the local design will reflect national history, especially a country's political and legal nuances. A broad anthropological view of cultural origins identifies six major legal traditions: Islamic, traditional, Talmudic, civil law, common law, and Asian (Glenn 2004). Each legal tradition brings its own approaches to land issues and to the concept of land. Colonization spread different legal systems throughout the world, shown graphically and indicatively in figure 2.12, each of which approached land administration design in different ways. The colonial experience of each area varied according to the absorption of the colonial framework amid the original legal traditions of the local people. Land administration was often a point of contention between imposed and original systems. However, some generalities are valid. As countries built LAS capable of supporting land markets, these different legal traditions and colonial experiences affected the design of cadastral models and land registration systems. Countries using the European or German approach and the Torrens title approach (except where it is used in parts of the United States) tended to merge cadastral and registration functions. Countries with socialist and Mediterranean influences did not (see chapter 5, "Modern land administration theory").

The influence of the legal origin is particularly evident in the relationship between a country's cadastre and its registration system, and in the type of its registration system. Basically, two types of registration system evolved: the deeds system and the title system. The differences between the two relate to the extent of involvement of the state and the cultural development and juridical setting of the country. The key difference is whether the transaction alone is recorded (deeds system) or the title itself is recorded and secured (title system). Deeds systems provide a register of owners, focusing on "who owns what," while title systems register properties representing "what is owned by whom." The cultural and juridical aspects relate to whether a country is based on Roman law (deeds systems), or Germanic or Anglo common law (title systems).

Deeds registration is rooted in Roman culture and is, therefore, common in Latin cultures in Europe (France, Spain, Italy, and Benelux—Belgium, the Netherlands, and Luxembourg), in South America, and parts of Asia and Africa that were influenced by these cultures. The concept is also used in most of the United States but was derived from English deeds conveyancing. Deeds registration systems in the United States are now diversified, locally managed, and supported by private title insurance. In the Eastern United States, deeds registration is

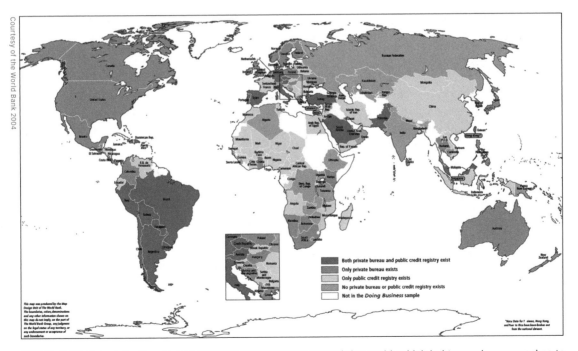

Figure 2.12 Colonialization spread different legal systems around the world, which led to varying approaches to land issues.

sometimes intermixed with a Torrens-style system. Internationally, deeds systems are found in different forms, with significant variations in the roles of cadastral identification and surveyors.

Title registration originated in the German culture and is found in central European countries (Germany, Austria, and Switzerland). Different versions of the German system are found in Eastern European and Nordic countries. The various versions relate to the use of the concept of property and the organization of the cadastral process, including the use and role of private licensed surveyors. A second variant, based on the original German concept (Raff 2003), is found in the Torrens system introduced in Australia during the mid–1800s to serve the need of securing land rights in the New World. The popularity of the Torrens system increased so that it was eventually considered best practice and spread to many jurisdictions in Asia; the Pacific; North America, particularly in Canada; Africa; and even South America. The United Kingdom replaced its deeds conveyancing system with a unique version of a title system, in which the concept of general boundaries is used to identify the land parcels on a large-scale topographic map series created through the Ordnance Survey, the national mapping agency of the United Kingdom. In reality, the systems with German, Torrens, and English origins have more in common from a title registration perspective than they have differences. For example, most if not all title registration systems accommodate general boundaries and require registration to complete legal assignment of land.

The land registration systems in each country (and indeed in most states, territories, or provinces of federated nations) are unique in detail. Even in Australia, the Torrens systems in the eight states and territories differ significantly. The sources of different land registration systems worldwide are shown indicatively in figure 2.13.

Within each unique system, deeds and title systems share general characteristics as well as differences, as shown more fully in table 12.3. While these characteristics are not definitive, they provide guidance as to the understanding of the foundation of the two systems.

HISTORICAL VARIATIONS IN ENGLISH AND EUROPEAN APPROACHES TO LAS

Europe and England developed the two formative models of LAS that provide the historical roots of modern systems, creating what Hernando de Soto (2000) called "integrated legal property systems" that are capable of managing a nation's assets. Europe focused on the cadastre and attendant land registration systems, and England focused on a deeds-based system, which stood apart from any cadastral surveying. Comparatively, registries in England are relatively young and officially go back to the Middlesex (1708) and Yorkshire (1884) deeds registries and before that to less formal systems of depositing deeds in churches and court-houses. England relied on metes and bounds descriptions and general boundaries. Eventually, the Ordnance Survey filled the need for organized parcel mapping by default by showing all boundary occupations. The English deeds system became unworkable in the late nineteenth century, and in 1925, after twenty-five years of refinement, substantial law reforms were finally enacted. Though land registration had existed in a formative system since 1862, the 1925 UK Land Registration Act set up the familiar process of government-run, centrally organized land records. The streamlined conveyancing processes eventually replaced the old deeds conveyancing and registration system, and by 1990, all land transactions were compulsorily registered.

In Western European countries, registries and organized land records are much older, with the prototype established in the Hamburg–Hanseatic land title registration system (Raff 2003). The Hanseatic experience showed clearly how organized land administration contributed to stability, longevity of land arrangements, and economic wealth of a series of city-states over hundreds of years. Titles to land allowed substantial population mobility, without threat of loss of property. Registration freed up land use from feudal clutches. By 1840, this land title registration was "highly sophisticated (and) rationally integrated, reflecting centuries of experience" (Raff 2003).

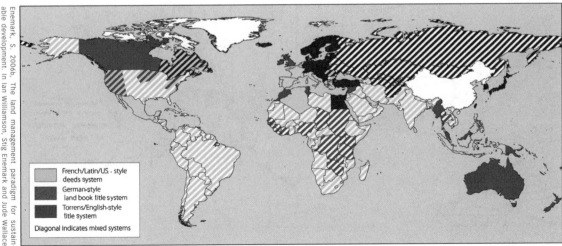

Enemark, S. 2006b. The land management paradigm for sustainable development. In Ian Williamson, Stig Enemark and Jude Wallace (eds.), Sustainability and Land Administration Systems. Geomatics, Melbourne, used with permission.

Figure 2.13 Land registration systems, though of three major types, are unique in detail across the world.

The simultaneous development of cadastres for taxation purposes, then for multiple purposes, also was important. Consequently, the Germans continue to be world leaders in land administration, along with other Western European nations, particularly the Danes (for cadastral development), the Dutch (for integrated systems), and the Swedes (for pioneering IT).

LAS IN THE USA

The English-style LAS took a different course in the United States, where the deeds system still persists. The federal nature of the government spread responsibilities for management of land records throughout the various states. A plethora of legal and administrative systems was created, generally reflecting the pre-Revolution, post-Revolution, antebellum, and postbellum settlement and the French heritage of particular states. The major land identification tool was a formal surveying of land according to national baselines and grids, independent of natural

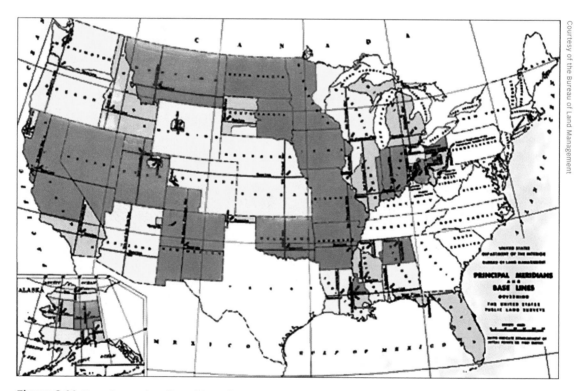

Figure 2.14 Formal surveying allowed the United States to develop a set of meridians and baselines that allowed identifying parcels, sight unseen.

topography. Figure 2.14 shows the historical development of U.S. meridians and baselines for the territory west of the Mississippi and Ohio rivers. In a sense, this allowed the United States to roll out an artificial cadastre that allowed it to identify parcels and subparcels, sight unseen, within the townships (rectangles) defined by the meridians and baselines, thereby allowing the U.S. government to grant land and open up the country to the west of the Mississippi River.

The localization of these standards continues to produce discontinuities, illustrated by figure 2.15, which shows the grid surveying system in the Western United States.

The influence of ideology, including limited government and expanded individual freedoms, was apparent in the American land administration infrastructure, especially in its arrangement of tenures and property rights. De Soto (2000) describes the acquisition of property rights, stressing the mercantilism that is still apparent in the American concepts of allodial titles, featuring an unfettered landowner, and "hands off" government. America also pioneered the constriction of state involvement in land record management and developed private-sector title insurance to shore up the fallible deeds-based conveyancing system inherited from England. The land markets of the United States are sufficiently robust to absorb the overhead of running private insurance systems in addition to deeds registration. Compared with the European systems, the U.S. tools of separate insurance, disbursed registration and cadastral

Figure 2.15 The rectangular survey system is used across the American West.

agencies, and privatized risk management systems are not considered best practice for developing countries where capacity and governance issues predominate. Given its relative economic resources, the United States is capable of addressing the disadvantage of having no national cadastre through alternative, though highly expensive, multiple information systems, provided cooperation among agencies is possible. Nonetheless, efficiency of government and the need for an authoritative national parcel information system are driving change (National Research Council 1980, 2005). The emerging U.S. vision recognizes the land parcel database or cadastre as an essential part of a national SDI (Cowen and Craig 2003).

DUALITIES IN POSTCOLONIAL COUNTRIES

Whatever the historical source, the designs of modern systems converged over time. The differences between the English general boundaries system and the European cadastral models faded as both systems improved. Now, they both provide scientifically identified land parcels and comprehensive land registration, which provides government-protected titles, universal registration of all transactions, and well-managed institutions supported by highly trained professional cadres of lawyers, notaries, surveyors, and administrators. Likewise, those countries (except the United States) that still register deeds provide the degree of certainty associated with land registration programs at large.

Meanwhile, the postcolonial experiences of many countries involve difficult challenges. Conquest and colonialism in the nineteenth and twentieth centuries led to the variable spread of the institutional approaches used in England and Europe across the world. Among the variations in colonial situations, a general pattern emerged. Typically, "old country" formal approaches were used to manage colonists' land, leaving local populations to continue their unique land management practices. This normative dualism in colonized countries remains a problem today and is often compounded by divergent national, regional, and local approaches to land administration. Dualism, or more often, pluralism, is capable of undermining efforts to formalize land administration (Fitzpatrick 1997) unless the alternative normative structures are recognized and their implications taken into account at the LAS design stage (Lavigne Delville 2002b; Chauveau et al. 2006).

LAND ADMINISTRATION IN SILOS

Despite these diverse institutional and political histories, early land administration theory concentrated on support for land markets and land taxation by the establishment of formal methods of parcel identification, legalistic identification of interests in land, and administrative infrastructure surrounding these tasks. Within their local differences, up to World War II, and even beyond, the formal institutions engaged in land administration throughout the world had one thing in common: They were run as independent agencies, called "silos" or "stovepipes." Generally, there was no reason for a particular agency in a country to deal with other related agencies. Land taxation, valuation, registration, mapping, and surveying were conducted as if their partner activities did not exist, though in some exceptional situations, multiple activities were serviced through a single agency. These silo arrangements were challenged after World War II, particularly once computers were introduced, but they remain in many countries. The need to reorganize these institutional arrangements under one roof was sufficiently obvious to drive land administration theory to its next stage of development in which the cadastre forms the connecting link between the silo agencies and their internalized processes.

IMPORTANCE OF THE CADASTRE

The cadastre is only one part of LAS, but its significance is profound. However, the international experience in designing and building cadastres is so variable that it is the most difficult and complex component to explain. The components (shown in table 2.2) can be supplied in both paper-based and digital systems. Variations reflect the diverse patterns of legal traditions, colonial histories, and parcel registration systems, drawn from each country's respective historical, administrative, and legal contexts (Kain and Baigent 1992).

TABLE 2.2 – CADASTRAL COMPONENTS (AFTER UNECE, WPLA GLOSSARY 2005)	
Cadastre	A type of land information system that records land parcels. The term includes **Juridical cadastre:** a register of ownership of parcels of land **Fiscal cadastre:** a register of properties recording their value **Land-use cadastre:** a register of land use **Multipurpose cadastre:** a register including many attributes of land parcels
Cadastral index map	A map showing the legal property framework of all land within an area, including property boundaries, administrative boundaries, parcel identifiers, sometimes the area of each parcel, road reserves, and administrative names
Cadastral map	A (detailed or technical) map showing land parcel boundaries. Cadastral maps may also show buildings.
Cadastral surveying	The surveying and mapping of land parcel boundaries in support of a country's land administration, conveyancing, or land registration system

Within this variability, international experience suggests commonalities in the design and historical development of the "cadastral engines" of each national LAS, suggesting three basic approaches. These approaches are based on countries grouped according to their similar background and legal contexts (German style, Torrens/English approach, and French/Latin style). While each system has its own unique characteristics, most cadastres can be grouped under one of these three approaches (see section 5.2, "The cadastre as an engine of LAS"). Just as there are three different styles of land registration systems, these translate to three different roles that the cadastre plays in each system. Again, while the role of the cadastre and the land

TABLE 2.3 – GENERAL RELATIONSHIPS BETWEEN LAND REGISTRIES AND CADASTRES		
STYLE OF SYSTEM	**LAND REGISTRATION**	**CADASTRE**
French/Latin/ U.S. style	Deeds system Registration of the transaction Titles are not guaranteed Notaries, registrars, lawyers, and insurance companies (U.S.) hold central positions Ministry of justice Interest in the deed is described in a description of metes and bounds and sometimes a sketch, which is not necessarily the same as in the cadastre	Land taxation purposes Spatial reference or map is used for taxation purposes only. It does not necessarily involve surveyors. Cadastral registration is (normally) a follow-up process after land registration (if at all) Ministry of finance or a tax authority
German style	Title system Land book maintained at local district courts Titles based on the cadastral identification Registered titles guaranteed by the state Neither boundaries nor areas guaranteed	Land and property identification Fixed boundaries determined by cadastral surveys carried out by licensed surveyors or government officers Cadastral registration is prior to land registration. Ministry of environment or similar
Torrens/ English style	Title system Land records maintained at the land registration office Registered titles usually guaranteed as to ownership Neither boundaries nor areas guaranteed	Property identification is an annex to the title • Fixed boundaries determined by cadastral surveys carried out by licensed surveyors (Torrens) • English system uses general boundaries identified in large-scale topographic maps Cadastral registration integrated in the land registration process

registration styles are not definitive, table 2.3 describes the three approaches in general terms. A more detailed account is in table 12.3, "Differences among registration systems."

Despite the importance of the cadastre as a multipurpose and essential tool in LAS, its underlying benefits still are not fully realized. Cadastres hold data that is verified by scientific surveying processes and held on a large scale. Whether manual or digital, cadastres reflect the unique arrangements communities create with land and record the arrangements on cadastral maps using scales large enough to contain detail relevant to a multitude of purposes. Businesses of all kinds need reliable information on a large enough scale to organize their activities, and land-use planning requires specific, accurate, timely information. Postal authorities, utility suppliers, census collectors, emergency managers, risk analysts, insurers, and dozens of other industries use land information on this scale of detail. Sometimes, they use it to build even larger scale maps for asset management, especially in the land servicing industries that provide water, power, gas, communications, and so on.

Cadastral information is also reliable in the sense that it generally relies on surveyors to create, verify, and re-create both the descriptive data and positions of parcels on the ground. The representation of the parcel on the map is therefore verified, even in countries with inadequate professional skills, to the best possible standard. While most cadastres are regarded as "capable of being made more accurate," they still represent the on-the-ground configuration of land arrangements according to engineering standards that are not capable of being matched by data from other sources.

A multipurpose cadastre capable of forming the engine of LAS and an SDI remained a mere vision until computer systems developed sufficiently to offer an implementation path. When cadastres are digitized, they become even more important, because they are capable of forming the basic layer in an SDI that provides easily understood identification of each significant space or place. Because the parcel configuration is dynamic, a well-maintained cadastral map stays much more up-to-date than many other spatial datasets. The most important engineering feature of digital cadastres is their enduring vitality for countries that build them once, build them well, and use them many times over.

The digital reorganization of land information systems stimulated new theoretical responses, principally the identification of the SDI as the means of visualizing land in digital systems. Coordination of land and spatial information became a major research focus. The scope of spatial information is, however, much larger. It has closely followed the development of the land management paradigm since 2000.

Part 2

A new theory

Part 2 introduces land administration as a new theory developed since the 1980s. However, its key components of the cadastre, land registration, and surveying and mapping have been a part of civilizations for millennia. The changing role of property with regard to land is explained and its influence on the design of the next generation of land administration systems (LAS) is described in chapter 3. Importantly, Part 2 discusses and describes this new theory of land administration in depth—in particular, highlighting key declarations and statements of the United Nations and other organizations on the evolving role of LAS. The central role that LAS play in supporting sustainable development objectives is explained in chapter 3. The growth of restrictions and responsibilities related to land, and better understanding of rights in land, are described, so they can be incorporated in LAS.

Part 2 next introduces land administration processes in chapter 4 as the central activities in LAS. LAS are not static entities but revolve around central tenure processes common to most nations:

◆ Formally titling land

◆ Transferring land by agreements (buying, selling, mortgaging, and leasing)

◆ Transferring land by social events (death, birth, marriage, divorce, and exclusion and inclusion among the managing group)

◆ Forming new interests in the cadastre, generally new land parcels or properties (subdivision and consolidation)

◆ Determining boundaries

The theory proposes that the key to successfully reforming LAS is to improve management of processes. Examples of land administration processes from a range of jurisdictions are presented in chapter 4.

Part 2 concludes by introducing a modern theory of land administration in chapter 5 that focuses on its role in managing land and resources from a sustainable development perspective. Central to this perspective is the land management paradigm that considers the land administration functions of land tenure, land use, land value, and land development in the context of a land policy framework and a land information infrastructure within a particular country context. The paradigm explains how these land administration functions interact to deliver efficient land markets and effective land management that collectively contribute to sustainable development. Central to the paradigm is the cadastre. Chapter 5 describes the cadastre in detail and examines the issue of land units within the cadastre. The use of the cadastre as the engine of modern LAS is presented in a new model, referred to as the "butterfly diagram."

Chapter 3

The discipline of land administration

3.1 Evolution of land administration as a discipline

3.2 Land administration and sustainable development

3.3 Incorporation of restrictions and responsibilities in LAS

3

3.1 Evolution of land administration as a discipline

GROWTH OF SCIENTIFIC METHODS OF LAND MANAGEMENT

Land administration as the name of the discipline first appeared in 1996 (UNECE 1996), though the intellectual roots of the discipline in the management of people-to-land relationships and the specialized tool of surveying are much older. Modern surveying, as a defined activity involving scientific and rigorous collection of land information through precise boundary and parcel identification, has a long history of more than 400 years.

The heyday of land surveying history began with the Napoleonic era, when what we know as modern Europe was surveyed according to precise standards. This delivered enduring benefits: coherent, adaptable land distribution systems, which formed the basis of efficient land taxation; formal land registration and transaction tracking; and eventually, effective land markets.

These functions helped stabilize landownership and manage any disagreements. European educational and technical institutions appointed professors in cadastre and land management, and surveyors became highly regarded professionals, whose activities were widely understood among their respective communities. In global terms, especially since the 1970s, surveyors have become the profession interested in developing a broad expertise in land management, as evidenced by the efforts of the International Federation of Surveyors (FIG). Surveyors' technical focus made them the primary users of new technology. When the digital world expanded, they became masters of new applications. When computer science joined land management, surveying became a spatial technology.

The reliance on scientific methodologies in applied systems gave land administration theory its primary focus on designing, building, and monitoring systems to achieve articulated goals. This, in turn, gave land administration its practical, hands-on approach of finding solutions to very difficult land management issues. It also gave the discipline a self-critical capacity to absorb and learn from unsuccessful efforts, since failures were clearly apparent. The tradition of trying to get things to work better helped produce a literature in which large-scale land administrative systems design is discussed vigorously (for example, FIG 1996, 1998; UNECE 1996, 2005a, b, and c; GTZ 1998, among the many contributions), and project evaluation is openly tracked.

The combination of critical evaluation and applied scientific methods, or the engineering approach, remains apparent in modern land administration theory and practice.

ADMINISTRATION OF LAND AFTER WORLD WAR II

From the end of World War II to the 1970s, administration of land continued its prewar configuration of institutions and ideas and carefully refined its core concepts of the cadastre and land registration for implementation of land markets. The focus on war repair and land markets fits well with traditional ideas of lawyers, surveyors, and economists. In general, the steady state of established institutions was undisturbed. Immediately after the war, Japan and Taiwan were stabilized. Then some postcolonial African countries, among them Kenya and Uganda, were the focus of land administration projects and land law reform, mainly geared toward stabilizing access of farmers to their land. Later, reform of Latin American infrastructure in land administration was also started, along with land reform and redistribution activities (Lindsay 2002). The former British Colonial Office (renamed the Ministry of Overseas Development, then the Department of Overseas Development, and now the Department for International Development (DFID) sponsored significant publications, including

Land Registration by Sir E. Dowson and V. L. O. Sheppard in 1952, the first text to analyze land registration systems for an audience wider than lawyers; *Land Law and Registration* by R. W. Simpson in 1976; and *Cadastral Surveys within the Commonwealth* by Peter Dale in 1976. Dale comprehensively examined international efforts to build land administration in Commonwealth countries in a major effort to facilitate information exchange. Elements of cadastral survey systems and their potential for multiple uses extending into valuation and taxation, and planning and development, together with surveying options in day-to-day use in land administration offices were described in detail.

The Food and Agriculture Organization (FAO) publication in 1953 of *Cadastral Surveys and Records of Rights in Land* by Sir Bernard O. Binns (revised in 1995 by Dale and republished) identified the importance of formal organization of land records in agricultural development. The Land Tenure Series of FAO began and remains a fundamental source in land administration theory and practice, particularly in relation to rural land, as does its journal, *Land Reform and Settlement and Cooperatives*. The Land Tenure Center was established in 1962 at the University of Wisconsin–Madison and began its forty-year-plus engagement in land research and documentation, with a focus on the land tenure issues in Latin America and later in Eastern and Central Europe.

In 1975, the World Bank board of directors articulated a land policy approach. J. W. Bruce (Bruce et al. 2006) suggested that the bank as an institution had no actual land policy, but for practical purposes, its global influence reflected its economic policies. Thus, the economic development paradigm was applied to land administration activities. This approach prevailed for the next thirty years and remains highly influential today. Stable land institutions, similar to those in Europe and the United States, were seen as essential for the economic capacity of nations. The formation of properties for land markets through provision of Western-style institutions (cadastres, registries, and property-based land rights) became the focus. The resulting paradigm of economic development as a focus for activities of institution building and reform in land administration produced mixed results, however.

The proceedings of the UN Regional Cartographic Conferences (UNRCC) led to meetings on cadastral surveying and mapping in 1973 and 1985, and later to a meeting on surveying and mapping legislation in 1997. The UNRCC is currently administered through the UN Department of Economic and Social Affairs (DESA) in New York.

The UN Center for Human Settlements (UN–HABITAT) was active in land issues after HABITAT 1 in Vancouver in 1976 in the areas of security of tenures, the formalization of informal settlements, and access to land. Its contribution to pro-poor land management, security of tenure, and multiagency conferences on the urban crisis provided substantial literature on urban issues. It pioneered new theoretical and practical land administration approaches over the next thirty years.

These theoretical developments eventually delivered general acceptance of the multipurpose model. In the English-speaking world, doctoral theses and scholarly work in the United Kingdom (Dale 1976), Canada (McLaughlin 1975), and Australia (Williamson 1983) built on the concepts of coordination of registration and the surveying and mapping systems that originated in Germany. Meanwhile, the United Nations worked on cadastral surveying and mapping (United Nations 1973, 1985). In the longer term, these efforts set the stage for design of national LAS to take up the challenges posed by the arrival of computers and supported Europe's adoption of multipurpose models.

THE 1980s

The realization of the importance of a well-defined and effective cadastral system gained momentum in the English-speaking world in the Maritime Provinces in Canada (McLaughlin 1975), then reached its peak with the vision of a multipurpose cadastre. The 1980 National Research Council study, "Need for a multipurpose cadastre," began a new era (McLaughlin 1998). The approach to land administration next entered the implementation phase, which centered on how to build multipurpose cadastres rather than on why they should be built.

Though a distant reality, the vision of the multipurpose cadastre functioned as a means of organizing and directing change in the context of very well-established, and even rigid, approaches to surveying and institutional arrangements.

Reform of administrative and technical support systems saw the replacement of paper records and large numbers of staff with computers and trained managers and technicians. Land information was central to Peter Dale and John McLaughlin's 1988 book, *Land Information Management*. The arrival of the computer extended the use of spatial information across a broad range of industries and professions—lawyers and surveyors, fiscal systems, local governments, utilities, land-use planning, and others (figure 3.1). A digital cadastral database (DCDB), linked to the national geodetic reference framework for scientific veracity, and supporting computerized land registration served as the foundation for implementing a land information vision to build capacity to deliver multipurpose uses. These interconnections contributed to the subsequent development of the SDI.

Leadership in land policy and related issues came from diverse organizations, including the FAO's Land Tenure Service, the University of Cambridge's Department of Land Economy, the Republic of China's International Center for Land Policy and Training in Taiwan, and the University of Wisconsin–Madison's Land Tenure Center.

From the mid-1980s on, another revolution occurred, particularly in the United States, where the role of the private sector, technical innovations, and wide access to land information expanded to fill the vacuum of an ineffective cadastral approach. The spatial-information revolution had begun. The conversion of centrally organized governments in Eastern Europe to market economies and the engagement of the European Union in redesigning entire national approaches to land stimulated comprehensive, and more successful, market-based LAS design and construction (Dale and Baldwin 2000; Bogaerts, Williamson, and Fendel 2002). Meanwhile, land projects in many other countries produced mixed results. Reevaluation of project aims and design broadened land administration theory.

Many developed countries began their major commitment to development assistance in land administration, particularly to the establishment and reform of land administration and cadastral systems in developing countries. Contributors include the Netherlands (through the Institute for Aerospace Survey and Earth Sciences at the International Institute for Geoinformation

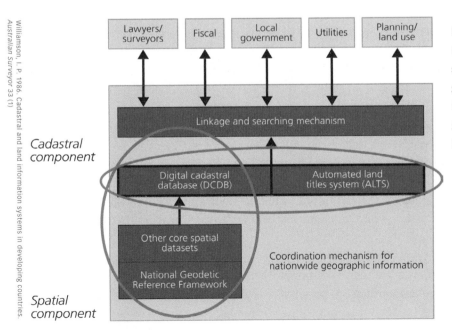

Figure 3.1 A parcel-based land information system uses the national geodetic reference framework to coordinate geographic information for government purposes.

Science and Earth Observation (ITC)), the United Kingdom, Sweden, Australia, Germany, France, Canada, the United States, and Spain. These activities added to knowledge about implementation of LAS, including "best practices" publications.

THE 1990s AND BEYOND

The fall of the Berlin Wall in 1989 saw the start of a Central and Eastern European realization, which also had significant impact globally, of the role of property in a market-based economy. This had a major impact on the rebuilding of facilities and development of LAS theory and practice (UNECE 1996).

A trend away from narrow, historically defined land administration tools of cadastres, registries, and property-based land rights to broader and adaptable tools capable of meeting the economic, social, and environmental issues raised by sustainable development policy can be seen in the work of the United Nations and nongovernmental organizations, such as FIG–Commission 7, responsible for cadastre and land management (FIG 2008a). Germany, through its aid agency, GTZ, encouraged the documentation of best practices with its notable publication *Land Tenure in Development Cooperation — Guiding Principles* (1998). The new focus consolidated the substantial groundwork done in the 1970s and 1980s. The challenge was taken up by FIG–Commission 7, which spent several years developing the Statement on the Cadastre (1995). The statement was designed to be used globally, was truly multipurpose, and was accepted by all FIG associations, representing more than eighty countries.

International surveying, land administration, and cadastral conferences produced a stream of policy, technical, and interdisciplinary literature circa 1996 and later (for example, Holstein 1996a; Burns et al. 1996; McGrath, MacNeill, and Ford 1996). Numerous conferences, workshops, and meetings also added to the literature — notably, the international cadastral reform conferences at the University of Melbourne, Australia, in the early 1990s, the International Land Policy Conference in Florida in 1996, the International Conference on Land Policy Reform in Jakarta in 2000, many conferences in Western and Central Europe, and numerous events sponsored by organizations such as the United Nations and FIG, especially through its Commission 7 on Cadastre and Land Management, which provided leadership throughout the 1990s and beyond.

The challenge of the emerging digital environment led to consideration by a working group, set up in 1994 by FIG–Commission 7, of what a cadastre might look like in 2014. The resulting vision, "Cadastre 2014: A vision for a future cadastral system" (1998), made a major contribution to the debate and discussion of where cadastres were heading. Various books provided comprehensive and timely overviews of land administration theory, such as *Land Registration and Cadastral Systems* (Larsson 1991) and *Land Administration* (Dale and McLaughlin 1999). Together with the earlier literature, these publications set the academic framework for development of the use of geospatial data in land administration and SDIs.

However, the greatest theoretical reorganization of the discipline from a technical to a multidiscipline endeavor was driven by another, overarching trend: delivery of sustainability policy.

THE INFLUENCE OF SUSTAINABLE DEVELOPMENT

The efforts to design a cadastral approach capable of incorporating environmental and social goals, following the Brundtland Commission in 1987, especially Agenda 21 and the United Nations meeting of UN-HABITAT II, began to consolidate in the Global Plan of Action. Together with the United Nations, FIG, through its Commission 7, developed the Bogor Declaration on Cadastral Reform (UN–FIG 1996) to stimulate efforts to build effective and efficient multipurpose cadastres in individual countries. This led to the development in 1999 of the joint UN–FIG Bathurst Declaration on Land Administration for Sustainable Development, which came in response to countries facing intractable poverty and environmental issues that demanded new inclusive approaches to land administration. Contributors to the text of the document included anthropologists, economists, land policy professionals, lawyers, surveyors, and spatial information experts from all the major organizations. A framework for multidisciplinary cooperation in land policy and administration was firmly established.

The Bathurst Declaration became the formative document in modern land administration theory. It established a strong link between land administration and sustainable development. The declaration identified evolving concepts and principles, which added to, and built on, the rich body of knowledge in land administration, particularly cadastral systems developed since World War II. This body of knowledge included a wide range of journal articles, books, reports, statements, policies, and declarations from international organizations, especially the United Nations and World Bank, individual country governments, and many individuals. These trends culminated in the clear theoretical articulation that cadastral activities in particular, and land administration in general, should focus on sustainable development.

EVOLUTION OF THE CONCEPT OF LAND ADMINISTRATION

FIG has been active in promoting discussion of cadastral and land management issues for almost 100 years. Availability of its material through the Internet contributes to the theoretical maturity of land administration, and FIG remains a major contributor to the electronic libraries of the world. FIG sponsored the International Office of Cadastre and Land Records (OICRF), supported by the Netherlands since 1958, as one of its permanent institutions. It provides access to an extensive electronic library at `www.oicrf.org`. In particular, FIG–Commission 7 produced formative publications including, in addition to those already mentioned, the 1997 report on "Benchmarking cadastral systems" (Steudler et al.). The FIG congresses, held every four years, and the annual FIG Working Week continue to provide rich sources (`www.fig.net`) of cadastral and land administration papers (figure 3.2).

When countries in Eastern and Central Europe changed from command economies to market economies in the early 1990s, the UN Economic Commission for Europe (UNECE) saw the need to establish the Meeting of Officials on Land Administration (MOLA). In 1996, MOLA produced Land Administration Guidelines (UNECE 1996) as one of its many initiatives.

MOLA's 1996 initiative was sensitive to there being too many strongly held views in Europe of what constituted a cadastre. Another term was needed to describe these land-related activities. MOLA also recognized that any initiatives that primarily focused on improving the operation of land markets had to take a broader perspective to include planning or land use as well as land tax and valuation issues. As a result, MOLA replaced "cadastre" with the term "land administration" in its guidelines. A structural reorganization followed. In 1999, MOLA became the UNECE Working Party on Land Administration (WPLA). Most WPLA activities still concern the traditional cadastral areas of land registration, cadastral surveying and mapping, and associated computerized land information systems (LIS).

Widening the concept of a cadastre to include land administration reflected its variety of uses throughout the world and established a globally inclusive framework for the discipline. WPLA reviewed land registration and published an "Inventory of land administration systems in Europe and North America" (UNECE) in several iterations, the most recent in 2005. The Working Party also analyzed the issues relating to real estate units and identifiers (UNECE 2004). In 2005, another formative policy document, "Land administration in the UNECE region: Development trends and main principles," updated the 1996 guidelines (2005a).

For the first time, efforts to reform developing countries, to assist countries in economic transition from a command to a market-driven economy, and to help developed countries improve LAS could all be approached from a single disciplinary standpoint, at least in theory. That is, to manage land and resources "from a broad perspective rather than to deal with the tenure, value, and use of land in isolation" (Dale and McLaughlin 1999, preface). The importance of land information in the formation of national land policies came out of the *Aguascalientes Statement* (FIG 2004). A key finding in this statement called for an integrated information strategy:

> *"There is a need to integrate land administration, cadastre, and land registration functions with topographic mapping programs within the context of a wider national strategy for spatial data infrastructures." (14)*

These efforts formed the basis for understanding the relationships among LAS institutions and processes involved in land tenure, valuation, and use. They established the principles of taking a holistic approach to these institutions and diagnosed the problem of historical silos, or

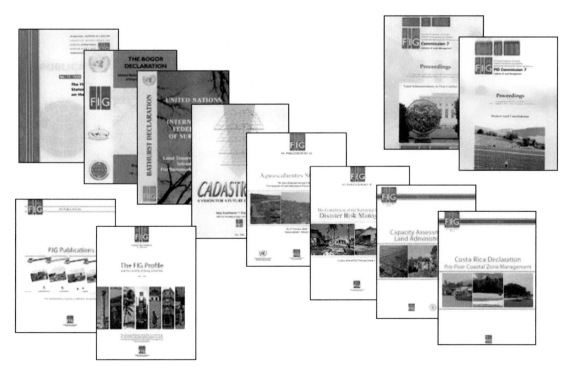

Figure 3.2 The International Federation of Surveyors has produced a variety of formative publications on land administration and the cadastre.

stovepipes, that managed each process from an internal perspective. They saw the need to integrate functions of registration and cadastral surveying. The usefulness of land information, and the need to collect it once and use it many times, was also identified. Most significantly, this development brought the cadastre to center stage.

INFLUENCE OF DEVELOPMENT AID PROJECTS

The broadening of the discipline was reflected in changes in land policy and the activities of major organizations over time, including the World Bank (Deininger and Binswanger 1999; Deininger 2003). Forerunners of this policy analysis included H. B. Dunkerley and C. M. E. Whitehead (1983), G. Feder et al. (1988), G. Feder and D. Feeney (1991), L. Holstein (1996b), and F. K. Byamugisha (1999). The World Bank's Land Policy Network at `www.worldbank.org/landpolicy` contains an extensive list of activities and publications. Noting that the Land Policy Network has a primarily rural focus, the World Bank also supports a complementary Land and Real Estate Network, which has an urban focus (see Razzaz and Galal 2000). Gender equity became a popular goal of land projects (Giovarelli 2006), as did the need to take a comprehensive approach to land issues, although paths toward implementation were not as well defined.

Particularly after 2000, land administration literature grew remarkably, through directed academic and land institute research and the activities of organizations such as FIG, UN–HABITAT, FAO, and many other multi- and bilateral aid agencies and professional organizations. Substantial development policy required widening the LAS vision and conducting intensive research on existing land practices to improve understanding of how people think about and manage land. The cumulative effect extended modern land administration, both in theory and in practice, as a multifaceted endeavor capable of having an influence on such newly articulated focus areas as good governance through LAS (Van der Molen 2006), population movements, emergencies and natural disasters, and difficulties in political and economic transition.

The role of land administration in foreign aid and government administration transformed public administration theory. It had an immediate impact on land administration institutions in successful economies and through foreign aid in developing countries. Government downsizing and privatization led to Structural Adjustment Programs (SAPs) that restricted the role of governments, raised the prominence of the private sector, and treated the poor as a target population, basically as aid recipients (McAuslan 2003). While registration and cadastral standard setting remained government functions, just about everything else was turned over to the private sector. The core functions of land administration institutions within government were clearly identified. Government was charged with the responsibilities of defining and protecting property rights for a number of reasons:

> *"First, the high fixed cost of the institutional infrastructure needed to establish and maintain land rights favors public provision, or at least regulation. Second, the benefits of being able to exchange land rights will be realized only in cases where such rights are standardized and can be easily and independently verified. Finally, without central provision, households and entrepreneurs will be forced to spend resources to defend their claims to property, for example, through guards, fences, etc., which is not only socially wasteful but also disproportionately disadvantages the poor, who will be the least able to afford such expenditures." (Bell 2006)*

Another equally important consideration is the need to link the performance of LAS with public confidence in government. If land administration is tied to democratic performance, enhanced civil peace and good governance in general will result.

INSTITUTIONALIZING INTERNATIONAL SUPPORT

Academic and professional research grew commensurately. Numerous articles appeared in technical journals such as *Survey Review* (United Kingdom), *Australian Surveyor* (Australia), and *Geomatica* (Canada); in land policy journals such as *Land Use Policy*; in more general planning journals such as *Computers, Environment and Urban Systems*; and in many others, including spatial science publications.

Various research groups, typically housed in universities within surveying, geomatics, geography, or law departments, investigated land administration issues, particularly cadastral topics in developing countries. Examples include the University of New Brunswick, Canada; the Technical University of Delft, the Netherlands; Aalborg University, Denmark; several German universities; the University of Florida; and the University of Melbourne, Australia. The ITC in the Netherlands is of particular importance because of the significant resources it provides and attracts for both education, and training and research. It now emphasizes land administration education and research in developing countries.

The Organization of American States, Latin America, and the newly established Commission for Legal Empowerment of the Poor (UNDP 2008) are also contributors to both activities and theory. Theoretical incorporation of informal tenures within LAS, new tenure analysis and tools, land management in Francophone African countries, pro-poor land management, betterment paths, and strategic use of possession as a source of tenure are only some of the emerging research results. The Global Land Tool Network (GLTN) is one of the initiatives dedicated to sharing information about these emerging concepts.

INFLUENCE OF COMPUTERS

Computer technology will continue to drive fundamental change in land administration for decades. Computers have already stimulated substantial administrative and institutional change. The conversion of LAS records from paper systems to digital began in the 1970s when integrated circuit technology reduced the cost of computers and personal computers came into vogue. In the 1980s, silo agencies involved in land administration converted their processes to digital systems, but each system was unique. Some agencies, even in developed countries, did not computerize: for example, some of the land registries in Swiss cantons still use paper-based systems. The arrival of computers challenged the managers of land administration agencies to improve their services. In the 1980s and later, the problem of integrating information, such as combining value and registration data, was conceived as involving two choices: Either each agency provided all its data to a main computer, which provided access, or it provided it to a hub system. Land administration agencies generally did not favor any solution to access or sharing that allowed another agency to handle their data. The concern about data sharing was generated by acknowledging the primary importance of data about land and the need to ensure its integrity. These concerns continue today.

Once desktop computers took over from mainframe computers, data-sharing issues became significant. First came the issue of sharing among all the computers in a business or agency via an intranet. Then when the Internet arrived, data could be shared with the world at large, expanding the need for security. Simultaneously, improvements in software systems took the capacity of computers far beyond 1980s expectations. The arrival of GIS, geodesy, imaging, layering, and entirely new object-oriented methods offered opportunities in land administration and commensurate challenges. The technical issues of how to relate cadastral surveying software with other spatial software systems became evident. Meanwhile, the private sector's inventiveness in developing spatial systems took computers into an entirely new dimension, where images, information, and access could all be mixed and matched according to a user's needs. The many land administration agencies that innovated and computerized in the 1980s and 1990s were struck with the anomaly of being held back by out-of-date or "legacy" systems that put these new opportunities beyond their reach. The barriers they faced included the need for fundamental institutional change, the huge capital cost involved in new systems, and lack of understanding by the political powers that be.

Technology is, of course, central to related disciplines in human and natural geography, resource management, and environmental research. These disciplines are especially active in building integrated Web-enabled information transfer and monitoring systems that use topographic and other GIS information, satellite imaging, and many other new applications. These applications potentially overlap with LAS and LIS, which rely on parcels as basic building blocks, and together, they form an SDI.

Worldwide, the speed of development in computer theory and capacity throws up a particular challenge for land administration as a discipline. Technology demands a future vision flexible enough to comprehend trends and directions—before they happen. The questions of how to manage land information and how to use the traditional land administration processes for broader public good and optimum business outcomes are now on the LAS agenda (see chapter 9, "SDIs and technology"). Computers exposed the potential for sharing land information, provided it was organized. This opportunity stimulated the creation of an SDI, as part of the land administration infrastructure, to manage the great, untapped resource of land information, and to merge the information layers in GIS and digital cadastral databases.

The arrival of the Internet required each land administration agency to enable at least some Web access to its data, and eventually to convert processes to digital, interactive systems, known globally as e-land. These changes also contributed to e-government, with its philosophy of greater accountability of agencies and more involvement of the public in government processes. Land administration institutions, however, remain structurally the same, though their use of technology revitalized their processes. Meanwhile, new spatial technologies offer even greater potential for using land information. The integration of land information with the institutional arrangements creating and sharing it are now crucial to the delivery of sustainable development. Just as geographers are beginning to add to the public knowledge base by explaining the relationships among water, soil types, salinity, and vegetation, land administrators need to provide information to government policy makers in a way that assists sustainability and protects public interests. Most of the relevant information comes from land administration processes.

3.2 Land administration and sustainable development

IMPLEMENTING SUSTAINABLE DEVELOPMENT THROUGH LAS

LAS evolve in response to changes in people-to-land relationships primarily driven by the development of land markets. But increasingly, these changes come about because of pressures on the environment caused by population increases, use and misuse of resources, reorganization of national, state, and local agencies, and advancements in technology. Especially in developing countries, supply of money and credit, labor, food, and other agricultural products requires government action, which in turn has an impact on land administration processes. Most nations are also beginning to make increasing demands of their administrative infrastructure as they seek to improve land management. While land markets remain the major driver, other pressures are now beginning to be absorbed by land administration institutions through the prism of sustainable development. As introduced earlier, sustainable development is now the major policy justification for LAS and related technical capacities in land information systems and GIS. Still, policy implementation remains a significant issue.

The international land policy literature observes three components within the broad goal of sustainability:

* Efficiency and promotion of economic development

* Equality and social justice

* Environmental preservation and a sustainable pattern of land use (GTZ 1998; Deininger 2003)

A fourth component of good governance is also recognized as essential for institutional and government capacity to deliver sustainable development.

Following the 1987 Brundtland Report, highlights of international efforts to promote sustainable development include the adoption of Agenda 21; the 1992 UN Rio Earth Summit and subsequent summits; the Copenhagen Declaration and Programme of Action of the World Summit for Social Development on empowering civil society; advocacy for women's and children's rights demonstrated at the fourth World Conference on Women's Rights in Beijing, China, in 1995; food security and sustainable rural development incentives delivered at the World Food Summit in Rome, in 1996; the UN City Summit in Istanbul, Turkey, in June 1996, instigating discussion that resulted in the UN–HABITAT Human Settlements campaigns for adequate shelter and tenure security for all (1999); the stream of activities of the Food and Agriculture Organization of the United Nations; and, most recently, the Millennium Development Goals adopted by UN member states in September 2000 in support of global human development among developing and developed countries (Feder et al.1988; Deininger and Feder 1999; Dalrymple 2005). These are described in figure 3.3.

These international efforts were the antecedents of one of the most significant land policy documents—the World Bank's research report, "Land policies for growth and poverty reduction" (2003a). The report reviewed World Bank activities since 1975 and made three significant conclusions (Van der Molen 2006). First, the previous focus on formal titling is no longer appropriate, and much greater attention should be paid to the legality and legitimacy of existing institutional arrangements. Second, an uncritical emphasis on land sales should be extended to include rental markets. Third, careful assessment of an intervention is needed for land redistribution efforts. Land related strategies need to be integrated with other strategies, especially to link land to broader economic development in a long-term strategy capable of gaining broad support. The World Bank, along with other international aid agencies, made it a priority that land titling processes would include enhancement of tenure security through innovative practices, allowing gradual upgrading over time and strengthening of government institutions.

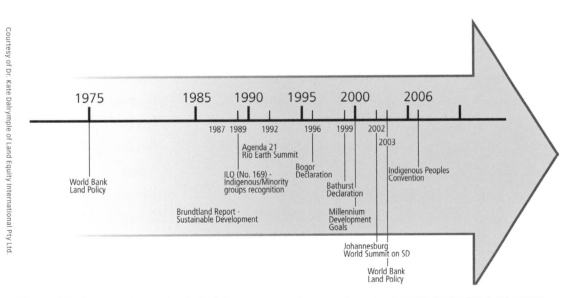

Courtesy of Dr. Kate Dalrymple of Land Equity International Pty Ltd.

Figure 3.3 Since 1975, international efforts have been on a trajectory toward sustainable land policy development.

Articulation of these imperatives by the World Bank followed similar trends in other UN and international organizations, particularly UN–HABITAT and development aid agencies. These policy changes stimulated a broadening of emphasis in land administration theory and practice, especially since 2000. Examination of national and subnational activities and identification of processes to upgrade them widened formal titling projects. What were previously called land administration projects were refocused and renamed land management and policy development projects. For example, Indonesia's Land Administration Project in 1995 became the Land Management and Policy Development Project in 2002.

The change in focus produced innovation in the theory of LAS design and tentative changes in system design and activities. It stimulated development of the land management paradigm (see chapter 5, "Modern land administration theory") and widened the theoretical capacity of the discipline to integrate its formal and familiar tools into the new realms of social tenures, marine environment administration, and complex commodities and restrictions management, among other innovations. These extensions provide challenges for LAS designers who seek to implement them.

TRANSLATING SUSTAINABILITY INTO CLEAR OPERATIONAL LAS STRATEGIES

While international land policy, and most articulated national land policies, revolve around sustainable development, it is not clear how land administration activities can be related to this broad goal. One almost universal approach is to reduce sustainable development into more explicit achievable outcomes, sometimes called strategies, implementation policies, or principles. While many versions of implementation policies exist, a durable set was designed by Germany's GTZ (ILC 2004) for developing countries. These policies were defined as

- ◆ Improvement of resource allocation by minimizing the land issue, especially for the benefit of small and midsize landholders
- ◆ Support of access to land for groups living in poverty
- ◆ Creation of higher legal security in the transfer and use of land, especially for women
- ◆ Design of sustainable land-use patterns
- ◆ Demand for education and training in the field of land tenure systems and land management

This list of policies, or principles as they are sometimes called, is generic and suitable for any country, though its focus is on issues faced in developing countries. UNECE (2005a) also provided a detailed blueprint of best practices and principles directed at countries seeking market-based economies in democratic political systems.

Especially since 1995, land administrators have tried to systematically relate sustainable development to the specific administrative processes they deliver. These efforts led to the realization that effective social, environmental, and economic management demands a holistic approach to land and resources across three distinct areas:

- The natural environment—land, resources, and related features

- The built environment—man's impact

- The virtual environment—computer technologies assisting management of the other two environments, specifically the digital systems used to reflect the natural and built environments

It is no longer considered best practice to treat land owned by the government (national parks, forests, riverbeds, and the like) separately from land owned in other ways. Nor is it best practice to manage land and water separately or to suspend administrative systems at the coastline. Seamless management of the entire terrestrial environment, coastal zone, and marine environment is essential. So far, efforts have been piecemeal, although the insights they have produced are invaluable (see section 3.1).

At the major policy level, land administration attempts to accommodate sustainable development through improved land management resulted in comprehensive analytical and comparative literature. One of the trends in this literature is improvement in the ability to measure sustainability outcomes delivered by land administration processes (see chapter 4, "Land administration processes"). Land administration processes are so closely related to the ways communities use, distribute, and organize land that they are crucial to land management. Land management encompasses the broader processes that control and organize human activity in relation to land. These land administration processes then are capable of delivering far more than what they were historically designed to do. When understood from land management perspectives rather than their narrow original purposes, land administration processes, individually and collectively, provide systematic feedback on sustainability policy. This two-way loop allows "sustainability accounting."

3.3 Incorporation of restrictions and responsibilities in LAS

GROWTH OF REGULATIONS AND RESPONSIBILITIES

An immediate challenge for sustainable development lies in extending the capacity of modern land administration beyond management of rights in property-based commodities in land and resources to managing rights, restrictions, and responsibilities (called RRRs in most of the literature; although RORs, referring to rights, obligations, and restrictions, is also used). The key to meeting this challenge is not through managing land itself, but in managing the business processes and administration systems affecting and influencing people's activities in relation to land. The analytical need is to move away from managing the physical assets and toward managing people's behavior in relation to these assets. This jump in philosophy mirrors the perception of Peter Drucker in 1946 in *The Concept of the Corporation* and his idea that the major resource of a company (indeed, of a country) is its people. In Drucker's approach, a company should facilitate decision making and agree with subordinates on objectives and goals, then get out of the way on how to achieve them. This model therefore directs attention to evaluation and monitoring at the back end, and to shared goal setting at the front end, lessons equally applicable to LAS design.

The first stage in applying this kind of people-based analysis to management of RRRs requires appreciation of the dual nature of these relationships. A right is not a relationship between an owner and land. It is a relationship between an owner and others in relation to land, backed up by the state in the case of legal rights. This duality of owners and others is also present in

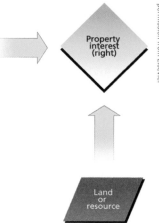

Figure 3.4 The original LAS approach to managing property in land simply dealt with the rights to ownership.

Reprinted from Publication *Land Use Policy*, Vol. 25 / 1, Rohan Bennett, Jude Wallace and Ian Williamson, Organizing land information for sustainable land administration, 13, Copyright (2008), with permission from Elsevier

restrictions and responsibilities affecting landowners and users. Each restriction/responsibility involves a duality that imposes obligations on owners in relation to the land for the benefit of others. An administrative framework is robust and successful when it takes this duality into account and also identifies the appropriate managing or implementing authority. The conceptual framework for managing property has therefore changed dramatically. In the earlier analyses of how the institution of property worked, rights were seen as property interests in terms of the owner or benefiting party and the owner's land or resource (figure 3.4). A catalog of the rights affecting the parcel therefore seemed a sufficient administrative framework for management of these people-to-land relationships.

This model is now outdated and not capable of servicing the needs of modern government concerned with delivering sustainable land uses. Modern analysis (figure 3.5) therefore exposes the duality of the arrangements created by RRRs and relates them to the institution of property. This model identifies both the parties benefited and the parties burdened for all RRRs or property interests. It offsets the theoretical wall built by the original analysis between the rights and opportunities of owners and their responsibilities and restrictions vis-à-vis stewardship, environmental planning, and other concerns. The tools in the land administration toolbox suggested for the management of restrictions and responsibilities take their dual nature into account.

Theory is only part of the answer to managing restrictions and responsibilities. The problem of systematic management or administration has become urgent. Restrictions and responsibilities attached to land and resources flow from the global, regional, and national attempts to address problems of land use, environmental degradation, and climate control. These attempts

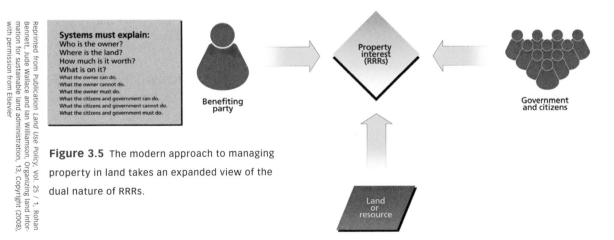

Figure 3.5 The modern approach to managing property in land takes an expanded view of the dual nature of RRRs.

are only a small part of the regulatory intervention and legalization of spheres of endeavor affecting the world at large and plots of land in particular.

The increase in the number and complexity of restrictions and responsibilities in Western democracies is phenomenal. It was initially spurred by the consumer movement and the need to regulate land transactions, but it expanded as a result of industrial and building standards emerging from land-use planning and quality controls affecting land-related behavior. Restrictions and responsibilities created by the actions of a state or nation multiplied along with the growth of government—e.g., taxation, pollution controls, environmental protection, land-use management, and so on. Add planning controls, and the picture of government-built regulatory systems affecting land takes on huge significance. Despite this, restrictions and responsibilities are largely ignored in existing institutional LAS, though some attempts are being made to integrate land-use planning information and processes.

This new regulatory environment has two main features: first, the massive growth in the number and variety of regulations, illustrated in the growth of statute books, regulations, codes, and standards throughout Western economies. The second, more fundamental feature is the complex nature of these new normative orders. Regulatory patterns have moved away from formal standard setting through strict laws and regulations to extended processes of incorporating monitoring and compliance processes and techniques of contract accounting and auditing, administered by a loose mixture of public authorities, independent agencies, and private individuals.

As a consequence, many significant restrictions and responsibilities are found outside the traditional legislative framework and even outside the framework of government.

CHANGING THE CONCEPT OF "PROPERTY" IN LAS

Though much analytical space is given to land rights, all nations must regulate and restrict land uses for a variety of reasons including environmental protection, building standards, social equity, provision of utilities and infrastructure, tax fairness, and cultural issues. Government restrictions on land differ in their nature and impact, depending on the stability or mobility of communities and their opportunities for collective action and land planning capacity (Webster and Wai-Chung Lai 2003). Traditional analysis of LAS reflects their historic role in defining private parcels and confirming ownership through registration systems, as well as the theoretical ascendancy of private property in land in Western economies (also see figure 2.6). The institution of property (using "institution" as the rules of the game) and the human constraints that shape social interaction (North 1990; Auzins 2004, 59) are the keys to management of land arrangements in successful economies. Historically, land administration focused on the process of capturing information about the land rights component of the property institution. By contrast, modern LAS are required to organize the large panoply of processes, activities, and information about land and to use best practices to facilitate integration of all information to better deliver sustainability. This includes information about restrictions and responsibilities affecting land (Lyons, Cotterell, and Davies 2002; Bennett et al. 2008). Given the number and variation of systems that apply restrictions, one of the challenges facing modern land administrators is how to incorporate this information effectively within the technical, informational, and administrative options available.

Changes in regulation theory and practice continue to challenge land administration theory and practice in fundamental ways, because most of the land administration infrastructure is designed to manage property as a rights-based institution. From a sustainability viewpoint of the land management paradigm, a structure or facility should be available to manage restrictions and responsibilities as well. Any land policy that focuses on opportunities of landowners and their land rights without considering their restrictions and responsibilities will fail to deliver sustainable development. Drains, roads, and utilities demand intrusion on landowners, while conferring obvious benefits. The complex arrangements supporting modern multioccupancy residential and multipurpose buildings in crowded cities also require extensive documentation of the obligations of residents, owners, and third parties. These restrictions are frequently instituted by governments, but other kinds of restrictions stem from arrangements among owners themselves or from private-sector systems. Some restrictions are derived from the cultural arrangements used in a country, particularly those still using social norms to manage land. Many of these arrangements are appropriately managed in informal systems. However, the complexity of modern arrangements affecting land is driving the demand for more formal management of restrictions and responsibilities within LAS, especially if they are imposed through public agencies and governments.

Historically, the concept of property was conceived as solely related to land distribution and exchange functions in a market-driven economy, in which definitive issues were between owners and third parties, mediated by dispute resolution and transaction systems provided by government. This theory of property is about the allocation of entitlements and the means for protecting entitlements against nonowners. If the property concept is to remain effective in the modern regulatory environment, it needs to incorporate the regulatory function in its various forms. Similarly, property theory needs to shift to the new role of arbitrating the relationship between citizens and government. The methods used to manage and administer restrictions and responsibilities will inevitably engage LAS. Whatever mechanisms are used, transparency is essential, a requirement sadly overlooked in many existing regulatory systems and processes.

A country that decides to organize restrictions and responsibilities related to land activities will find that there is no theoretical framework or common language to assist this process, even among countries that share a legal heritage. The highly refined and theoretically clear framework of using tenures to organize rights to land stands in stark contrast. There are a number of reasons why this happened. The history of restrictions and responsibilities is much younger than the history of rights. Restrictions increased in number and significance as governments set up controls over land-related activities and attempted to deliver land polices. In addition, many of the activities involved in economic regulation resulted in ad hoc creation of restrictions and responsibilities, only some of which relate to land. Restrictions and responsibilities were therefore seen as the analytical realm of administrative lawyers, bureaucrats, and political scientists, not land administrators. Until now, no one saw the need to create a metatheory and ontology of restrictions and responsibilities equivalent to a tenure system relating to land (Bennett, Wallace, and Williamson 2006).

A LAND ADMINISTRATION RESPONSE

In the context of worldwide LAS, the absence of records of encumbrances and restrictions created by public law is a major problem. Government measures can restrict the right of disposal and use of land to a certain and, on occasion, substantial degree. The restrictions can vary from mild (such as the obligation to paint heritage colors) to severe (such as requiring a specific use of the land or even government acquisition of that land).

These recommendations raise the issue of how records of restrictions and responsibilities should be maintained. The recurrent suggestion is to use land registries (sometimes in conjunction with the cadastre), rather than specific databases, as the supply chain for information. The deeper concern, however, lies with the overall diagnosis of the problem. Most people think the solution lies in government creation of systems to reveal everything about land. The sheer effort of determining what land, rather than what citizens, is affected by government policies, strategies, plans, and other documents as they change over time is profound, even when a restriction relates to a defined group of land parcels. Nor is it feasible to include all restrictions and responsibilities within the realm of orderly administration. Rights of entry, intermittent controls relating to noise emissions, housing subdivision rules and restrictions, and myriad other abridgments of opportunities gain no better coherence or impact from inclusion in a management system beyond their source agency. The questions are what should be included in separate and additional management or information systems and how inclusion should be effectively achieved. Thus, on reflection, the disorganized nature of restrictions and responsibilities, in addition to lack of information, exacerbates management issues.

A functional approach to management of restrictions and responsibilities immediately raises questions of whether creation and access should be organized around land transactions or around the regulatory activities themselves. Most governments in modern democracies see a need to ensure that a person buying, leasing, or mortgaging land has a reasonable chance of discovering what restrictions and responsibilities affect that land (UNECE 2005a). A typical approach, however, is to place the responsibility for providing this information on vendors, lessees, and mortgagors, apart from the agencies that create and manage restrictions and responsibilities. The alternative would require the agencies that create restrictions, grant permits, establish warranties, grant licenses, and so forth to provide that information through easily accessible, centralized, networked systems.

In any estimation, transaction-based provision of information on restrictions and responsibilities is only a microcosm of the general problem of finding out about decisions that affect land: Where? What? Who? When? These particulars take their point of reference from the making and enforcing of the relevant restriction or responsibility. From the point of view of the originating agency, answers to these questions must be available as a corollary of their day-to-day functions. Queries about particular parcels of land by the parties to land transactions are merely a small subset of occasions where information about restrictions and responsibilities is needed. Thus, a broader LIS is needed.

The functional approach used at the empirical level would have the agencies (public as well as private) involved in the creation of restrictions and responsibilities assume a general obligation to make their decisions and activities publicly known, especially when they affect land and resources. How an agency makes regulatory information available thus depends on its significance as land information and the stage of development of its land administration system and SDI.

Chapter 4
Land administration processes

4.1 Importance of land administration processes

4.2 Core land administration processes

4.3 Examples of tenure processes

4.4 Reforming LAS by improving process management

4

4.1 Importance of land administration processes

Three kinds of land-administration tasks are undertaken in all settled societies: identifying land, defining interests in land, and organizing information or inventories. Land administration theory encompasses the variety of processes countries use to undertake these tasks, but the discipline focuses on the way these tasks are undertaken in market economies where they are now associated with the core functions of tenure, use, valuation, and development.

Land administration is basically about processes, not institutions. An examination of the processes a nation uses for tenure, use, valuation, and development, not its institutions and agencies, reveals its administrative approach. Simply put, land administration systems cannot be understood, built, or reformed unless the core processes are understood. If the processes are well organized and integrated, the structure of agencies and institutions that manage them is much less important. Once they are broken down, processes tend to take on similar characteristics, even though the institutions and agencies are highly variable.

The focus on processes is underscored by the United Nations in "Land administration in the UNECE region: Development trends and main principles," where land administration is defined as "the processes of recording and disseminating information about the ownership, value, and use of land and its associated resources" (UNECE 2005a). Similarly, the UN-FIG Bogor Declaration on cadastral reform states, "Cadastral reform or improvement should focus on the functions of the cadastre and in particular the key *processes* that are associated with adjudicating, transferring, and subdividing land rights" (1996; emphasis added).

The importance of processes to cadastral reform is often highlighted in the literature; for instance, "Cadastral reform must focus on the key *processes* which are associated with adjudicating, transferring, and subdividing land rights, not just the concept of a cadastre or the individual activities of title registration or cadastral surveying" (Williamson 1996; emphasis added).

Each process in the core areas of land administration needs to first be designed and built, then managed. Whatever the level of a country's development, each process needs to incorporate and integrate all the land administration processes, rather than approach an individual activity, such as cadastral surveying, land registration, subdivision, or valuation, in isolation. The design of each process must also account for issues of capacity and the social environment. For example, design of processes to manage changes in ownership in countries with land markets will vary depending on their land registration program, the formalities associated with land transfer, the education of professionals and their respective skills, and the ways people use documents and think about land.

The focus on processes as a basis for understanding and improving systems is not new. Corporate management and governance studies can rely on a large body of literature on process design, improvement, and management and on the related activity of reengineering. In the land administration context, process management refers to the activities of planning and implementing a process, then monitoring performance. Key land administration processes are clearly business processes, though conducted mainly by government institutions. Reform or change involves application of knowledge, skills, tools, techniques, and systems to define, control, and improve processes for the purpose of building customer satisfaction.

Contrast reengineering, which involves radical redesign and reorganization not only of business processes, but also of the organizations and institutions that use them. The classic example of reengineering in land administration is the conversion of paper-based systems to digital during the 1980s and beyond by each of the silo agencies operating in successful market economies. A pending example is potentially much more significant. It involves the isolated institutions of land registries, cadastral authorities, and valuation agencies absorbing the new tools in spatial technology to release the inherent power of land information for use by governments, business, and communities. While reform of isolated land administration agencies is often achieved within process management, coordinated reengineering of all the related core processes in land administration is much more difficult.

This problem of cooperation is not new. A. F. Hall wrote about land administration in New South Wales, Australia, in 1895. His "object was to put before the public the system of survey, to expose its faults, conspicuously *the want of unanimity amongst the various branches of the Civil Service charged with its administration*, and to show in what direction there is room for improvement, and where, without impairing its efficiency, the administrative cost might be considerably lessened" (149; emphasis added). Though the comments were made more than a century ago, they apply just as clearly to land administration institutions and processes in the twenty-first century.

Applying land administration theory to the design of a new system, or to reform or reengineering of an existing system, therefore involves examining the particular approaches and existing local processes used to organize land. While each system is different, general themes play out based on human experience. For example, if a society needs a rule about animal ownership, it will almost always attribute ownership of a baby animal to the owner of its mother. In the context of land, the most generalized processes are related to tenures, which reflect the same human needs for certainty. Thus, those who sow can expect to reap, those who build can expect to benefit, and so on. Amid the vast variety of tenures, a trend toward family and individual ownership is manifest, accompanied by appropriate formalization of processes (see chapter 6, "Building land markets").

4.2 Core land administration processes

Each social system that manages land allocation, from the earliest attempts to provide stable housing, hunting areas, or croplands to recent efforts to organize modern cities, undertakes core processes associated with tenures. These core processes vary from country to country. Understanding the core tenure processes in any nation or group reveals how their land administration system works as well as its effectiveness. Five core tenure processes common to most nations are

◆ Formally titling land

◆ Transferring land by agreements (buying, selling, mortgaging, and leasing)

◆ Transferring land by social events (death, birth, marriage, divorce, and exclusion and inclusion among the managing group)

◆ Forming new interests in the cadastre, generally new land parcels or properties (subdivision and consolidation)

◆ Determining boundaries

The last four processes are historic and widespread, and the subprocesses vary according to the stage of LAS development. Most market-based systems, however, have common characteristics because they are derived from Western models. By contrast, the first process is more recent and essentially technically oriented.

These five tenure processes cannot be understood in isolation but must be related to parallel processes in land development, planning, and valuation. They are also influenced by national land policy, and social and economic systems. In mature systems, each process is highly developed into subprocesses and attracts administrative support, the work of dedicated professionals, and legal recognition.

The Western models for tenure processes are robust, supported by technology, and flexible enough to meet local conditions and policies. The objective of the design and build component in these land tenure processes is to create the initial data—that is, build the ownership and cadastral records from a land titling process. Reform of the processes typically involves projects to automate a land title register or to move a hard-copy basemap into a digital environment. Similarly, design and build processes to create a new valuation system involve establishing its legal and institutional framework, improving local capacity to run a valuation system, and undertaking individual property valuations. Equally important are establishing infrastructure to manage the valuation data and creating systems that permit access, interoperability, and multipurpose use of the data along with maintenance.

Land titling also involves equally important processes for systems maintenance. For example, the resilience of any land transfer or subdivision process depends on maintenance of the processes after they are established. The key to maintenance is for the formal system to capture derivative or postregistration changes in ownership and use patterns. Maintenance of a land titling system and, indeed, all core systems in land administration, is critical. Without maintenance, a system loses relevance and will be replaced by an informal system.

At the same time, land administration processes continually evolve in response to economic, social, and institutional pressures. Improvement in the capacity of a jurisdiction (at the country, institutional, and individual level) to operate an effective land administration system leads to increased wealth as the system generates income from transaction taxes, wealth taxes, and so on, thus improving the economic yield from land.

4.3 Examples of tenure processes

STANDARD PROCESSES FOR SYSTEMATIC INDIVIDUAL LAND TITLING

The first tenure process of bringing land into a formal registration system is relatively modern. It contrasts with extensions of registration through applications made sporadically by individual owners. Given the comprehensive coverage of LAS in developed countries, this process is mostly used in developing countries, often through large-scale land administration projects (LAPs). The process involves a state or country identifying areas to be systematically adjudicated, surveyed, and titled, then using subprocesses to implement a systematic land titling program. The design assumes appropriate legislation and regulations are in place, and that the jurisdiction has the capacity to undertake the titling, with adequate basemaps, orthophoto maps, and geodetic control.

Subprocesses include legal identification, adjudication, demarcation, surveying, and registration. The establishment of geodetic control and the provision of basemaps, including rectified aerial photomaps or orthophoto maps, are technical functions that can be quite time-consuming components (see chapter 12, "The land administration toolbox"). Setting up appropriate basemaps can often take a couple of years, or even longer, given the necessity to award contracts, the restrictions weather can impose on aerial photography, and the need to build local capacity to produce the maps.

The engagement of the community is essential and involves awareness programs (television, radio, town hall meetings, newspapers, posters, leaflets, and so on). Typically, a land titling team is set up in a town or village or region, using local advisers, adjudicators, surveyors, administrators, and computer support personnel. Each land parcel is systematically identified on the ground in the presence of an owner or lessee, and adjoining owners or neighbors, usually with some independent local person, like a village chief or mayor, overseeing the process. The parcel boundaries are physically marked, and all land parcel and ownership data for the parcel is collected, including copies of any documentation that confirms the interests of owners or lessees.

The parcel boundaries are surveyed in formal systems or marked on photomaps, such as orthophoto maps or rectified photomaps. Surveyors or technical staff adds land parcel measurements to a cadastral map for the area, including a unique identifier for each land parcel, while administrative or technical staff updates cadastral indexes with physical and ownership information about the parcel. The appropriate certificates of ownership, or land titles, are prepared by administrative staff and are usually handed over to landowners or lessees at a ceremony. Often, a small fee is paid to receive the certificate. Cadastral maps, a cadastral index, and copies of land titles are transferred to a local land registry so that subsequent land transfers or subdivisions can be recorded.

One of the most successful examples of systematic land registration was achieved in Thailand, which formed a model for many other registration programs, though these were frequently less successful (Angus-Leppan and Williamson 1985). The surveying process used for systematic land titling in the Thailand Land Titling Project (TLTP) is shown in table 4.1. Using this process, one survey field party, consisting of two surveyors and one adjudicator, could survey 150 parcels a month for seven to eight months a year. These basic surveys comprised about 90 percent of the surveys for land titles at the early stages of the 1983 land titling project. Some 20 percent of the areas were surveyed with traverse and tape surveys at 1:1000 in village and urban areas. The remaining 80 percent done in rural areas relied on rectified photomaps at 1:4000. The process outlined in the table shows how elementary and old technologies can build reliable land information systems, especially where the land is predominantly flat.

TRANSFERRING LAND BY AGREEMENT

All LAS operating where land trading is permitted implement transaction management processes. Processes differ according to the level of literacy, degree of professionalism, standardization of paperwork, and other formalities. Production of evidence of the transaction for third parties is the primary subprocess. Even in premarket systems, some early transfer processes involved elaborate ceremonies. "Feoffment by livery of seisin" was a dominant means of transfer until 1536 in England and finally abolished in 1925. It involved a ceremony in which the transferor handed the land, represented by a clod of earth or other remnant, to the transferee before significant witnesses. It was generally replaced by deeds, and then by transfers registered in the land registry. Many villages based on traditional systems still use ceremonial means of transfer, relying either on general witnesses or the approval of the village head as a means of evidence.

	TABLE 4.1 – PROCESSES FOR SECOND-CLASS SURVEYS IN SYSTEMATIC LAND TITLING USED BY DEPARTMENT OF LANDS, THAILAND, CIRCA 1983
OUTPUT	**PROCESS**
Flyovers	Aerial photography was taken at a scale of 1:15,000, with 2 km between flight lines. No signalization of boundaries or other control points was carried out in the field.
Control points	Four horizontal control points are required for rectification. These were obtained via ground methods by a Mapping Control Division.
Photogrammetric measurements	Technical measurements for aerotriangulation involved two Wild A8 Autographs and one Zeiss C8 Steroplanigraph.
Rectification	Rectified photomaps were prepared at 1:4,000, in a 500 x 500 mm format representing 2 x 2 km on the ground.
Title delivery	Photomaps were used for issuing land titles where physical boundaries were visible in the photograph. Where boundaries were not visible, surveys were based on the coordinated traverse control using tapes and optical squares, or sometimes only tapes.
Adjudication	Boundaries were identified by an adjudication process involving the owners and officials, and signed by all present. Surveyors placed numbered circular concrete blocks at each corner. The lengths of the boundaries, but not angles, were measured.
Boundary overlay	In the presence of all adjoining owners, the boundaries were marked on the photomap and transparent overlay. Corners and corner numbers were also marked on the photomap. The owner's name, boundary distances, land parcel numbers, and road names were marked on the overlay. The overlay included a table showing the parcel number, group or adjudication number, and approximate area.
Field cadastral map	A cadastral map was prepared in the field showing the parcel numbers, adjudication numbers, and areas, determined graphically. The final cadastral map showed all parcel boundaries, numbers, corner numbers, and road names.
Titles	Titles were prepared and issued.

The general trend requiring evidence of a transfer exists because ownership has social and legal consequences beyond the immediate parties, especially if land markets operate. Third parties need to identify interests of the true owners of each parcel of land. This is reinforced in many systems by the execution of the document, which creates an interest in land for the transferee, or even makes him the owner. Modern systems require explicit evidence in the form of standard documents or deeds, accompanied by registration. In a Torrens system, only registration itself transfers the interest.

Another typical process involves the buyer physically investigating the land himself or choosing to take the risk that anything adverse to his interests would be discovered prior to sale. Modern systems require far more investigation of peripheral information to discover the status of rates, taxes, and charges, building information, and other conditions of the land. Generally in market

TABLE 4.2 – SIMPLE LAND TRANSFER PROCESS USED BY VICTORIA, AUSTRALIA, CIRCA 2009	
AGENCY	**ACTIVITY**
Vendor or conveyancer	Prepare statement of details of property, title, rates, zoning, restrictions, and service information for marketing and disclosure statement
Buyer	Investigate land
Real estate agent, conveyancer, buyer, and seller	Sell by private treaty or public auction, paying 10% deposit and using standard written contract including terms of purchase, price, and property
Buyer and conveyancer	Search title at land registry and confirm the seller is the same as the last recorded official owner who holds the guaranteed and indefeasible title
Seller, conveyancer, buyer, and bank	Wait contract period during which buyer organizes financing and seller arranges land for handover
Buyer and conveyancer	Prepare transfer of land and submit to seller
Buyer and seller	Settle contract. Buyer pays balance of price to seller and seller's lenders; seller hands over document of title and signed transfer to buyer. Buyer takes possession.
Buyer	Pay stamp duty
Buyer and bank	Lodge transfer with document of title and transfer for registration
Buyer or conveyancer	Notify council, water, body corporate, and tax offices

systems, management of risks in the transaction process is the responsibility of the buyer, who is dependent on public registries working well and other essential land information being readily available. Consumer protection trends since the 1980s have partially reversed the assumption of risk through systems of vendor disclosure. Success in legislating transparency has varied, though the transaction process has become more consistent.

While Torrens-type systems offer simplicity in transaction processes, transaction efficiency in most modern systems is improving dramatically (World Bank, Doing Business reports, 2004, 2005, 2006, 2007). A simple transfer process for Torrens registered land is shown in table 4.2 (also see Dalrymple, Williamson, and Wallace 2003).

Similarly, a modern and effective mortgage process in a developing country is shown in table 4.3 (also see Smith et al. 2007).

TABLE 4.3 – SIMPLE MORTGAGE PROCESS USED BY VIETNAM, CIRCA 2004	
AGENCY	**ACTIVITY**
Household	Borrower collects application form from district bank branch.
	Borrower requests Commune People's Committee to certify the land-use right certificate (LURC) or other document verifying land-use right for the mortgage.
	Borrower compiles documents: business plan, ID card, LURC or other legal document, and permanent residency certificate.
District bank branch	Head of credit department and branch director approve the loan and return the file to the credit officer.
	Credit officer submits the credit appraisal form to the head of the credit department and branch director for approval.
	Credit officer visits borrower to appraise land and assets and completes the assets examination form.
	Bank staff assists the borrower to fill in the mortgage contract, business plan, and request for registration of the mortgage.
Commune	After appraisal, the borrower sends the application file to the commune cadastral officer for certification that the land is not already mortgaged.
	Commune People's Committee chairman approves the mortgage, signs and stamps the application form, and the commune cadastral official records the mortgage in the Mortgaged LURC book.
District bank branch	Credit officer sets the loan amount, term, and interest rate, and notifies the borrower of the date of payment.
Household	Borrower travels to district bank branch and signs two copies of the mortgage contract. Borrower receives the loan and retains a copy of the contract and a loan book. The bank retains the LURC.
	When the mortgage is repaid, the bank returns the LURC to the borrower.

Land transactions in developing countries are frequently informal, as is the case in Indonesia. About 70 percent of land is held in forest tenures rather than land tenures, and transactions in "land rights" rely on informal systems. The majority of nonforest land is outside the Badan Pertanahan Nasional (BPN) land registry jurisdiction and awaits conversion. Typical transaction flows are shown in table 4.4.

TRANSFERRING LAND THROUGH SOCIAL EVENTS

The management of changes to landownership or entitlement related to social processes—marriage, divorce, birth, death, or entry to and exclusion from the land holding group—is a neglected aspect of land administration. Transition following death involves inevitable tensions, compounded by the vagaries of inheritance systems. There are two general kinds of

TABLE 4.4 – FORMAL AND INFORMAL TRANSFERS OF LAND USED BY INDONESIA, CIRCA 1998	
AGENCY	**ACTIVITY**
INFORMAL	
Buyer and seller	After working out the terms, the buyer and seller make an agreement in the form of a *perjanjian jual beli*, or sales purchase agreement. This can be verbal and is based on some form of cash transaction.
Village head	Typically, a village head or other authority figure in the group observes the transaction.
FORMAL	
Buyer and seller	The parties sign a formal *akte jual beli*, or deed of sale, to ratify the sale purchase agreement.
Official	The deed is notarized by a *pejabat pembuat akte tanah* (PPAT) or land deed official employed by the BPN; a *notaris*, or notary public; or a *camat*, a civil servant who is head of a subdistrict and responsible to the regent (in a regency) or to the mayor (in a city).
Badan Pertanahan Nasional (BPN) National land agency	BPN uses the deed of sale as evidence to register and record the transaction and the right it creates in the *buku tanah*, or land book. This may be accompanied by some form of on-site adjudication and formal survey by the BPN. The office creates a *sertifikat tanah*, or land title deed, with a *surat ukur*, survey certificate to record the transaction.

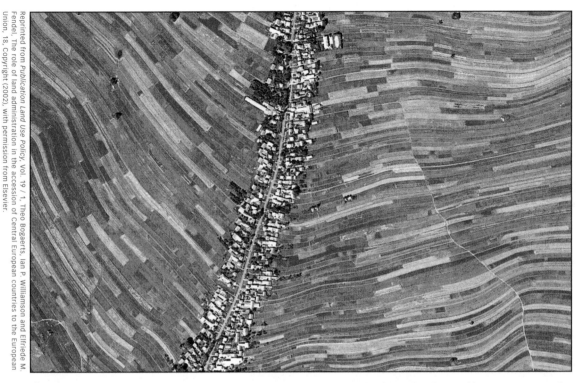

Figure 4.1 Polish cadastral parcels are continuously subdivided, resulting in an abundance of long, narrow parcels.

inheritance systems: testamentary systems (wills or other testamentary wishes) and bloodline inheritance. The first involves literacy and a system for giving effect to the written or proven instructions of a deceased owner and incorporating these into land title records. The second involves identification of the successors of deceased owners by a sociolegal system. Blood inheritance systems tend to follow one of two general models: an English model, in which bloodline inheritance identifies a single recipient, typically the firstborn male heir, or an Islamic model, which involves sharing among all members of the next generational group. The English model of primogeniture is by far the simplest to include in land records.

In Islamic and some European systems, inheritance of land historically involves sharing the land among all blood-related descendants, though allocations are variable. This sharing or fragmentation is typically undertaken in one of two general ways depending on the culture and legal system. Both create problems for LAS maintenance. Either the properties are continually subdivided with new parcels distributed to the descendants (sometimes with strips of land a meter or less wide and many hundreds of meters long as in figure 4.1), or the heirs are added as co-owners of the parcel (sometimes yielding hundreds of joint owners for a small parcel). The latter process was historically prevalent in Hawaii. Where land is physically reassigned, population increases induce parcel shrinkage. For example, the desirable economic size for a Javanese farm plot is judged to be about 0.8 ha. The agricultural census of 1993 showed the plot size had fallen to 0.17 ha per household from 0.26 ha per household in 1983. The sustainability of these small individual farms, apart from the sustainability of LAS, is problematic.

Fragmentation causes major problems resulting in uneconomic and unsustainable land-use patterns and unsalable land with too many owners to permit reliable transfer of title. Popular solutions involve land consolidation programs where the small parcels are combined to form usable parcels or the interests in land held by many co-owners are consolidated back to holdings by one or two representative owners. Land consolidation continues to this day in many countries, including in Europe and Japan, and even jurisdictions like the state of Hawaii.

Land markets also use "overreaching" as an alternative solution to fragmentation. It was used in England to reassign land following the agricultural recession of the 1880s via the Settled Land Act of 1882. The land of the aristocracy was tied up in strict settlements that kept land in the titled families for generations making it unsalable. Given the drop in produce prices, funds were unavailable to maintain the land, so the solution was a statutory scheme giving the power of sale to a person closely associated with the land that overreached the owner's interests. While the formal machinery varied, the settled land process transferred the interests out of the land and into a fund of money collected out of the proceeds of sale. The fund was then held in trust in a bank account for the owners according to their respective interests in the land. Claims and disputes were shifted from the land to the money fund. Thus, market processes, rather than a government-sponsored consolidation, were used to return the land to economic use. Other options include the use of adverse possession of whole parcels even in a Torrens-type system such as in New South Wales, Australia. Such an approach can return land to productive use after an economic downturn where owners have walked away from their land decades previously and cannot be contacted.

The failure to incorporate processes for tracking social changes is common in LAS project design. Many of the social changes involve court orders or state-sanctioned and recorded ceremonies (marriages, funerals, divorces) or religious ceremonies. Decisions of the different institutions and agencies engaged in social processes need to be reflected in LAS. Processes of registration must be geared to timely tracking of decisions of non-land agencies and the extent they affect land entitlement. If they are not, LAS inevitably fail, even in a single generation.

In transitional economies, landownership is affected by a whole new problem of absentee owners who leave impoverished rural villages or community farms to work in urban areas, or even in other countries, for example in rural areas of the Philippines, Pakistan, Vietnam, and many parts of Africa. The mobility of labor distorts the processes of identifying people presently entitled to or capable of inheriting a share in land. Dealing with absentee claims is an essential part of many land reform programs in developing countries and their supporting LAS.

Whatever the complications associated with accommodating social processes in land administration, they are comparatively simple when contrasted with the processes used to absorb changes resulting from commercial activities in highly developed land market systems. Debt failure, bankruptcy, corporation failure, and competition among land security holders and others, including equity holders, company security holders, asset security holders, and holders of court orders in general, can all result in land transition. The design of LAS therefore needs to manage changes derived from commercial processes.

FORMING NEW INTERESTS AND PROPERTIES (SUBDIVISION PROCESSES)

Land-use patterns change in tandem with cultural and market conditions. The boundaries associated with a particular use may need modification, either to consolidate pieces of land into one usable parcel or to break up parcels into smaller plots. In market systems, generally an owner will prepare the design and employ a surveyor to lay it out and draw up the plans. The processes are overseen by local authorities. These authorities are generally required to consult interested stakeholders, including owners of adjoining parcels as well as water, electricity, sewerage, telecommunications, and gas authorities. The local authority approves the subdivision, usually with conditions such as roads or other construction carried out to their satisfaction. A final survey is undertaken usually by a professional surveyor, and a subdivision plan is prepared and submitted to a local authority for approval. The final subdivision plan and existing title or ownership documents are submitted to the land registry. New titles for each new parcel are issued in the name of the owner of the original parcel (figure 4.2).

DETERMINING BOUNDARY PROCESSES

Boundary identification normally raises no difficulties, provided the boundaries are well documented and monumented or the adjoining owners agree on their mutual dividing line. Establishing boundaries in LAS involves a series of subprocesses: marking boundaries on the ground, including boundaries in the cadastre or cadastral map, and maintaining consistency

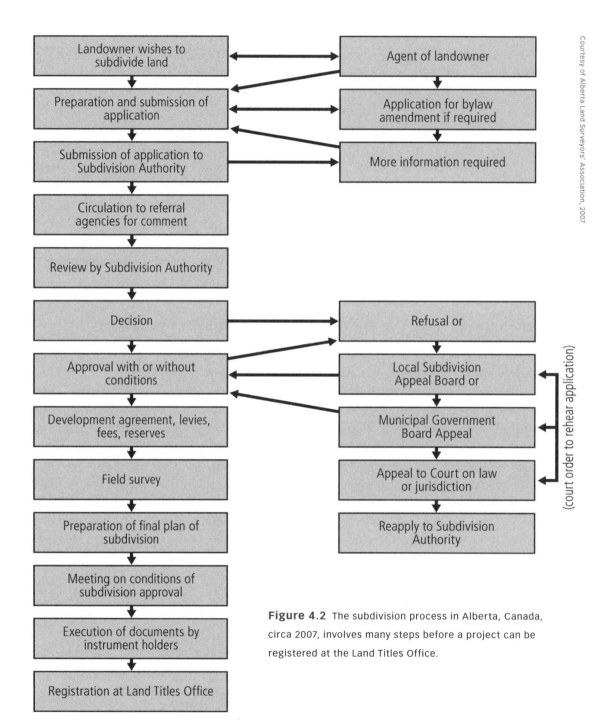

Figure 4.2 The subdivision process in Alberta, Canada, circa 2007, involves many steps before a project can be registered at the Land Titles Office.

between the on-the-ground and recorded boundaries. If there is a disagreement, the system needs another subprocess for determining the boundaries according to a range of criteria, including the history of occupation, the legal standing of boundaries, the physical evidence of a boundary, the title and cadastral information, and availability of skilled surveyors. Table 4.5 illustrates a Danish solution.

The Korean example, in table 4.6, illustrates a simple cadastral surveying process. When sales are negotiated and completed, the new parcels are transferred to the new owners.

The processes are similar in countries that use general boundaries systems, such as Zambia, shown in figure 4.3.

Boundary identification in a cadastral system raises the overriding issue of achieving and maintaining consistency between boundaries on the ground and boundaries in the cadastral record base. The options vary, usually reflecting the different legal solutions adopted to regularize occupation irregularities and adverse possession, on the one hand, and the legal status of boundaries, on the other. In some extremes, incongruity can be eliminated by forcing

TABLE 4.5 – BOUNDARY DETERMINATION PROCESSES USED IN DENMARK, CIRCA 2009	
AGENCY	**ACTIVITY**
Owner	Instructs surveyor to determine boundary
Surveyor	Compares cadastral information to the conditions on the ground. Three situations may occur: • If the field conditions are consistent with the recorded cadastral information, the boundary is final • If there is a prescriptive right acquired by time (20 years), the cadastre must be updated to reflect the new boundaries • If the position of the boundary has changed because of an unrecorded agreement between the neighboring parties, the cadastre must change to reflect the agreed boundaries If the neighbors disagree, the surveyor acts as a judge following a formal procedure to determine the legal boundary and establishes a temporary boundary. The temporary boundary becomes the final boundary, if the court option is not invoked and the parties agree.
Owner and neighbors	An interested party can bring the case to court for official designation of the boundary, but this occurs very rarely
Court	Decides final legal boundary

physical boundaries to agree with title or cadastral boundaries by continuously realigning fences, buildings, and boundary markers in accord with the records. Even though no cadastral survey system is this demanding, rare and highly developed systems, such as in Hamburg, Germany, and the Australian Capital Territory, tend to approach this level of rigor in a coordinated cadastre. On the other hand, in most systems, boundaries may move according to established prescriptive rights for statutory parcels. LAS usually choose solutions between these extremes. Any solution can work with land registration and cadastral recording, provided it is understood by the community, consistently applied, and integrated with other core processes.

The best solutions, however, are the ones that work to reduce boundary disputes and encourage placement of boundary markers on the official boundaries, so that over time, congruity is improved. The combination of practice, understanding, official recording, and boundary recognition rules, rather than any particular principle in itself serves to control these disputes.

TABLE 4.6 – SIMPLE CADASTRAL SURVEYING PROCESS USED IN KOREA, CIRCA 2000		
CLIENT	**KOREAN CADASTRAL SURVEY CORPORATION (SURVEYOR)**	**AUTHORITY**
Apply for cadastral survey	Accept the application	
	Prepare the survey	Approve copying and reading the map and attribute data
Attend the survey	Undertake the field survey	
	Produce the survey	Inspect the survey plan
Receive the survey	Deliver the survey map to client	Deliver the survey result to the Korean Cadastral Survey Corporation
Apply to authority for arrangement of cadastral record		Arrange the cadastral record
Apply to authority for copying of cadastral record		Deliver copy of the cadastral record

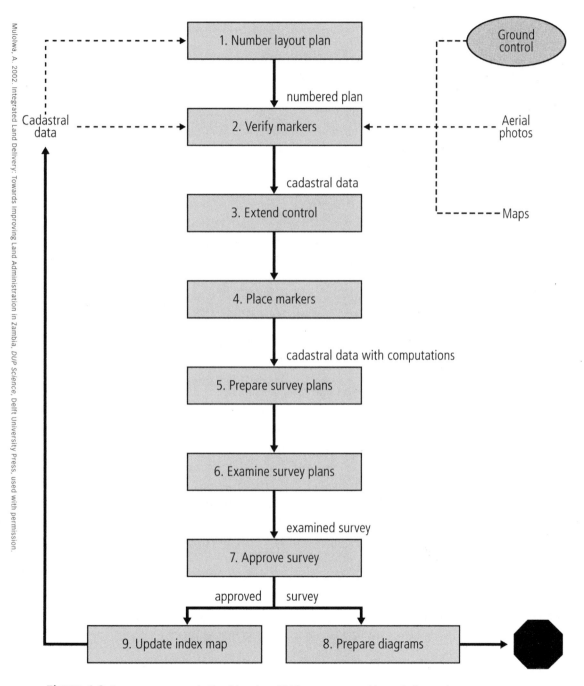

Figure 4.3 Survey processes in Zambia, circa 2002, use a general boundaries system.

4.4 Reforming LAS by improving process management

Analysis of how LAS work led to a renewed interest in reforming particular processes as a means of rebuilding and reengineering systems. The documentation of processes related to land administration or cadastral systems is a common strategy to understand and reform these systems. I. P. Williamson and L. Ting (2001) designed a framework for reengineering land administration and cadastral systems, and D. Steudler, A. Rajabifard, and I. P. Williamson (2004) used the framework as part of an evaluation of LAS.

Much of the recent information about land administration and cadastral dataflows and analyses of their importance form a background for reform efforts. W. W. Effenberg, S. Enemark, and I. P. Williamson (1999) investigated the process related to digital spatial cadastral data. K. Dalrymple, I. P. Williamson, and J. Wallace (2003) highlighted the key land administration processes of land transfer and subdivision. Williamson and C. Fourie (1998) analyzed cadastral processes in the context of understanding cadastral systems from a case study perspective. The most extensive and influential documentation of processes that relate to land activities was done by Hernando de Soto (2000), also as a basis for land administration reform. De Soto's approach to documenting land market processes was used in Vietnam to better understand and reform LAS in support of the rural land market (Smith et al. 2007).

These and other sources of information about LAS processes indicate a high level of volatility in the specific processes and tensions between keeping formal processes geared to business and social needs and driving informal processes into formal systems. These tensions are constant. Others are also evident. The capture of a process by a professional group, whether in a bureaucracy or the private sector, creates opportunities for rent seeking—that is, the extraction of fees and creation of arbitrary power for unproductive activities. An assumption that land processes should include elaborate formalities needs to be replaced by minimum formalities consistent with adequate third-party evidentiary proofs and reliable public records. Fees and charges, including government taxes on land transactions, need to be compatible with the capacity and willingness of participants to pay.

While processes vary, the increased world interest in best practices is drawing much more standardization of systems and sharing of ideas, evidenced by the extensive interest in reengineering of conveyancing and registration processes to achieve e-conveyancing.

Chapter 5

Modern land administration theory

5.1 Designing LAS to manage land and resources

5.2 The cadastre as an engine of LAS

5

5.1 Designing LAS to manage land and resources

THE LAND MANAGEMENT PARADIGM

The cornerstone of modern land administration theory is the land management paradigm in which land tenure, value, use, and development are considered holistically as essential and omnipresent functions performed by organized societies. Within this paradigm, each country delivers its land policy goals by using a variety of techniques and tools to manage land and resources. What is defined as land administration within these management techniques and tools is specific to each jurisdiction, but the core ingredients—cadastres or parcel maps and registration systems—remain foundational to the discipline. These ingredients are the focus of modern land administration, but they are recognized as only part of a society's land management components.

Consolidation of land administration as a discipline in the 1990s as described earlier reflected the introduction of computers and their capacity to reorganize land information. UNECE viewed land administration as referring to "the *processes of determining, recording, and disseminating information* about the ownership, value, and use of land, when implementing land management policies" (1996; emphasis added). The emphasis on information management served to focus LAS design on information for policy makers, reflecting the computerization of land administration agencies after the 1970s. The focus on information remains, but the type and quality of information needed for modern circumstances has changed dramatically. Thus, the need to address land management issues systematically pushes the design of LAS toward an enabling infrastructure for implementing land policies and land management strategies in support of sustainable development. In simple terms, the information approach needs to be replaced by a model capable of assisting design of new or reorganized LAS to perform the broader and integrated functions now required.

This new land management paradigm is described in figure 5.1. The paradigm provides the reason to reengineer agencies and their processes to deliver policy outcomes through more integrated task and information management, rather than merely managing land information for internal purposes. The paradigm enables LAS designers to manage changes in institutional arrangements and processes to implement better land policies and good land governance by identifying a conceptual framework for understanding each system. In theoretical terms, the paradigm identifies the principles and processes that define land management as an endeavor. It recognizes that in practice, the organizational structures for land management differ widely among countries and regions throughout the world and reflect the local cultural and judicial settings of a country. Within the country context, land management activities may be described by three components: land policies, land information infrastructure, and land administration functions that support sustainable development.

The paradigm invites LAS designers to build systems capable of undertaking the core functions of tenure, value, use, and development for the purpose of specifically delivering sustainable development, in addition to implementing national land policy and producing land information. The key tenet of the paradigm is that proper design of the land management components and their interaction will lead to sustainable development. While sustainability goals are fairly loose, the paradigm insists that all core LAS functions are considered as a whole, and not as separate, stand-alone exercises.

Land management is broader than land administration. It covers all activities associated with the management of land and natural resources that are required to fulfill political objectives and achieve sustainable development. Land management is then simply the processes by which a country's resources are put to good effect (UNECE 1996). Land management requires interdisciplinary skills based on the technical, natural, and social sciences. It is about land policies, land rights, property, economics, land-use control, regulation, monitoring, implementation, and development. The concept of land includes properties, utilities, and natural resources and encompasses the total natural and built environments within a national jurisdiction, including marine areas.

Land management activities reflect the development agents of globalization and technology. They stimulate the establishment of multifunctional information systems, incorporating diverse land rights, land-use regulations, and other useful data. But a third force for change is sustainable development. It stimulates demand for comprehensive information about environmental, social, economic, and governance conditions in combination with other land-related data.

Land policy is part of the national policy on promoting objectives such as economic development, social justice and equity, and political stability. Land policies vary, but in most countries, they include poverty reduction, sustainable agriculture, sustainable settlement, economic

Enemark, S., I. Williamson, and J. Wallace. 2005, Building Modern Land Administration Systems in Developed Economies, *Journal of Spatial Sciences*, Perth, Australia, Vol. 50, No. 2, pp 51–68, used with permission.

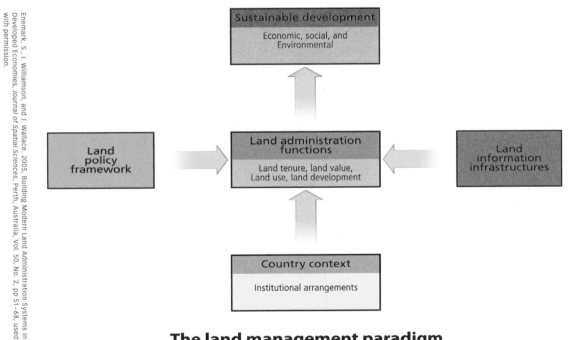

The land management paradigm

Figure 5.1 Within the country context, land management activities may be described by three components: land policy, land information infrastructure, and land administration functions in support of sustainable development.

development, and equity among various groups within society. Policy implementation depends on how access to land and land-related opportunities is allocated. Governments regulate land-related activities, including holding rights in land, supporting the economic aspects of land, and controlling the use of land and its development. Administration systems surrounding these regulatory patterns facilitate the implementation of land policy in the broadest sense, and in well-organized systems, they deliver sensible land management, good governance, and sustainability.

The operational component of the land management paradigm is the range of land administration functions that ensure proper management of rights, restrictions, responsibilities, and risks in relation to property, land, and natural resources. These functions include the processes related to land tenure (securing and transferring rights in land and natural resources); land value (valuation and taxation of land and properties); land use (planning and control of the use of land and natural resources); and, increasingly important, land development (implementing utilities, infrastructure, and construction planning). These functions interact to deliver overall policy objectives and are facilitated by appropriate land information infrastructure that includes cadastral and topographic datasets.

Sound land management requires operational processes to implement land policies in comprehensive and sustainable ways. Many countries, however, tend to separate land tenure rights from land-use opportunities, undermining their capacity to link planning and land-use controls with land values and the operation of the land market. These problems are often compounded by poor administrative and management procedures that fail to deliver required services. Investment in new technology will only go a small way toward solving a much deeper problem: the failure to treat land and natural resources as a coherent whole.

All nations have to deal with the management of land. They have to deal with the four land administration functions of land tenure, land value, land use, and land development in some way or another. A country's capacity may be advanced and combine all the activities in one conceptual framework supported by sophisticated information and communications technology models. More likely, however, capacity will involve very fragmented and basically analog approaches. Different countries will also put varying emphasis on each of the four functions, depending on their cultural bias and level of economic development.

In conclusion, modern land administration theory requires implementation of the land management paradigm to guide systems dealing with land rights, restrictions, and responsibilities to support sustainable development. It also requires taking a holistic approach to management of land as the key asset of any jurisdiction.

A GLOBAL LAND ADMINISTRATION PERSPECTIVE

A global perspective is needed to share experiences in designing LAS and diagnose trends in design and implementation of local systems. According to this view, LAS ideally sit within the land management paradigm as the core infrastructure for achieving sustainable land management. The global land administration perspective is shown by enlarging the role of the land administration functions at the center of the paradigm, then linking them with each other to support efficient land markets and effective land-use management. In turn, market and management activities must work to promote sustainable development.

The four land administration functions (land tenure, land value, land use, and land development) are different in their professional focus, and are normally undertaken by a mix of professionals, including surveyors, engineers, lawyers, valuers, land economists, planners, and developers. Furthermore, the actual processes of land valuation and taxation, as well as the actual land-use planning processes, are often not considered part of land administration activities. However, even if land administration is traditionally centered on cadastral activities in relation to land tenure and land information management, modern LAS designed as described in figure 5.2 deliver an essential infrastructure and encourage integration of the four functions:

◆ **Land tenure:** the processes and institutions related to securing access to land and inventing commodities in land and their allocation, recording, and security; cadastral mapping and legal surveys to determine parcel boundaries; creation of new

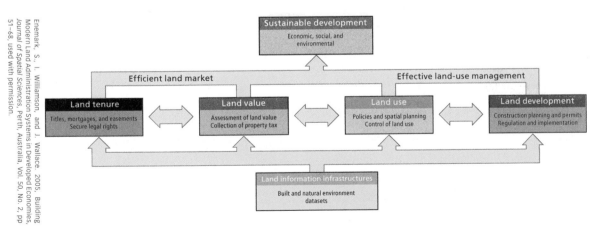

Figure 5.2 A global land administration perspective promotes sustainable development through efficient land markets and effective land management.

properties or alteration of existing properties; the transfer of property or use from one party to another through sale, lease, or credit security; and the management and adjudication of doubts and disputes regarding land rights and parcel boundaries.

◆ **Land value:** the processes and institutions related to assessment of the value of land and properties; the calculation and gathering of revenues through taxation; and the management and adjudication of land valuation and taxation disputes.

◆ **Land use:** the processes and institutions related to control of land use through adoption of planning policies and land-use regulations at the national, regional, and local level; the enforcement of land-use regulations; and the management and adjudication of land-use conflicts.

◆ **Land development:** the processes and institutions related to building new physical infrastructure and utilities; the implementation of construction planning; public acquisition of land; expropriation; change of land use through granting of planning permissions, and building and land-use permits; and the distribution of development costs.

Sustainable development policy requires the four functions to be integrated. This is achieved in four general ways:

1. In theory, the functions are approached as four parts of a coherent whole, not as independent activities. This means that each function is not an end in itself, but all four together are the means to support sustainable development.

2. The processes used to perform the functions must pursue sustainable development, ideally within a broad framework of monitoring and evaluation of performance against sustainability outcomes.

3. Information and outputs generated by processes need to be mutually shared and made widely accessible.

4. All functions must be built on core cadastral knowledge.

Inevitably, all four functions are interrelated. The interrelations appear because the conceptual, economic, and physical uses of land and properties serve as an influence on land values. Land values are also influenced by the possible future use of land determined through zoning, land-use planning regulations, and permit-granting processes. And land-use planning and policies will, of course, determine and regulate future land development.

Land information should be organized to combine cadastral and topographic data and to link the built environment (including legal and social land rights) with the natural environment (including topographical, environmental, and natural resource issues). Land information should, in this way, be organized through an SDI at the national, regional, federal, and local level, based on relevant policies for data sharing, cost recovery, access to data, data models, and standards (see chapter 9, "SDIs and technology").

Ultimately, the design of adequate systems of land tenure and land value should support efficient land markets capable of supporting trading in simple and complex commodities (see chapter 6, "Building land markets"). The design of adequate systems to deliver land-use control and land development should lead to effective land-use management (see chapter 7, "Managing the use of land"). The combination of efficient land markets and effective land-use management should support economic, social, and environmental sustainable development.

From this global perspective, LAS act within adopted land policies that define the legal regulatory pattern for dealing with land issues. LAS also act within an institutional framework that imposes mandates and responsibilities on various agencies and organizations. LAS should service the needs of individuals, businesses, and the community at large. Benefits arise through the LAS guarantee of ownership, security of tenure, and credit; facilitation of efficient land transfers and land markets; support of management of assets; and provision of basic information and efficient administrative processes in valuation, land-use planning, land development, and environmental protection. LAS designed in this way form a backbone for society and are essential for good governance, because they deliver detailed information and reliable administration of land from the basic level of individual land parcels (figure 5.3) to the national level of policy implementation.

CADASTRAL SYSTEMS

Modern land administration theory acknowledges the history of the cadastre as a central tool of government infrastructure and highlights its central role in implementing the land management paradigm. However, given the difficulty of finding a definition that suits every version (see section 2.2, "Historical evolution"), it makes sense to talk about cadastral systems rather than just cadastres (figure 5.4). These systems incorporate both the identification of land parcels and the registration of land rights. They support the valuation and taxation of land and property, as well as the administration of present and possible future uses of land. Multipurpose cadastral systems support the four functions of land tenure, value, use, and development to deliver sustainable development.

By around 2000, cadastral systems were seen as a multipurpose engine of government operating best when they served and integrated administrative functions in land tenure, value, use, and development and focused on delivering sustainable land management. A mature multipurpose cadastral system could even be considered a land administration system in itself. This multipurpose design is the touchstone of best practices, sought by many LAS designers and managers. Achieving it, however, is another story, because each unique existing system needs a different group of strategies to implement the proposed multipurpose design.

The way forward can, nonetheless, be rationalized. The dominance of market economic theory and the influence of colonialism suggest three general formal approaches have historically influenced the design of cadastral systems as described under the subheading "Importance of the cadastre" in section 2.2, "Historical evolution" (see table 2.3). The German and Torrens approaches are able to include a spatial cadastre directly within a state or national SDI, thus delivering actual and potential advantages to countries that use either of these two approaches.

Most countries using the French/Latin approach make only a loose connection between the cadastre and deeds registries, and in many, the two activities are completely distinct and separate. Therefore, in practice, countries relying on this approach often have great difficulty including a spatial cadastre within an SDI and commensurate difficulty in supporting effective LAS, especially where the registry functions are carried out in the private sector.

Figure 5.3 Land-use pattern divides land into minor parcels for separate and individual use in the Philippines.

Courtesy of Land Administration and Management Project, DENR, Philippines

The German and Torrens approaches often create confusion in the minds of people seeking to understand or design cadastres and LAS that work together, since the cadastral engine of each model plays only one, or predominantly one, of two different roles. Historically, in the German approach, the concept of a complete and multipurpose cadastre has been adopted, in many cases for more than a century. Thus, the cadastre always supported separate activities in land tenure, value, use, and development and sometimes other functions, including, for example, showing buildings that are linked to insurance identifiers (Switzerland).

In the multipurpose German approach, operation of the land market generally stood aside from the cadastre, leaving land market activity primarily to the *Grundbuch*, or land registry, an institution still often found in a Ministry of Justice but based on the cadastral identification.

The Torrens approach has a shorter history and only evolved in the most advanced countries to include a complete spatial cadastre after about 1970, or even 1980. The original focus was on building a land registry with a dual function of supporting titles, deeds, and tenure, as well as

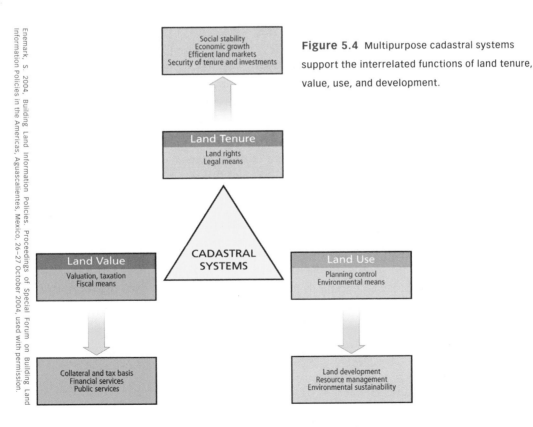

Figure 5.4 Multipurpose cadastral systems support the interrelated functions of land tenure, value, use, and development.

legal surveys, then eventually, cadastral mapping. Historically, the function of the land registry in Torrens jurisdictions was to support a land market. The system used individual isolated surveys to identify parcels and supported transfers and other essential transactions as they occurred. Indexes were created and referenced to charting maps of various accuracy and currency. The focus was usually not on the charting maps but on the individual surveys—the charting maps simply helped to locate isolated cadastral surveys. In the 1970s and 1980s, jurisdictions using this approach upgraded their charting maps to show all land parcels. In many developed systems, these maps achieved a high degree of accuracy and currency to the point that they equal the attributes of the spatial cadastres used in the German approach. However, even today, many land registries in these jurisdictions still focus on their land market function, with the spatial cadastre a secondary objective, if at all. In practical terms, they usually do not achieve the multipurpose results of the central European approach and bear the inefficiencies of duplication. A LAS model that integrates the cadastre with land tenure activities is now considered best practice.

So, in summary, land market activity under the German approach is separated from the primary objective of creating and maintaining a cadastre. By contrast, in the Torrens approach, the land market is the main focus of the land registry with tenure and cadastral survey activities closely integrated and the spatial cadastre eventually developing as an additional benefit. The international trend to amalgamate the cadastre (cadastral surveying and mapping) and tenure activities in the land registry is more and more evident. In the last decade, this occurred in the Netherlands and Sweden, for example; both countries run world-class LAS. So, in one sense, the first two approaches are coming closer together. The systems using the German approach are moving to adopt the principles inherent in the most sophisticated systems used in the Torrens approach where the cadastre and tenure activities are fully integrated.

An even more important principle for countries using the German and Torrens approach (and some using the French approach, such as in France and Spain) is that a complete spatial cadastre is produced to form a key layer(s) in their national SDI. However, for any jurisdiction, this goal raises the major challenge of integrating cadastral (or built environmental) data with other topographic (or natural resource) data. It arises in part because most countries historically separate their cadastral and land registry activities from national mapping activities. These separate and silo administration systems have historically different data models and different cultures surrounding the creation and maintenance of the two types of data. This issue is explored in more depth in chapter 9, "SDIs and technology."

In the array of situations found throughout the world, many countries lack the capacity to build even rudimentary systems, and others use relatively partial and even informal systems, including ad hoc and messy systems in response to massive and uncontrolled urban expansion. Cities like Jakarta, Indonesia; Lagos, Nigeria; Kabul, Afghanistan; Manila, the Philippines; Mexico City, Mexico; São Paulo, Brazil; and others illustrate situations where administrative capacity lags well behind need. For all situations however, the land management paradigm is capable of assisting stakeholders in developing LAS that improve national options. For countries with limited capacity, the paradigm defines a path toward improving land management capacity and building robust administrative systems.

CADASTRAL UNITS—PROPERTIES, PARCELS, AND ENTITIES

The most important component of any cadastral system is the land parcel. A land parcel is normally understood as a single piece of land that is determined geographically by its boundaries and held under relatively homogeneous property rights. The UNECE "Guidelines on real property units and identifiers" (2004) provides a framework within which appropriate identifiers can be developed (also see section 12.3, "Professional tools"). The guidelines show that terminology varies widely across Europe, for example. Just as there is no unique cadastral solution that fits all countries, so there is no unique land parcel or address system.

In land administration theory, a property is normally understood as a legally defined term for ownership of land units. A property may consist of one or several land parcels, and each parcel may consist of several plots, where a plot is something that can be plotted on a map and is often equivalent to the way in which the land is used or managed. Each parcel needs a unique identifier so that data concerning the parcel can be given an exclusive reference. The form of this reference varies from country to country. Within the land book, or land register, and cadastral systems, the identifiers currently used generally reflect historical practice rather than contemporary need (UNECE 2004).

The relationship between properties and parcels is often problematic because "land parcel" has different meanings in different countries, and its use in conjunction with the term "property" is also variable. The Cadastral Template (see section 10.3, "The Worldwide Cadastral Template Project") illustrates the problem of identifying land units in terms of land parcels or properties. Three scenarios, illustrated in figure 5.5, are presented to distinguish simple differences in the ways the two terms are used. While the surveyed or registered units (in thicker lines) may be different for each of the scenarios described as follows, the number of the smallest uniquely identified units in each case would be fifteen:

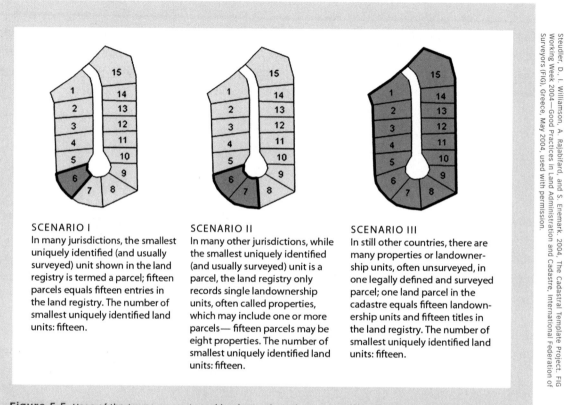

SCENARIO I
In many jurisdictions, the smallest uniquely identified (and usually surveyed) unit shown in the land registry is termed a parcel; fifteen parcels equals fifteen entries in the land registry. The number of smallest uniquely identified land units: fifteen.

SCENARIO II
In many other jurisdictions, while the smallest uniquely identified (and usually surveyed) unit is a parcel, the land registry only records single landownership units, often called properties, which may include one or more parcels— fifteen parcels may be eight properties. The number of smallest uniquely identified land units: fifteen.

SCENARIO III
In still other countries, there are many properties or landowner-ship units, often unsurveyed, in one legally defined and surveyed parcel; one land parcel in the cadastre equals fifteen landown-ership units and fifteen titles in the land registry. The number of smallest uniquely identified land units: fifteen.

Figure 5.5 Uses of the terms property and land parcel can have very different meanings: In scenario I, a surveyed land parcel is shown. In scenario II, a surveyed property containing two land parcels is shown. And in scenario III, a surveyed parcel consisting of fifteen landownership units is shown.

Steudler, D., I. Williamson, A. Rajabifard, and S. Enemark. 2004. The Cadastral Template Project. FIG Working Week 2004—Good Practices in Land Administration and Cadastre, International Federation of Surveyors (FIG), Greece, May 2004, used with permission.

Despite the variable meanings of the terms property, land parcel, and landownership unit in jurisdictions throughout the world, the land unit—normally understood as the land parcel identified in the cadastre—is the key object in LAS. The systematic treatment of these key objects requires well-designed cadastral systems.

5.2 The cadastre as an engine of LAS

IMPLEMENTING THE PARADIGM

The land management paradigm makes a national cadastre the engine of LAS, underpinning a country's capacity to deliver sustainable development. Though the paradigm is neutral to how a country's cadastre developed, systems based on the German and Torrens approaches are much more easily focused on land management than systems based on the French/Latin approach.

The cadastre as an engine of LAS is shown in figure 5.6. The diagram highlights the usefulness of the large-scale cadastral map as a tool by exposing its power as the representation of the human scale of land use and how people are connected to land. The digital cadastral representation of the human scale of the built environment and the cognitive understanding of the

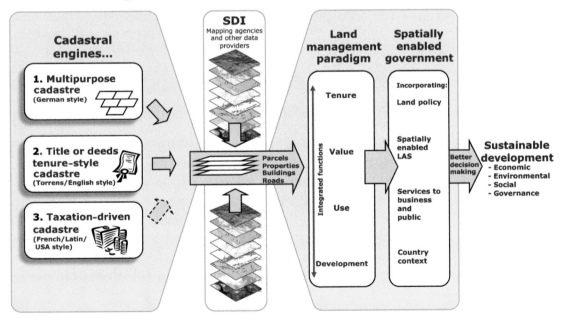

Figure 5.6 The "butterfly" diagram shows the cadastre as the engine of LAS and the means to implement the land management paradigm. The cadastral information forms a key component within the SDI as it supports each of the four land administration functions for delivery of sustainable development.

land-use patterns seen in people's farms, businesses, homes, and other developments then form the core information that enables a country to build an overall administrative framework to deliver sustainable development.

The neutrality of the paradigm in terms of any actual national cadastral approach is emphasized by showing how all three formal approaches used throughout the globe are capable of feeding into a national SDI and then into sustainable development. Wherever the cadastre sits in a national land administration system, ideally it should assist the functions of tenure, value, use, and development. In this way, the cadastre or cadastral system becomes the technical engine of LAS, delivering the capacity to control and manage land through the four land administration functions. The cadastre supports business processes of tenure and value, depending on how it is locally built. It identifies legal rights, where they are, the units that form the commodities, and the economy in relation to property. These cadastres are much more than a layer of information in a national SDI.

While these connections are usually thought of as computer generated, even in manual systems, cadastral information about parcel attributes and their unique identifiers can be used throughout the four land administration functions to implement the land management paradigm and to deliver efficiencies for government services and businesses. The requirement that this vital information should be created once and used many times underscores the identification of the cadastre as the authoritative register of parcel information—an idea appropriate for any formal system, whether digitized or not. In this way, the paradigm provides a foundation for eventual digital conversion of emerging LAS processes for countries about to embark on upgrading their system.

The diagram demonstrates that the cadastral information layer cannot be replaced by a different spatial information layer derived from GIS. The unique cadastral capacity is to identify a parcel of land both on the ground and in the system in terms that all stakeholders can relate to—typically an address, plus a systematically generated identifier (given that addresses are often duplicated or are otherwise imprecise). The core cadastral information of parcels, properties, sometimes buildings, and in many cases, legal roads, thus becomes the core of SDI information, feeding into utility infrastructure, hydrology, vegetation, topography, imagery, and dozens of other datasets.

The diagram is a virtual butterfly: One wing represents the cadastral processes, and the other the outcome of using the processes to implement the land management paradigm. Once the cadastral data (cadastral or legal parcels, properties, parcel identifiers, buildings, legal roads, etc.) is integrated within the SDI, the full multipurpose benefit of LAS, so essential for sustainability, can be achieved.

The body of the butterfly is the SDI, with the core cadastral information acting as the connecting mechanism. This additional feature of cadastral information is an additional role, adding to the traditional purposes of servicing the four land administration functions. This new function takes the importance of cadastral information beyond the land administration framework, enlarging its capacity to service other essential functions of government, including emergency management, economic management, administration, community services, and many others. In advanced systems, integrated cadastral layers within a jurisdiction's SDI ideally deliver spatially enabled LAS to support the four functions of land tenure, land value, land use, and land development. However, building this kind of interaction among the four functions is not easy. The historic institutional silos, separate databases, separate identifiers, and separate legal frameworks need to be reorganized. For most countries, this presents another major land administration challenge.

Since 2000, and especially since 2005, new spatial technologies raise entirely new possibilities for using the cadastre and cadastral information to service government and business (see chapter 14, "Future trends"). Even though cadastral systems around the world are clearly different in terms of structure, processes, and actors, they are increasingly merging into a unified global model in which the multipurpose cadastre takes on increased importance. Globalization and technology development support establishment of multifunctional information systems with regard to land rights and land-use regulations in combination with comprehensive information about environmental conditions. As a result, the traditional surveying, mapping, and land registration focus of LAS has moved away from being primarily provider driven to now being clearly user driven. Thus, the land management paradigm offers a means of adapting the cadastral engine in ways that were not available a decade earlier to serve open-ended functions essential to modern governments. From this perspective, the butterfly diagram is a key theoretical graphic in this book.

A LAND MANAGEMENT VISION

New opportunities present an emerging challenge for LAS design: using the cadastre to incorporate LAS into the land management paradigm. The success of a cadastral system depends on how well it internalizes these new influences while achieving broader social, economic, and environmental objectives. One of the ways the cadastre performs these wider functions is by institutionalizing spatial enablement—that is, facilitating the use of spatial information.

There are many forms of spatial information, ranging from coordinated positioning data using GPS to much wider uses of the concept of position or location to spatially enable information. These wider uses open up a world of information types whose importance lies in their capacity to provide spatial enablement by relating the information to a specific place. For instance, land boundaries, cadastres, topographic information, demographic information, natural resource data, and many other forms of information are now spatially enabled. From the perspective of building and reengineering LAS, information that potentially offers the significant benefits delivered by new spatially enabling technologies includes

- **Land administration information** generated by cadastral, land recording, and sometimes valuation activities

- **Land information** about land use, land planning, and some rights records

- **Geographic information** about terrain, natural resources, and infrastructure that relates to them

Spatial enablement is just one form of interoperability stemming from the capacity of a computer to identify "where" something is. It is, however, far more versatile than a mere organizational tool and offers opportunities for visualization, scalability, and user functionality. The capacity of computers to place information in on-screen maps and to allow users to make their own inquiries has raised the profile of spatial enablement. This is further underpinned by the "open systems" of service-oriented IT architecture that allows governments, enterprises, organizations, and citizens to build their own applications on top of authentic registers and maps and their connected data services (see chapter 9, "SDIs and technology").

Spatial enablement of LAS will increase the usefulness of the information they generate. When the interaction between the four key functions is made operational through spatial enablement, LAS themselves are spatially enabled and can play a central role within the land management vision that in turn will support sustainable development. Wider opportunities for spatial enablement throughout government also arise. A spatially enabled government organizes its business and processes around "place"-based technologies—as distinct from using two-dimensional maps and visuals—and Web enablement.

While the paradigm conceptually unites land management arrangements, a vision for modern LAS within the paradigm is needed to generate potential dynamic responses to contemporary developments. A vision was developed in an Expert Group Meeting on Incorporating Sustainable Development Objectives into ICT–Enabled LAS held in Melbourne, Australia, in November 2005 (Williamson, Enemark, and Wallace 2006). Compared with the paradigm, the vision recognizes that land management activities must include a strong focus on benefits for people and businesses. Feedback is encouraged to aid ongoing adaptation and innovation. The vision also aims to integrate land information infrastructure with land administration functions to form what is called spatially enabled land administration (figure 5.7).

Through spatially enabled land administration, other opportunities also open up. LAS in developed economies can promote sustainable development of the built and natural environments through public participation alongside informed and accountable government decision making. The interface between the land administration infrastructure and professions and the public will expand as ICT helps implement e-government and e-citizenship. While e-citizenship mobilizes society to engage in planning, use, and allocation of resources, using technology to facilitate participation, e-government involves a government agency putting government information and processes online and using digital systems to assist public access and service. Ultimately, e-government is e-democracy—allowing government of, by, and for the people through the use of the Web.

The land management vision presents another major challenge for LAS designers—that is, for a jurisdiction to understand and accept the vision as well as the operation and interaction of the key components as being the cadastre, the SDI, and spatial enablement of LAS. Sustainable development objectives will then be easier to achieve and evaluate. Adaptability and usability of modern spatial systems will encourage more information to be collected and made available.

Williamson, I. P., S. Enemark, and J. Wallace. 2006. Incorporating sustainable development objectives into land administration, Proceedings FIG XXIII Congress, Shaping the Change, Munich, 2006, used with permission.

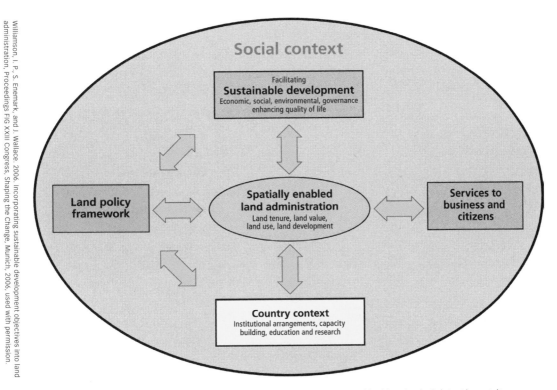

Figure 5.7 The central land management vision has spatially enabled land administration at its core.

Improved information chains will help governments develop and implement a suitable land policy framework. The services available to the private and public sectors, and to community organizations, should then commensurately improve. Ideally, these processes are interlinked: Modern ICT, the engagement of users in the design of suitable services, and the adaptability of new applications should all increase and have a mutually positive influence on each other.

HIERARCHY OF LAND ISSUES

The motivation to respond to change in any particular jurisdiction will depend on how local leaders and decision makers understand the land management vision. While the larger theoretical framework described here is futuristic for many countries, LAS must still be designed around the land management paradigm. A simple entry point showing how to do this (figure 5.8) uses a hierarchy of land issues to illustrate how the concepts involved in the paradigm and the vision for spatially enabled land administration fit together, building on the land parcel:

- ◆ **Land policy** determines values, objectives, and the legal regulatory framework for management of a society's major asset—its land.

- ◆ The **land management paradigm** drives a holistic approach to LAS and forces their land administration processes to contribute to sustainable development. The paradigm allows LAS to facilitate overall land management. Land management activities

Figure 5.8 The hierarchy of land issues forms an inverted pyramid with land policy at the top and the land parcel at the bottom.

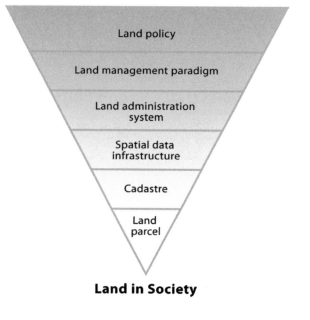

Land policy

Land management paradigm

Land administration system

Spatial data infrastructure

Cadastre

Land parcel

Land in Society

include the core land administration functions of land tenure, value, use, and development while encompassing all activities associated with the management of land and natural resources that are required to achieve sustainable development.

◆ The **land administration system** provides the infrastructure for implementation of land policies and land management strategies and underpins the operation of efficient land markets and effective land-use management. The cadastre is at the core of a land administration system.

◆ The **SDI** provides access to and interoperability of cadastral and other land-related information.

◆ The **cadastre** provides the spatial integrity and unique identification of every land parcel, usually through a cadastral map updated by cadastral surveys. The parcel identification provides the link for securing rights in land, controlling the use of land, and connecting the ways people use land with their understanding of land.

◆ The **land parcel** is the foundation of the hierarchy, because it reflects the way people use land in their daily lives. It is the key object for identification of land rights and administration of restrictions and responsibilities in the use of land. The land parcel links the system with the people.

The hierarchy illustrates the complexity of organizing policies, institutions, processes, and information for the purposes of dealing with land in society. But it also illustrates an orderly approach represented by the six levels. This conceptual understanding provides the overall guidance for building a land administration system in any society, no matter its level of development. The hierarchy also provides a framework for adjustment or reengineering of existing LAS. This process of adjustment should be based on constant monitoring of the results of land administration and land management activities. Land policies may then be revised and adapted to meet the changing needs of society. The change of land policies will require adjustment of LAS processes and practices that, in turn, will affect the way land parcels are held, assessed, used, or developed.

Part 3

Building modern systems

Part 3 is the core of the book and explains all the dimensions of building modern land administration systems (LAS). It starts in chapter 6 with a detailed investigation of a land administration view of land markets and how to build effective land markets. Chapter 6 includes an important discussion of how to build infrastructure to support the evolutionary stages of formal markets. Recognizing that land rights are covered in depth in other parts of the book, the three other land administration functions of land use, land value, and land development are examined.

The book does not purport to cover these functions in depth as it does with land rights but introduces them as part of the land management paradigm. However, chapter 7 explores managing the use of land more in depth because of its central role in land management and sustainable development. As such, chapter 7 explores planning control systems, urban land-use planning and regulations, rural planning and sectoral land-use regulations, land consolidation and readjustment, and integrated land-use management.

Part 3 includes an introduction to marine administration in chapter 8 in recognition of the fact that land administration does not stop at the water's edge. This chapter introduces the concept of marine administration and the challenges in building effective systems. It looks at existing systems and introduces the marine cadastre concept as well as the key components of marine registers and marine SDIs.

A detailed review of SDIs and technologies used in LAS is presented in chapter 9. This chapter explains why we need SDIs to support land management, and it introduces SDI concepts. It explores the importance of effectively managing information about the natural and built environments and how to make appropriate ICT choices. A new approach to cadastral data modeling as part of modern LAS is examined.

Part 3 concludes by presenting an overview of worldwide land administration activities in chapter 10. It highlights the importance of land projects in LAS activities and gives an insight to recent land administration activities that draw upon the concepts and activities presented in the book. The Worldwide Cadastral Template Project is introduced. This is a joint initiative of the United Nations-supported Permanent Committee on GIS Infrastructure for Asia and the Pacific (PCGIAP) and the International Federation of Surveyors (FIG).

Chapter 6
Building land markets

6.1 A land administration view of land markets

6.2 Building infrastructure to support formal markets

6.3 Land valuation and taxation

6

6.1 A land administration view of land markets

MANAGING LAND MARKETS

Land administration as a discipline relies principally on engineering methodology to design, build, and manage effective institutional infrastructure to achieve established policy goals. Creating and managing dynamic land markets are the most common reasons why governments invest in LAS. Countries wanting an effective land market need to bring land into a market distribution system. This involves identifying both the land and the commodities related to that land through suitable infrastructure. When infrastructure (including core land administration institutions and processes related to tenure, value, use, and development) is built to support the land management paradigm, daily functions of the market are capable of delivering sustainable development, including social and environmental goals, not just economic goals. The land management paradigm allows detailed examination and understanding of land markets and suggests opportunities for substantial improvement of LAS design. Practically speaking, however, relating markets to a land management paradigm is a remote vision for most countries, and achievable by very few. In the short term, our understanding of how LAS works with land markets needs improvement.

Most existing LAS treat land markets only as simple land trading; the land itself is perceived as the commodity. Descriptive and analytical literature about land markets generally comes from the discipline of economics and focuses on the activities of buying and selling, leasing, developing, using capital, raising credit, and so on. The business end of land markets also receives a great deal of attention, because it is the public face of local, and even global, land markets. Comparative economic analyses track relative levels of market activity, pricing, and investment patterns. New approaches are evident within the framework of economics. Under "new institutional economics," economists can use a multidisciplinary approach to examine the relationships between the institutions of property rights and the economic activities involved in land use, particularly those promoting sustainability (Auzins 2004; North 1990). Institutional economics shows the need for comprehensive and integrated institutions that incorporate all aspects of land management. Indeed, the lack of integrated institutions is recognized as a reason for many of the difficulties experienced in converting centralized systems of land management in ex-Soviet command economies into land markets (Dale and Baldwin 1998; Auzins 2004). The new institutional economics approach is compatible with land administration theory in this book. Together, they identify a generic and pressing problem: Modern land markets are now multilayered and complex, while formal LAS, even in developed economies, still manages only the simple trading of land.

RELATIONSHIPS BETWEEN FORMAL AND INFORMAL MARKETS

Land markets can be formal or informal, but all markets require an administrative system and established rules of the game. In the land administration discipline, a market is more or less formal according to the level that its activities are serviced by public, authorized systems provided by, or at least organized through, government. There are, of course, many markets that operate beyond government, under the auspices of some local system; some are even illegal. Globally, markets in land and land-related commodities are more likely to be informal than formal. For land administration as a discipline, the art is to formalize systems as much as possible when governments and communities decide to build effective markets. Standard processes of formalization involve creating infrastructure to manage processes to deliver registration, valuation and taxation, and planning and development.

Of the 227 world nations and discrete jurisdictions, only about forty or so, depending on the criteria used in the count, can claim they run effective formal, comprehensive, national land markets. Arguably, these include most of the thirty countries that ratified the Convention on the Organization for Economic Cooperation and Development (OECD) — Australia, Austria, Belgium, Canada, the Czech Republic, Denmark, Finland, France, Germany, Greece, Hungary, Iceland, Ireland, Italy, Japan, Korea, Luxembourg, Mexico, the Netherlands, New Zealand, Norway, Poland, Portugal, Slovakia, Spain, Sweden, Switzerland, Turkey, the United Kingdom, and the United States (OECD 2005). Some aspiring members, and others with broad-based economies, might also be included. In the remaining countries, informal land markets make up an important, and sometimes the only, system operating in the land economy. For example, in 2001, 92 percent of apartments in Egypt were not registered (Galal and Razzaz 2001, 2). In Hanoi in 2005, only 15 percent of land was subject to land-use right certificates. In both these environments, and in their equivalents worldwide, land markets operated informally.

Informal land markets organize and permit simple land transactions and social transitions of entitlement. They are sometimes very successful, at least in terms of the level of local trading they support. However, they have major limitations. They lack the infrastructure used in developed countries to deliver public confidence or to attract the participation of formal financial institutions in the trading processes. Their rules are not apparent, and therefore, the interests in land are frequently unrefined, irregular, happenstance, or, worse, insecure. These informal markets therefore cannot attract formal, institutional credit at competitive rates, develop into complex commodity markets, or support secondary levels of trading at standards comfortable for global investment. Their success in organizing the processes of trading land among participants depends on local systems of enforcement that are often far from transparent. To the extent these are effective, local markets will allow land to be bought and sold, leased, and shared successfully among selected players, typically only between members of the group and insiders, and often with severe constraints. These informal markets sometimes operate in countries that provide parallel, legalized formal market systems in order to reduce the human and financial overhead of doing business, and because some people prefer local and informal practices over expensive formalities. In countries with parallel markets using varying degrees of formalization, large-scale developments and transactions in high-value land tend to engage the most formal processes.

The distinction between formal and informal land markets is not black and white. Both kinds of markets can operate simultaneously (figure 6.1), transitional processes are frequently ad hoc, and the differentiation involves the degrees to which markets are formalized rather than definitive separations. The most successful markets have converted virtually the entire realm of land-related activities to formal processes managed through official systems. In many cases, building an effective infrastructure took hundreds of years and countless human and financial resources.

Informality is not synonymous with simplicity. Informal markets can feature complicated processes of trading and inheriting land. The evidentiary practices can also be complicated. They tend to lack transparency to strangers, and they reinforce the exclusionary functions of their beneficial group. They sometimes involve highly refined systems of microcredit. The design of any LAS or development project needs to account for the features of, and practices used in, local informal land markets and offer appropriate and attractive transitional processes that lead to a more formal market.

Informal land markets might escape official organization, but they can sometimes provide comfortable levels of tenure security nonetheless. They can also involve high-value transactions. The informal urban land market of Hanoi generated high unofficial land values, comparable with other, more organized Asian cities, such as Singapore and Hong Kong. The Hanoi market in 2004 was lively, expensive, and managed by its participants according to local systems. Jakarta, Indonesia, likewise, experienced growth in land market activities outside the purview of its national land agency, Badan Pertanahan Nasional (BPN). Similar high-activity informal land markets exist in other nations. These situations raise significant issues, including loss of government revenue (especially transaction taxes), lack of formal credit systems, difficulties in providing land for commercial development and social housing systems, and ad hoc provision of services.

Successful formal land markets do not require that all land interests and commodities be included in formal processes. Indeed, all countries, even those with complex land markets, allow informal activities and trade in commodities beyond government purview. In common law countries, most of the trusts used to organize land are "off the register" and not formally accounted for. Most countries do not register domestic or residential leases: It is too much trouble for little return. Nor do successful land markets generate universal approval. In all countries where they operate, markets spawn some opposition. Also, in many countries, nonmarket relationships with land are used by groups alongside successful formal land markets. Canada, the United States, New Zealand, Japan, Sweden, and Australia, among others, support indigenous groups who reject land markets. Thus, the extent of formal and informal systems is both variable and changeable, and in many countries, the processes of transition between the two kinds of markets are sufficiently complicated as to require multidisciplinary description and approaches.

FORMAL LAND MARKETS

Successful formal land markets require institutions organized by government. Institutions include the agencies and organizations (land registries and cadastral authorities) and, most important, the institution of property. LAS design and performance is central to all of these. In addition to land administration infrastructure, they require well-balanced legal systems, dispute management systems, and financial systems of international standing. These systems underpin trading in land and land-related commodities and, in the most developed systems, in complex commodities. Most successful LAS provide the confidence and public face of land trading that, in turn, supports highly geared trading processes that accelerate creation of national wealth. One of the major potential reforms of LAS in developed countries lies in extending their capacity to support trading in complex commodities.

Three key disciplines are involved in highly formalized land markets: economics, law, and land administration. The nature of land is quite different in economic theory than it is in the discipline of land administration. An analysis of land in terms of economic market theory sees its special characteristics (Galal and Razzaz 2001, 16) as being fixed in location and heterogeneous (generating positive and negative externalities), as a bulky investment, and subject to derived demand. These characteristics of land expand in response to the creation of specific property rights, which allow recognized owners to retrieve the benefits of developing and using the land and to absorb the detriments or losses.

Courtesy of Land Administration and Management Project, DENR, Philippines

Figure 6.1 Formal and informal markets exist side by side in Manila, the Philippines.

Figure 6.2 The Mexican landscape reveals dense urbanization that can only exist with some form of land market, either formal or informal.

LAS manage property rights. Legal systems define them. By converting land rights into tradable assets, legal and administrative systems start the process of commoditization (called "commodification" in some countries). The formalization of property rights into tradable commodities involves identifying robust land rights and restrictions within existing cultural norms, managing disputes, establishing priorities among conflicting rights, and layering different opportunities within a single parcel (figure 6.2).

Central and Eastern European countries trying to access the European Union (EU) created a flurry of interest in converting centralist land delivery systems to open-market systems (Le Moule 2004), with mixed results. Understanding how they did this provides insight into the problems faced by other countries that want to adopt similar conversions.

LAS are critical to the organization and effectiveness of formal land markets. The most successful management of land markets is delivered by seamless and integrated management of all land and associated resources within the jurisdiction. Ideally, then, land administration covers all land, not merely land available for commodification. Government assets and public land are common instances. Roads are another. These nonproperty assets must be managed in a manner compatible with markets, especially in countries engaged in transferring land out of public and into private ownership. The most coherent LAS therefore provide support for public asset management and include all land within the cadastre or parcel map system. LAS should also extend to commodities in the resource and marine environments.

Given the integral relationship between private ownership and land markets, communal or community ownership by traditionally organized groups tends to take land out of the market in favor of preserving spiritual and social relationships with land. Traditional land should nevertheless be contained within LAS, despite the inherent difficulties involved in identifying both the land and its owners.

LAND MARKETS AND NATIONAL LAND POLICY

Formation of a national policy to pursue land markets and to extend markets into new areas are common global processes in land administration. This reflects anticipation that markets will release the value inherent in land into the general economy and raise overall living standards. The capacity of Western governments to extract revenue streams from land is a powerful model for other countries. Consider the funds generated by local rating systems, land taxation, transaction taxes and duties, capital gains taxes, goods and services taxes, income and corporate taxes, and so on. The value of land in private hands also delivers significant wealth to landowners. The ability of governments to provide housing, retail and business areas, industrial facilities, and essential infrastructure of roads, drains, and utilities, is limited: Engagement of the private sector in these activities offsets government obligations. Thus, turning land into an economic engine is a goal shared by many. The question, therefore, is not why, but how, this should be done.

A cautionary approach is appropriate. Grand claims that individualized property rights are the crux of Western capitalism and are transportable to developing countries need to be balanced by appreciation of other methods of distributing land access, particularly communitarian and social tenure systems that generate human comfort and provide food for millions. These claims also need to be contrasted with other methods available to communities for growing capital. Geoffrey Payne (2001, 58) compared British low levels of ownership (in 1914, at the apex of Britain's economic power, a mere 10 percent of its population owned property), with high ownership levels in German, Swedish, and Swiss history, along with 55 percent ownership in Jakarta and 53 percent in New Delhi, India—all accompanied by disastrously low per capita incomes. Payne suggested that no one has demonstrated a causal relationship between development of property rights and affluence in the West, and he pleaded for diversified and localized approaches to tenure in lieu of a single-minded market-driven approach. A transitional economy may get more immediate economic improvement by making its credit, labor, and product markets more effective, while it delivers tenure security through alternatives such as legal recognition of traditional and informal land arrangements. Additionally, formal markets can have negative economic consequences (Payne 2001). They immobilize housing-dependent work forces and the rural poor. Land speculation and rent-seeking behavior can appear. Land transactions tend toward rigidity, formality, and complexity. They often generate more informal and less formal market activity, especially if pricing policies are inappropriate. Local land market policy therefore needs to anticipate and counteract negative market effects and avoid a "one size fits all" approach.

CONTROLS ON LAND MARKETS

Land markets are managed according to national land policy. Apart from countries with little or no capacity for governance, most countries control where and how land markets work. All countries remove part of their national estate, or land and resource assets, from markets, but the decisions are highly variable. Countries with centrally organized economies tend to discourage land markets. This policy is changing rapidly, however, in countries like China and Vietnam. Certain groups in many countries prefer to use land for traditional, or collective, rather than economic, purposes. Some land is therefore not available for land markets. In free-market economies, virtually all land is held in freehold or leasehold tenures and distributed through markets, except for national parks and the like, roads and perhaps physical infrastructure, and public buildings. While this land is sequestered from trading activities, the land register and cadastre ideally still identify both the land and its managing authority or department.

Even in these free-market economies, substantial controls over land markets exist. Setting conditions for national land markets involves complex policy making. In many situations, land administration involves formalizing processes in existing informal markets. In others, processes of transition from nonmarkets to markets, and from informal to formal markets, occur through spontaneous, case-by-case decisions made by owners or groups of owners rather than being managed systematically. Governments seeking large-scale conversion must provide the infrastructures for implementation, typically through land titling or land administration projects. In land administration theory, a key to successful and managed transition is to engage the intended beneficiaries in the processes of change.

In free-market economies, market controls typically operate indirectly. Direct controls that define what and when the owner can undertake an activity are disavowed on the assumption that choices among individual owners will move the key economic resource of land toward its most economically efficient uses. Nevertheless, operations of free land markets are subject to extensive indirect controls. Among them are taxation of transactions; compulsory (or near-compulsory) registration of transactions; macroeconomic controls over money supply, including supply-side credit controls; land-use planning restrictions; extensive consultation processes and compliance standards for land development; environmental protections; provision of infrastructure of roads, drains, and utilities; regulation of professionals; transaction and construction standards; and so on. In free-market systems, LAS underpin implementation of these controls, particularly by providing information and facilitating transparent processes.

By contrast, centralist economies use direct controls over parties, prices, and timing of activities, even to the extent of forbidding or setting the terms of private arrangements. Developing countries tend to use cautionary controls aimed at reducing the accumulation of land in the hands of a few. For example, "privatization oligarchs" in some Eastern European countries and large accumulations by land speculators in developing countries are clearly undesirable outcomes. Common regulatory patterns therefore include limiting the amount of land owned, setting the minimum size of parcels (to eliminate noneconomic farms), controlling land uses (through tenure and planning systems), controlling change of use, anti-speculation provisions, moratoria on land transfer (especially for newly titled, traditionally held land), price controls to assist acquisition by the poor, and credit ceilings on use of land as collateral to avoid foreclosures and forced sales (van der Molen and Mishra 2006). Controls on foreign landownership and investment and ownership by corporations are also very common. From the land administration viewpoint, these controls tend to fail, either because their intended beneficiaries do not cooperate, and, in some cases, even oppose the controls, or because the government infrastructure supporting land market activities is inadequate to meet the regulatory challenges or is corrupt. Controls over land markets are only viable to the extent that governments have the capacity for, and willingness to, implement them consistently and transparently, without fear or favor. Moreover, implementation of the controls must be generally supported by the public.

WHY FORMAL LAND MARKETS ARE HARD TO ESTABLISH

Creation of land markets separates the haves from the have-nots. Markets require defined land tenures and titles. Formalization cannot achieve national coverage in developing countries quickly: The processes are incremental. Partial formalization creates correlative informalization of land occupation for people outside the system. Pursuit of markets without addressing the comparative disadvantage of those unable to participate is foolhardy. So is pursuit of markets in land traditionally or communally held, or in state-owned land, at the expense of inhabitants and their traditional associations with the land involved. Persistent land disputes are politically corrosive and, at the extreme, induce state failure.

Land market infrastructure is expensive. A country has to be relatively rich in economic and social capacities before it can develop formal land markets, even when substantial foreign aid is available. The experiences in Eastern European countries are illustrative (Dale and Baldwin 2000). Introduction of formal markets requires high-quality anticipatory planning and diverse sources of financial and human capital to build the necessary infrastructure. Technical support for developed land markets devours human and economic resources. In addition, land markets demand high levels of cognitive capacity—i.e., knowledge, shared understanding, abstract thinking, preparedness to participate transparently, inventiveness, ingenuity, and acceptance—in beneficiaries and participants. These aspects of market operations and the sociopolitical tools available to build capacity are only beginning to be explored (see chapter 11, "Capacity building and institutional development").

Land markets are surprisingly variable in their operations. Every market has its own momentum. The rental arrangements distributing beds in Calcutta, India, are not comparable to the office-space rental market in New York or house rentals in Sydney, Australia. A focus on the sales market should not be allowed to overshadow the rental market where very different processes must be used, particularly systems to give secure possession to tenants and prevent arbitrary eviction. Making generalizations about land markets, or borrowing tools from other markets, must be counterbalanced by grounding research in local contexts.

Land markets cannot be built in isolation from markets for labor, money, and agricultural products. All must be examined holistically and the results integrated into LAP design (Smith et al. 2007). Successful markets depend on credit. As a generalization, informal credit systems need to rely more on informal, and even predatory, tactics to protect loans. Even in formal systems, degrees of informal credit activity will remain and require policing to ensure associated practices do not undermine predictability and reliability. Institutionalized credit systems drawing on international capital require developed land tenures for proprietary ownership and security interests and other land rights, as well as well-regarded institutional and technical support systems.

LAND MARKETS AND LAND PROJECTS

During the late 1970s and early 1980s, the World Bank and other agencies commenced large LAPs with the intention of delivering prosperity, peace, and poverty alleviation to developing countries. Project designs emphasizing technical solutions and a rapid delivery of market options for economic growth were implemented. They focused on delivering straightforward individual private land rights as an investment incentive. Generally, the operative assumption was that titling would deliver effective land markets and economic improvement.

Later, the relationships among registration, titling, and land markets were examined more critically, especially as land projects failed to deliver anticipated benefits. The Thailand Land Titling Project indicated a positive relationship between economic improvement and registration

(Feder et al, 1988). David Atwood (1990) argued that there was not. But economists took a great deal of convincing that these assumptions needed to be modified. A short economics-oriented review of land titling projects expressed cautious optimism (Enterprise Research Institute for Latin America 1997). Many other people contributed to the debate, with anthropologists and sociologists expressing skepticism about titling as a universal means of delivering economic benefits to the intended beneficiaries. The debate resulted in documentation of the failure of titling programs to increase tenure security and reduce conflict, reflecting a widespread concern to find durable solutions:

> *"Failed titling programs are reported to have allowed wealthier and more powerful groups to acquire rights at the expense of the poor, displaced or female land occupiers (Binswanger, Deininger, and Feder 1993; Lastaria-Cornhiel 1997; Platteau 2000; Toulmin and Quan 2000); increased conflict by imposing simplistic legal systems on complex interrelationships (Fitzpatrick 1997; Knetsch and Trebilcock 1981; Lavigne Delville 2000; Simpson 1976; Toulmin and Quan 2000); and increased insecurity by overlaying formal institutional arrangements with informal arrangements (Bruce 1998; McAuslan 1998; Platteau 1996; Toulmin, Lavigne Delville, and Traore 2002)." (Dalrymple 2005)*

Philippe Lavigne Delville (2002a) explored the connection between survival of customary systems embedded in local social life and the failure of land registers. For professional land administrators, the problems were identified as lying not with the theory of capitalism, but with flawed project design and a narrow choice of tools used to transfer capacity. Titling was therefore identified as providing an answer for particular situations, while others required different solutions to secure tenure. Even economists realized there were considerable difficulties in building robust systems and thought the design of reforms and the tools chosen should be more comprehensive. Ahmed Galal and Omar Razzaz (2001) argued for simultaneous attention to institutional reform and property rights, capital markets, and market reforms to reduce distortions in prices in any analysis of real estate markets.

The critical literature also revealed another issue: lack of information. Africa remained an especially difficult case because of a fundamental lack of information to support sustained analysis:

> *"Economists' contributions have been essentially theoretical and deductive and are not based on solid empirical studies. Empirical studies of the economics of land tenure changes under the impact of demographic pressure have been conducted elsewhere and not yet repeated in Africa. The result is that there has been practically no empirical assessment of the economic benefits from land registration in Africa." (Lavigne Delville 2002a, 10)*

Too little analysis was given to the function of titling as a means of improving land management. Consequently, the "one size fits all" approach of titling individuals as owners in a registration program was finally abandoned in favor of more adaptable approaches (Deininger 2003). New research networks appeared, most prominently the Global Land Tools Network; multidiscipline comparative analyses were published (Torhonen 2001); and case study material increasingly became available (for instance, the Cadastral Template). These efforts will continue to refine project design and theoretical analysis. For land administration as a theory and its supporting discipline, the debate shifted away from ideological contests between promarket and antimarket factions. Instead, multidiscipline approaches, new technical solutions, and practical tools refined for local situations were identified. These tools include the participatory approach to building LAS, pro-poor land management processes, and slum-upgrade systems. These initiatives were set by an overarching policy framework of the land management paradigm—the globally adaptable framework within which particular tools are adopted according to conditions in local situations. Thus, titling systems were not only to be redesigned to ensure they continued after project experts left the scene, but they also were seen as but one aspect among broader solutions to land management problems (Burns 2006). Figure 6.3 shows a flourishing rural landscape in Greece that is only achievable with workable land management practices, where land administration is only one, albeit an important, component.

Land administration reform via titling projects, especially for the sake of economic advancement and poverty reduction, currently requires consideration of all aspects of sustainable development—i.e., environmental, social, economic, and good governance. A sociological and anthropological understanding of perceptions of people and of the importance of their local cultures is now part of betterment strategies, and indeed universally recognized (Harrison and Huntington 2000). Thus, the success of land markets depends not just on titling. They require three basic assumptions: public enthusiasm for material advantage, belief in and capacity for democracy, and belief in the sanctity of property. If these Western ideas can be transferred, well-managed, effective land markets can follow, as Thailand, Malaysia, Japan, and Korea demonstrate. But in Africa, Timor-Leste, Tonga, the Solomon Islands, and, indeed, most countries, any transitional process needs to start more with people's attitudes than with building GIS and titling programs.

THE NEW ROLE OF "PASSPORTING" PROPERTY

Hernando de Soto's influential book, *The Mystery of Capital* (2000), identified a much greater role for "passporting" assets than mere security of tenure: He viewed the passport, or official title for an asset, as having both an identifying role and a capital formation role. De Soto proposed to title land held by the poor and to create development and financial opportunities that

Figure 6.3 Simple rural land markets flourish in Greece.

would release the value of land. Titling would identify capital tied up in land and permit the land to be used as security, giving the poor access to credit.

These ideas are now applied to pro-poor empowerment, following more inclusive models for design of betterment paths rather than "title at all costs" interventions. This broadening is reflected in the UN-sponsored Commission on Legal Empowerment of the Poor, which links poverty and the inability of the poor to access acceptable legal structures to protect their economic assets. The commission's unique mission is built on the conviction that poverty can only be eradicated if governments give all citizens, especially the poor, a legitimate stake in the economy by extending access to property rights and other legal protections to populations and areas currently not covered by the rule of law. The commission wants the poor to have more access to property rights, assuming they are the right kind of land rights for their situation, but recognizes that, on their own, property rights are not enough.

The enigma faced by both de Soto and his critics is that titling the land of the poor sometimes makes little difference to their lives (Gilbert 2002), despite the observable truth that land titling delivers immense wealth to successful democracies. For land administration as a discipline then, the starting point is that the successful economies of the world are masters of land management, comparatively speaking, and provide expensive infrastructure to deliver tenure, value, use, and development processes. For the multilateral agencies and less successful governments, the problem is how to transplant these institutions and processes more successfully. The solution to the enigma lies in better project design, especially in the selection of tools for LAS (see chapter 12, "The land administration toolbox") and engagement of the intended beneficiaries.

The de Soto proposal, that is, of commoditization of land rights to defeat poverty, involves fundamental political questions for national governments. The debates about whether markets would successfully offset poverty and how best to create them clarify the tasks faced by any nation that decides to develop formal land markets or to better manage its informal markets. In approaching these market-building tasks, LAS design must respect the findings of the last decade:

- Ownership is not a single concept. Its content varies greatly among successful democracies, especially according to their civil law and common law origins, and each nation can afford to invest its operative concept of ownership in the unique texture of a national land administration system.

- Land is not just something that people walk on. In land administration theory, the fundamental aspect of land is the way people think about it, especially the construction of abstract rights and interests. No mere registration program, or LAP, can change the way people think about their land or the intrinsic value it has for them as members of social groups. Successful projects therefore are designed within the context created by the intended beneficiaries and seek to reflect this status quo in their design, build in change management paths, and allow the processes to adapt to the cognitive realities, and vice versa.

- People value land for spiritual, social, and economic reasons. Therefore, not all land rights relate to the economic institution of property. Many express other values. Attachments, however formed, remain even after people are forcibly removed and survive in later generations. Dispossession breeds disputation. Likewise, success in delivering stability in land, whether it is small plots for growing food in slum housing or high-density, complicated modern cities, adds to good governance. Thus, LAS should encompass all land rights, more than just all the physical land, in a country.

6.2 Building infrastructure to support formal markets

STAGES OF MARKET DEVELOPMENT

To better describe the kind of land administration institutions and tools needed to create and manage land markets, the stages of formal land market development are shown in figure 6.4.

Assuming formal markets are the goal, LAS is needed to manage market processes at the evolutionary stages: land trading, land markets, and complex commodities markets.

The five evolutionary stages do not represent discrete empirical experiences of how formal markets actually evolve. They are designed to show how LAS needs to be developed to assist the actual and potential economic development of a country. Given the predominance of informal markets, most nations will experience more than one stage at a time, and find that smooth transition from simple to complex markets is difficult to manage (Wallace and Williamson 2006). Moreover, the paths between the stages are essentially based on local experience. Without these detailed studies of on-the-ground reality or prior detailed knowledge, projects aimed at setting up or improving markets are likely to fail. A simple explanation of the characteristics of each stage is given in table 6.1.

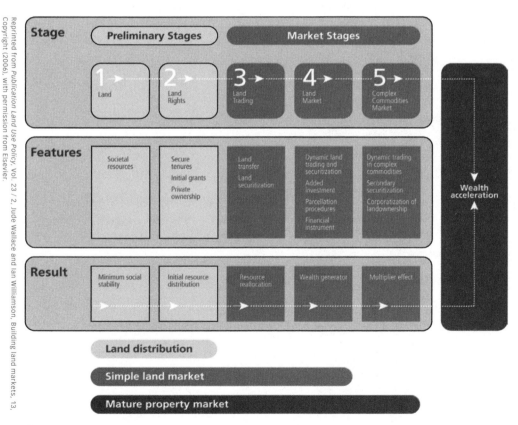

Figure 6.4 The evolutionary stages in development of formal land markets include land trading, land markets, and complex commodities markets.

While markets depend on the capacity to define commodities in the form of rights that are recognized as property, processes involved are typically mixed up with land trading and marketing. For a country to achieve a land market, its policy makers must obtain public commitment to the primary functions of property rights in land—stabilizing land distribution and generating capital. While land rights can exist without a market, markets cannot exist without land rights. Tradable land rights as commodities are the outcome of the institution of property.

TABLE 6.1 – SIMPLIFIED CHARACTERISTICS OF EVOLUTIONARY STAGES OF LAND MARKETS	
STAGES	**CHARACTERISTICS**
1. Land	A group or country establishes a defined location with territorial security. The securing of spatial relationships in land arrangements among competing groups is fundamental to all later developments.
2. Land rights	Within the group, regularities of access create expectations, which mature into rights. In formalized systems, the rights are reflected in the legal order. In some of these, the legal order is further embedded in a formal infrastructure of LAS. The crucial element of cognitive capacity of the participants starts with "my land" and "not my land" and matures into everyone appreciating "your land." The power derived from landownership is also managed and restricted by taxation and other systems.
3. Land trading	Virtually at any time in stage 2, a process of trading land between members of the group will develop. The rights in land traded evolve into property, the basic legal and economic institution in formal land markets. As economies become more complex, the trading will include strangers and depend on objective systems of evidence, eventually becoming a well-run program of recording of property rights. Processes of inheritance tracking will also develop.
	The commoditization processes will involve public capacity to view land as offering a wide range of rights, powers, and opportunities. The better these are organized and understood, the better the market will operate.
4. Land market	Now, the trading gets serious and increases in scale and complexity until it develops into a property market in which rights are converted with ease into tradable commodities. Significant government infrastructure supporting the market activities in land stabilizes commoditization and trading. Land is used extensively as security, multiplying the opportunities to derive capital. Capacity to invent and market new commodities emerges and gains strength.
5. Complex commodities market	The stability of the market allows spontaneous invention of complex and derivative commodities and "unbundling" of land into separate commodities of timber, water, carbon, planning permissions, and so on. This involves imagination and globalization. Typical machinery includes corporatization, securitization, and separation. The system relies heavily on the cognitive capability of society to understand and use tradable commodities, the rule of law, government capacity, and national ability to compete for capital in international marketplaces.

Robust land rights and effective LAS are necessary, though not sufficient, for success in the later market stages.

A functioning society always needs to rationalize the relationships between people and land; trading in commoditized land is one of the easiest methods of rationalization, especially when compared with bureaucratic or centralized allocation. Instances of commoditization of land were recorded in the earliest of human writings. Recent scholarship suggests that land was commoditized 4,000 years ago. Two conditions are regarded as essential: literacy and scarcity. Industrial capitalism is not a condition, though it is the engine of a complex market (Epstein 1993).

The message for designers of LAS is to manage the transitions through the evolutionary stages in a way that anticipates the complexities of a fully developed formal market. Whatever the process of change, the evolutionary stages in market development operate like building blocks: LAS capacity must be developed to manage each stage before the next is possible, and all earlier stages must operate successfully to support, or be subsumed within, the more complex stages. This is quite different from saying that every country must actually go through all stages. In fact, many countries attempt to collapse the evolution of formal land markets into a couple of decades: Their success depends on their ability to build robust administration to support stable land trading systems, attractive commodities, and cognitive capacity before they move on to the high-end property market sophistications of secondary mortgage markets and property trusts.

In countries with successful simple land markets, rights are based on secure and clear tenures, which give broad decision-making capacities to owners and allow others limited opportunities to restrict these capacities. A system of evidence of ownership, usually including land registration, exists to provide confidence in trading. The beneficiaries of the tenure system are willing participants and have a social and cognitive capacity to think of land as a commodity. They recognize that landowners can organize other people, enjoy a larger realm of decision impact, and can influence the lives of other people (Denman 1978, 46). The shared understanding of rights among beneficiaries is hard to build and maintain, because allocation of land to particular individuals and groups is, in fact, a state-sanctioned distribution of power. The array of infrastructure and tools is described in table 6.2.

TABLE 6.2 EVOLUTION OF INFRASTRUCTURE AND TOOLS IN LAS

STAGES	INFRASTRUCTURE AND TOOLS
1. Land	Territorial recognition
2. Land rights	Capacity to understand land as a series of rights A legal system to manage coherent fit of the various rights Basic administrative system to document rights: where, what, who, and when
3. Land trading	Public understanding and acceptance of the trading system A theory of property allowing private, individually owned land rights Formal transaction arrangements Trading between strangers Mature evidentiary systems relying first on paper trails, then ultimately on digital systems Objective identification of boundaries Inheritance tracking through the inventory system Government infrastructure supporting core LAS activities
4. Land market	Extensive trading and management of trading risks Flexibility in LAS to recognize new commodities Growth in separation of land, minerals, soils and gravel, and of trees, crops, and produce as unique commodities Extensive capacity to support supply and maintenance of utilities and services, and multioccupancy and multipurpose buildings Participation by corporations to spread risk, organize management of interests, and extend opportunities for participation Complex layering among interests in land, resources, and commodities Growth in human skills and administrative systems, particularly inventory systems High investment in government infrastructure, especially in technology
5. Complex commodities market	Investment in technology to maximize speed and range of services provided by government and private sector in core LAS processes "Unbundled" interests in land that are traded separately Highly geared systems capable of managing mass transactions Extensive participation in land-based activities by corporations Extensive, accountable, and transparent administrative systems; highly reliable inventories with clearly defined functions that operate simultaneously without conflict Public and private administrative systems operating in key areas Organized controls over land to deliver planning, environment protection, contamination and risk management, and more High level of inherent flexibility in creation of new commodities Participation opportunities other than outright ownership, especially through pension funds, superannuation schemes, trusts, and corporations

COMMODITIZATION SYSTEMS

The point of differentiation between simple land trading (stage 3) and land markets (stage 4) lies in understanding that land is not the only, or even the most basic, ingredient of land markets. Successful land markets are capable of inventing and commoditizing abstractions. Their vitality comes from the capacity of their administrators and participants to create and market abstract land rights and complex commodities, in addition to the land itself. Once abstractions are understood, the view that land is a commodity with limited availability ceases to be an overriding constraint on the market. "Land" in this sense is unlimited.

Property rights are the engine of a land market. They carry opportunities to exclude others, profit from use of the land, give away or sell the land, and create subordinate interests, especially leases and mortgages. Property rights in land share these opportunities in common with rights in other kinds of property—for instance, copyright, debts, shares, and interests in resources. Well-defined and formally tradable rights presuppose governmental capacity to announce and implement legal rules, especially laws about property in general, transactions, and disputes. These rules and their routine administration are necessary but not sufficient to turn the bundles of opportunities specified by the rights into marketable commodities.

If government institutions are stable enough, and land administration and land rights are established, market activities evolve into more complex products, typically by adopting an initiative tried out in another jurisdiction. Examples of more complex products include titles in multioccupancy buildings, secondary-mortgage market products created out of securitized mortgages in the secondary mortgage market, build/own/transfer arrangements, development trusts, property trusts, and so on. Some of these commodities are closely related to simple land market commodities and their related activities. Others require substantial legislative and administrative changes to expand private-property rights and registration schemes and to apply them to new commodities, for example, in New York (figure 6.5), where opportunities to construct high-rise buildings were created and traded.

Since the mid–1990s, new, radical processes of commoditization "unbundle" land into separate tradable assets. In this process, opportunities related to the land itself, and to minerals and petroleum, water, fauna, flora, tradable permits, carbon credits, wildlife credits, dryland salinity credits, planning opportunities, waste management, and so on, are repackaged and made tradable independently of ownership of the land. The idea comes from using market-based instruments (MBI) or incentive instruments for environment and resource management (Panayotou 1994). These initiatives borrow heavily from property theory and the main characteristics of

Figure 6.5 Complex
commodities such as high-rise
buildings abound in the New York
property market.

Western property: exclusivity, duration, quality of title, transferability, divisibility, and flexibility. They all require an administrative infrastructure, frequently incorporated into LAS but sometimes built separately. So far, analysis of the infrastructure needed to manage these commodities concentrates on registration, indefeasible or guaranteed title, suitability for securities and mortgages, and compensation for acquisition. However, these developments potentially challenge the capacity for holistic land management, unless the design of the administrative arrangements and the information generated are incorporated in LAS and treated within the land management paradigm. Moreover, little theoretical or practical research is available on how to incorporate social and stewardship values into these unbundled commodities or how to handle the public goods protected by the substantial restrictions affecting land.

COGNITIVE CAPACITY AND LAS EVOLUTION

The significance of land to capitalism is now better understood. In the theory so far, land is a potential market asset and source of capital. If a country cannot produce capital out of land, its population will remain poorer to the extent of unrealized opportunity. Unless other sources of wealth are readily available, its people will observe expansion of the gap between their economy and the economies of successful countries (De Soto 2000, 4–5).

This theory, however, oversimplifies land markets (stage 4) and the transition to complex commodities markets (stage 5). The "land market" label differentiates the earlier stage where mere land trading appears among group members and eventually between members and strangers. In land markets, the scale of activities is fundamentally larger; market management demands multiple and objective sources of integrity and reliability beyond mere group verification; and the state is, and must be, more involved. The result is a highly organized matrix of commodities, competencies, and participants. This mix makes the market work and forms the basis for moving to complex commodities markets (stage 5).

Despite their sophistication, most land markets grew without direction or design. Many informal markets relied on intuitive development of the three easily identifiable and essential activities for running a market:

♦ they invented diverse land-based commodities;

♦ they perfected capacity to use land as a security;

♦ they managed a huge increase in the scale of land trading.

Dynamism lies not just in the scale of trading. Increasing formalization allows more proprietary separations and reconstructions, derived from tenures that allow an owner to reduce his rights by creating derivative interests to permit actual use by owners of lesser rights, to recast his activity from actual land use to take profits from land use by others, and to reduce his activity on the land while increasing his gains — generally, to fragment the way land is used.

Successful commoditization in stage 4, land markets, and stage 5, complex markets, thus depends on an administrative system capable of building the capacity of participants to understand the nature of the commodities. Because land markets commoditize abstractions and make them tradable, LAS provide the necessary framework for reliable identification of and trading in commodities. Once LAS are built, the capacity to create new commodities out of land is open-ended, limited only by human imagination and capacity to invent appropriate administrative structures. This creativity allows land markets to constantly create new, and retire old, commodities, provided the underlying administrative infrastructure is reliable and flexible. Commodities are developed through three waves of creativity, each a little different from the other, but generally relying on an entrepreneurial response to perceived issues, including sustainable development. These waves are

♦ **Creativity in commodities reflecting changes in land use:** time shares, strata titles, community titles, utility infrastructure titles, and so on. These combine the surface land and complicated built arrangements, add a range of access opportunities, and provide for a wide variety of uses to suit specific needs.

◆ **Creativity in derivative interests:** This builds new commodities on top of activities in the simple land market. They include products for security tenures, secondary-mortgage markets, risk markets, and new financial interests. These commodities do not involve physical access to land, though it might become available in situations of individual and even structural breakdown, such as the multiple debt failures behind the subprime mortgage crisis in the United States in 2007–08. These secondary and derivative interests extend opportunities for participation in land markets exponentially and globally and require new systems of management and regulation, as well as understandable trading processes. For these developments to be sustainable, administrative frameworks similar to the LAS structure for management of RRRs are required. Given the global nature of trading and the dependence of these commodities on financial markets, provision of suitable frameworks was an elusive goal. The lack of appropriate frameworks to define the commodities, provide transparent trading opportunities, and apply sensible regulation is a major factor in the subprime mortgage problem turning into a global financial crisis in 2008–09.

◆ **Creativity in environmental protection instruments and unbundling of land and resources:** This concentrates on unbundling and separating land from resources to allow market forces to create and distribute property separated out of opportunities previously tied to landownership, such as water, timber, minerals, and MBI.

All these creative activities depend on LAS having well-developed processes for layering, separating, and defining. The capacity of a system to support creativity depends on its ability to set up a reliable basic system as a foundation that can incorporate the ideas of entrepreneurs.

The core ingredient of a complex property market is the cognitive capacity of its participants, who manage complicated sets of interrelated activities and outcomes. A fourth pillar added to P. F. Dale and R. Baldwin's Three Pillars diagram (2000) illustrates this point (figure 6.6). Mature cognitive capacity is both the incentive for and the outcome of LAS infrastructure (and other administrative systems), which specifies and enforces layers of conceptual, not physical, "reality" to support property rights in land and complex trading activities. Cognitive capacity cannot develop without the infrastructure of LAS to manage the commodities. Cognitive capacity involves society understanding the need for conceptual thinking and the ability to imagine opportunities and articulate a broadly accepted philosophy and set of values to undergird the entire system. The most important message for LAS designers is the necessity to build transparency in the system to encourage vigorous participation and thus support society's cognitive capacity.

When all these functions are established, institutional support for new commodities develops. For example, the opportunity to "own" land through membership in a corporate or trust vehicle is open-ended and available to individuals with even minimum capital. Opportunities to trade "land" through transactions involving shares, units, and pension fund investments are similarly opened up. The capacity of land to generate value can be mixed in dynamic and flexible ways with other economic opportunities for production and investment. Secondary markets flourish. More importantly, national trading attracts international investment. The basis of the market remains land, but what is now tradable is limited only by imagination and creativity. Figure 6.7 shows a selection of the interrelated new commodities drawn out of land.

Complex markets require and benefit from competent government infrastructure, and especially from technology. They also require substantial levels of formalization and commitment to publicly responsive systems. Additionally, management systems need to create predictable, reliable transaction patterns, particularly dealing with rent seeking as well as corruption, fraud, and forgery.

Complex markets benefit from remarkable improvements in technical support systems. The technical tools now in use are unrecognizable from their antecedents. GIS (Longley and Batty 2003), land registration systems, parcel definitions (UNECE 2004), information coherence and interoperability, SDIs, LAS, and computerized access in general are vastly different given new management, technology, and the changing roles of government. These developments were partly in response to improved technical capacity for creation and transfer of data (generated by computers and the

Figure 6.6 A fourth pillar (in red) of "cognitive capacity" is added to the quintessential Three Pillars diagram of land market activities.

Internet), new management styles, and devolution of the roles of government to public–private partnerships and the private sector. Thus, improvement in information integrity and standardization of rights fueled significant improvements in land markets, wealth acceleration, and opportunities for sustainable development. However, the largest contributor to the vitality of the marketplace remains the creativity of its participants. Still, nurturing this vitality is far from easy.

Like other complex social and economic systems, land markets generate their own myths and shared understandings. The significant difference between undeveloped and developed economies does not lie solely with the lack of records. Sometimes, even with records, the first group lacks the ability to systematically conceptualize land sufficiently to run an effective market, as the Indonesian example of idiosyncratic land rights illustrates. Recording of rights alone does not invite the next stage. It is not records, but the ability to work with abstractions that allows developed countries to accelerate wealth through creation and marketing of complex commodities.

Western countries allow landowners to remain attached physically to land, to think and talk about the characteristics of an individual parcel or building, and to regard the area within boundaries as "mine" and "yours," but they also do something far more important. They build concepts

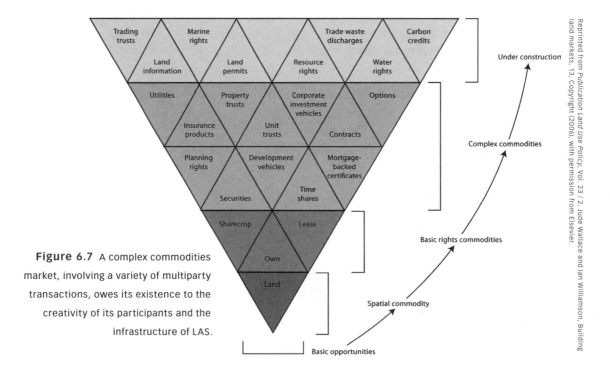

Figure 6.7 A complex commodities market, involving a variety of multiparty transactions, owes its existence to the creativity of its participants and the infrastructure of LAS.

Reprinted from *Publication Land Use Policy*, Vol. 23 / 2, J. Jude Wallace and Ian Williamson, Building land markets, 13, Copyright (2006), with permission from Elsevier.

in relation to land; embed these concepts in social behavior, language, and the economy; and then "trade" these concepts. Administrative systems provide the objective regularities that facilitate development, ownership, management, and trading of conceptual or intangible commodities. By contrast, if a country focuses on land simply as land, it cannot develop the functional processes required for wealth acceleration through commodification of land rights and complex commodities related to land.

A major function of LAS is to maintain the sociopolitical commitment to the commodities within the ancillary processes of securitization, corporatization, and separation functions associated with land markets (figure 6.8).

Securitization: In the banking sector, securitization involves repackaging financial instruments into new generic and more marketable commodities. Mechanisms include acquisition, distribution, classification, collateralization, composition, and pooling of commodities. These arrangements facilitate complex corporate-level borrowing and international investment. The much simpler activity of creating security by charging land with repayment of debts, thereby converting land value into spendable capital, is a primary activity supporting some securitization packages. From the economic point of view, multilayered opportunities for converting future yields into present capital are created. For developing economies, the lessons are simple. The vitality and reliability of the secondary systems depend on the strength of the primary assets of credit securities. In the realm of land securities, the connection between lending and recovery of the security on default is vital to the economic growth of the land and money markets. For developed countries, the global credit crisis reveals another lesson. The creditability of the connection between lending

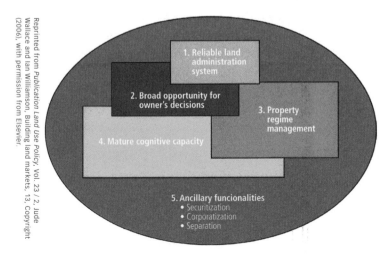

1. Reliable land administration system

2. Broad opportunity for owner's decisions

3. Property regime management

4. Mature cognitive capacity

5. Ancillary funcionalities
• Securitization
• Corporatization
• Separation

Figure 6.8 A complex property market involves securitization, corporatization, and separation of opportunities and responsibilities.

and recovering the security is the foundation of the secondary-mortgage market and other derivative trading. The subprime mortgage crisis in the United States in 2007–08 arose because this connection was broken. The United States thus provided a sobering reminder of the need for strong primary land interests to support the secondary market and demonstrated how a lack of confidence in land-related commodities markets can spread to global financial markets.

Corporatization: The process of allowing people to create a new legal entity of a company must exist to control risks, pool capital, divorce ownership from management, and increase opportunities for participation. For land, the company introduces a single owner and potentially open-ended numbers of benefit takers. It can provide professional management of complicated investments. Countries (such as Indonesia) where full ownership of land is limited to natural persons in effect deprive their population of the basic engine of capital raising and land management in the context in which it is most economically effective.

Separation: Dividing ownership, management capacities, and profit and benefit taking is especially necessary to manage commodities in this new, complicated environment. Companies remain a principal commercial separation mechanism. In countries with a common law background, trusts are an equally important commercial tool. Transportability of trusts is improved through international instruments; for instance, The Hague Convention on the Law Applicable to Trusts and on their Recognition (signed October 20, 1984) lays down the terms for recognition of trusts.

When commoditization, securitization, corporatization, and separation are combined with land, land rights, and land-based activities, wealth is created by carving out of land simultaneous, multiple trading opportunities. At the same time, ownership is separated from use, utilities from buildings, ownership from access, buildings from rights to build, possession from yield, development opportunities from risk, securitization from possession, ownership from management, yield from responsibility, risks from profit, benefit streams from management responsibility, corporatized owners from risk, control from sharing in benefits, securitization from direct capital raising, and so on. An open-ended range of opportunities capable of being converted into commodities opens up and accelerates wealth creation.

LAND MARKET CHALLENGES

From the viewpoint of land administration, the tasks in organizing land access and management remain constant, whether the country's underlying economic philosophy is capitalist or centralist. The distinction in the economies lies in the detail of the relationship between the state and citizens in regard to land: Capitalist economies balance power and responsibility in favor of individuals, while communist or centralist countries move that balance more toward the state. Except for property itself, the range of institutions and instruments (tenures, titles, approvals, controls, bureaucracies) used in each kind of economy is the same, though their operation, tools, and organization will be remarkably variable. These variations, and the variations in the ways people relate to land, explain the difficulties experienced by land policy makers who seek to copy a system in use in one nation to another. For most countries seeking to build markets, the key issues include

◆ understanding local features of the relationship between formal or managed activities and informal activities;

◆ designing a land market infrastructure capable of supporting each of the five evolutionary stages of land market development;

◆ managing the choice between using highly local rights and more generic, globally accepted rights.

Countries at the upper end of land markets need to rebuild LAS to integrate, or at least to service, trading in complex commodities. All countries need to use LAS to provide the information to form effective policy and the structures to implement it. By using the land management paradigm to direct LAS design, countries can ensure market activities feed into the delivery of sustainable development.

6.3 Land valuation and taxation

RETRIEVING VALUE FROM LAND FOR GOVERNMENT PURPOSES

Only with an effective, formalized land market can land valuation and taxation really work, such as in the move from a centralized economy to a market economy. On the other hand, a land market can benefit from an effective land valuation system to ensure transparency and efficiency, especially in developing countries. Trading activities set the price or value of land in the selling, renting, and credit markets. The formal processes used to manage transactions inform governments, from national to local, about pricing patterns and their vagaries. The best LAS deliver real-time or nearly real-time transaction information that feeds into the datasets of owners, parcels, transactions, trading patterns and so on. This book is about building this kind of interaction, rather than the technical, specialized, and professional activities of land valuation.

While the aim of integrating data to support land valuation and taxation systems is easy to understand, achieving integration is difficult. Most agencies that rely on collecting taxes, rates, duties, and contributions from landowners and users build internal systems to support their activities. The cadastral approach, as originally devised in Europe, was the first time a concerted effort was made to organize land information to streamline processes of tax estimation and collection. Since then, the means of extracting value from land have multiplied. Most countries use rates paid to local governments to fund the roads, drains, and infrastructure needed to service land. In some countries, notably the United States, local taxes and fees bear the costs of education, basic welfare, and policing services. The power of a local rating system to generate benefits for landowners and users is clearly a driver for many local authorities to build collection systems: The income stream available to them is frequently large enough to support high-technology solutions.

A similar process is found in systems that operate on a state or national scale. A state-based land taxation system will typically rely on owner parcel files. The European cadastral model that originated in the eighteenth and nineteenth centuries fits this pattern perfectly. It was driven by the physiocrat movement that influenced the Napoleonic cadastre. This philosophy believed that land was the basis of wealth and as such should be taxed. The European model used fiscal cadastres that recorded either the quality of the soil or the actual land use of the individual parcel as the basis for levying taxes that reflected the production capacity or the actual type of land use. Land parcels were then merged into properties to form the basis for land transfers to be recorded in the land book.

Today, property taxes are normally paid as a percentage of the market value, often confirmed by public valuation. This use of market value as a tax base has many consequences. Taxation of land at its improved (or assumed market) value will include the land and improvements. This might seem fair and reasonable where land is transacted with sufficient frequency to ensure that the property valuation is close to market value. But charging an annual tax on improved capital values has consequences on people's behavior that can operate as a disincentive to improving land. Many valuation systems underpinning annual land taxes therefore rely on an assumed or calculated "unimproved value" system.

The professionalization and objectivity of the profession of valuers in countries using these mass-valuation systems is fundamental to the maintenance of public confidence. So is the reliability and accuracy of the information in public records. People's willingness to pay taxes is therefore directly related to the efficiency and transparency of a country's LAS.

Even though this book is not about the technical, specialized, and professional activities of land valuation and taxation, a few main principles follow.

BASIC VALUATION PRINCIPLES

Valuation of land and property can be carried out by using two different approaches that are normally referred to as individual and mass valuation. Both approaches aim to assess the market value of the land or property. Market value means the price that a reasonable buyer would pay for the land or property — or "an estimated amount for which the property should exchange on the date of valuation between a willing buyer and a willing seller … wherein the parties had each acted knowledgeably, prudently, and without compulsion" (International Valuation Standards 2001).

Individual valuation is normally carried out on the request of the landowner for various reasons, such as an intended sale, a social event such as a divorce or inheritance, or an application for mortgage or property insurance. The valuation will normally be carried out by a recognized valuation professional. The estimation of the value will take into account all relevant issues to assess the actual market value. However, valuations may be carried out differently for different purposes, such as sale, mortgage, insurance, and so on.

Mass valuation is undertaken mainly for the purposes of taxation imposed by government. Mass valuation should ensure that land and property taxes are levied according to the actual market price, or in proportion to that price, so that similar properties pay similar taxes. Mass valuations are normally based on standard valuation models that include a range of components, such as property area, building area, building quality, materials and year of construction, building improvements, location, and possible use and restrictions according to planning regulations.

Mass valuation is normally carried out every four to five years, while updating may be carried out annually. The basis for such updating is normally the recording of actual sales prices that will enable the calculation of increasing values for the various kinds of properties, including dwellings, condominiums, summer cottages, and the like. Importantly, mass valuations may not be the same as market value and are, indeed, often lower, but the differences between the official and market valuations of different properties should be roughly equivalent to ensure equitable taxation.

In modern systems, property values recorded in valuation registers are normally maintained at the local government level but sometimes at the state level (in federal systems). This register is normally based on cadastral information showing the location of individual properties (cadastral maps). Landowners are informed annually about the actual valuation of their property and will normally have the opportunity to object to this assessment in a valuation appeals court.

While the mass valuation system primarily serves as a basis for taxation, it has a range of other functions, such as supporting an efficient land market, facilitating fair compensation in situations of compulsory purchase, and more generally underpinning the role of land and property as a basic asset of the national economy.

Although the market value assessment is the most common approach used for mass valuation, it may have some shortcomings—e.g., when the number of transactions in a given area or a specific kind of properties is very limited. The market value approach is therefore mainly used for the housing sector. Other approaches such as income capitalization or calculation of building costs can be used to support the assessment where sufficient market value evidence is not available.

BASIC TAXATION PRINCIPLES

Property taxes are usually levied as a small percentage of the estimated market value of the property, which is provided through the public mass-valuation system. In some countries, however, taxes are levied mainly on wealth rather than on land and improvements (UNECE 2001). Typical taxes include

- ◆ **A land tax**, normally levied as a percentage of the assessed marked value of the land, without buildings but including site improvements, such as road access, sewerage, and so on, though often the home or residence of the taxpayer is excluded or is "tax free."

- ◆ **A property tax**, levied as a percentage of the assessed market value of the total property including buildings and other improvements. And again, the home or residence of the taxpayer is often excluded or "tax free."

Other kinds of taxes include

- ◆ **A service tax**, levied on buildings for private businesses and also public buildings that are outside the general land market to cover the general public service provided.

- ◆ **A property transfer tax**, often referred to as stamp duties that may be paid as a percentage of the sales price or estimated property value when a property is transferred to another owner.

- ◆ **A development gains/betterment tax**, levied as a percentage of the profit gained through development opportunities provided through planning regulations.

◆ **A capital gains tax**, applied to businesses or private properties on sale as a percentage of the difference between the prices of buying and selling.

No matter what means of taxation a country uses, the objectives should be clearly defined and administered in a way that is transparent, well understood, and accepted by the public. Property taxes are relatively easy to introduce as long as there is adequate legislation and there are sufficient educated professional valuers. The procedures to be followed involve the identification and mapping of all properties to be taxed; the classification and valuation of each property in accordance with agreed procedures; identification of who will be responsible for paying the tax; preparation of the valuation roll; notification of the individual property taxpayer of the amount to be paid; the collection of taxes; and an appeals procedure for taxpayers who dispute their assessment (UNECE 2005c).

Taxation of land and property has advantages in that it comprises a broad tax base, making it easy to administer and inexpensive to introduce and maintain. These taxes are difficult to evade, and, provided that a country maintains good cadastral records, the collection rate can approach 100 percent. However, the valuation records must be integrated with the cadastral and land registration records; otherwise, tax evasion can occur. Unfortunately, in many countries, valuation and land registration records are in different "silos," using different land parcel and property databases. These records should provide a stable and predictable source of revenue that is transparent in the way it is calculated and collected. This encourages efficient use of land and property and discourages land speculation. It recognizes public claims on private property while allowing the development of private property.

EQUITABLE TAX BURDENS

The land taxation model dominated in the seventeenth and eighteenth centuries in Europe and was the wellspring of its cadastral enterprises. It enabled Europe to tax land so effectively that it formed the basis for building national wealth and durable infrastructure, especially for its cities. The land administration focus in taxation systems remains evident today with the majority of countries in a 2001 survey of mass land valuation for tax purposes showing reliance on land administration support (UNECE 2001).

When European land diminished in economic importance as a consequence of the Industrial Revolution, income taxes became the preferred national revenue stream. National and state land taxes diminished and sometimes even disappeared. Rating systems supporting the small-scale activities of municipalities and councils were what was left.

After the 1950s, many developed countries started to reintroduce the land tax. These taxes continue to be irritants to the extent that they are now a major factor in the viability of businesses and one of the major tools governments use to influence land use. Houses, farms, and charitable land are differentially taxed, so that conversion of land among categories of taxable and nontaxable land needs careful judgment.

Modern economies now depend on highly sophisticated information about property assets for taxation purposes. Graduated personal and corporate income taxes, value-added taxes, and capital gains taxes depend on information streams about land, land-based activities, and land transactions. The cost of building information streams outside a country's LAS to service taxation activities is enormous, if not prohibitive. Thus, the benefits of integrating LAS information with modern tax collection processes speak for themselves.

UNDEVELOPED COUNTRIES

The major problems facing collection of taxes, even land taxes, in undeveloped countries are differential collection and corruption. The paucity of records and their fragmentation breed situations that allow taxpayers to pay less, or even no, tax. This occurs in many ways, the three predominant methods being failure to include parcels in the tax base, concealment of the true owner's identity, especially where the value of all parcels is aggregated for tax assessments, and declaration of a sales price lower than the price actually paid. The result is typically an increase in the taxes charged, with consequent encouragement of tax avoidance behavior. Rudimentary systems built around internal nontransparent records or reliant on "briefcase" collections via personal visits of the tax collector are particularly fallible.

A valuation system based on a good cadastral map will help expose properties and ownership outside the system. If satellite imagery can be superimposed, the accuracy of systems is further improved.

A message for initial development of rudimentary LAS is the importance of using cadastral records to underpin an equitable taxation system: Not only is the taxation system more transparent and inclusive, but an income stream derived from taxation is delivered to maintain the cadastre. Unlike land registry-driven cadastres, a very low parcel tax, supported by an inclusive fiscal cadastre, is a great starting point for national LAS.

Chapter 7
Managing the use of land

7.1 Land use

7.2 Planning control systems

7.3 Urban land-use planning and regulations

7.4 Rural planning and sectoral land-use regulations

7.5 Land consolidation and readjustment

7.6 Integrated land-use management

7.7 Land development

7

7.1 Land use

Managing the use of land is an essential part of land administration systems. However, the means of land-use control varies throughout the world. In some developing countries, the means may be very basic covering only the allocation of land rights or approval of building construction. In more developed countries, the means may include advanced systems of planning control based on an integrated approach to land-use management.

Even if land-use planning is normally considered a separate discipline, the processes of land-use control should be considered a coherent part of LAS in any country. As argued in chapter 5, "Modern land administration theory," the four functions of land tenure, land value, land use, and land development are interrelated, and land should be treated as a coherent whole.

Some degree of land-use planning and regulation is essential to control development and prevent unregulated settlements, to protect natural values, and to manage environmental impact. How planning is done is a major policy decision for any jurisdiction. The ways countries undertake planning and regulation vary according to their historical experience, economic values, competency in building systems, social needs, legal framework, and many other factors. The design of any national or local system will be influenced by the level of maturity in the overall LAS.

Very few countries (perhaps about twenty) have successful citizen-supported integrated planning systems. These, however, provide models for many others. In highly organized systems, land administration logically includes the administrative aspects of planning and development controls (planning regulations and restrictions and sectoral land-use laws), but the planning processes themselves usually fall within the domain of highly trained planning professionals.

This chapter is not about the methods and means of land-use planning. It is about the institutional role of planning and land-use regulation within the context of LAS and the need for parcel-based information to execute that role. Therefore, only the administrative aspects of land-use planning and development controls are considered. The details of how planners work, consultation processes, dispute handling, development approvals, and the tasks involved in designing specific planning systems are left to texts dedicated to urban and rural land-use planning. Likewise, the land administration role in development lies in the bureaucratic and official means of controlling development, not in the way builders and developers actually build, the materials they use, or the processes they engage. Administration of both planning and development processes is essentially political. Unlike many other aspects of land administration, especially land registration and cadastral surveying in modern democracies, planning and development systems are often contested.

Land-use management includes the control of land use in both urban and rural areas as well as management of natural resources. Control of land use may be executed through spatial planning at various administrative levels and is often supported by land-use regulations within the various sectors such as agriculture, environmental protection, water catchments, transportation, and so on.

Effective land-use management should also ensure sustainable land development that includes, for example, design of new urban areas, distribution of hazardous and polluting facilities, and design and implementation of infrastructure such as roads, railways, and electricity lines. Proper land-use management should also prevent unauthorized or informal development that may complicate appropriate development at a later stage and impose huge collateral costs on society.

In the planning context, land policies may be seen as the set of aims and objectives set by government for dealing with land issues relating to how access to land and land-related opportunities is allocated. The land management paradigm drives systems dealing with land rights, restrictions, and responsibilities in support of sustainable development. By integrating land policies, land administration functions, and the land information base, the paradigm ensures that any new development or change of land use is consistent with adopted land policies, and current and updated land information, thus promoting sustainable development. This holistic approach to land management is the key asset of any jurisdiction and represents a huge political challenge for those setting up planning systems.

Arguably, establishment of mature systems that are trusted by the public is also the key to preventing and legalizing informal urban development. This goes, at least, for the developed part of the world. In developing countries, this approach must be supplemented by measures that address the issues of poverty, health, education, economic growth, and tenure security.

LAND-USE RIGHTS

Ownership and long-term leaseholds are the most important rights in land. The actual content of these rights may vary among countries and jurisdictions, but in general the content is well understood. Rights in land also include the right of use. This right may be limited through public land-use regulations and restrictions, sectoral land-use provisions, and also various kinds of private land-use regulations such as easements and covenants. Many land-use rights are therefore restrictions that control the possible future use of land.

Land-use planning and restrictions are becoming increasingly important as a means to ensure effective management of land use, provide infrastructure and services, protect and improve the urban and rural environment, prevent pollution, and pursue sustainable development. Planning and regulation of land activities crosscut tenures and the land rights they support. How these intersect is best explained by describing two conflicting points of view on land-use planning: the free-market approach and the central planning approach.

THE FREE-MARKET APPROACH

Property rights activists, most of them influenced by the private ownership viewpoint, argue that landowners should be obligated to no one and have complete domain of their land. In this extreme position, the government opportunity to take land (eminent domain), or restrict its use (by planning regulations), or even regulate how it is used (building controls), should be nonexistent or highly limited. Proponents argue that planning restrictions should only be imposed after compensation for lost development opportunities is paid (Jacobs 2007). Sometimes, they even argue that land should not be taxed (the State of Nevada, for example, passed a law allowing owners to buy out their land tax liability in perpetuity by paying a low up-front capital contribution to the government). These and similar views have become popular in the United States as evident in a 2004 ballot initiative, Measure 37, that was passed in Oregon. The measure forced local and state government either to remove the requirements of a thirty-year-old planning law on properties owned by people before the law was created and who have owned them continuously since then, or to compensate the owners for the burden of the law.

Not every state uses the free-market approach. U.S. law constrains the relationship between landowners and government by interpretation of the "takings" phrase in the Fifth Amendment, contained in the Bill of Rights, to the U.S. Constitution. This provides that "private property shall not be taken for public use, without just compensation." A similar provision appears in most formal constitutions, yet in the United States, its meaning is intensely debated and litigated. Generally, U.S. courts interpret this clause as allowing government to both plan the use of land and to condemn derelict or even nonderelict land to further redevelopment. In *Kelo v. City of New London* (125 S.Ct.2655 (2005)) in Connecticut, the U.S. Supreme Court ruled in favor of eminent domain for the transfer of land from one private owner to another for the pursuit of economic development.

While land was in excess and there were few situations for deleterious land uses to negatively affect neighbors, private-rights activists did not make much impact on the way land uses were regulated. However, population growth, industrialization, and urbanization brought new problems. In 1920, the U.S. Census officially recorded the shift from a rural to an urban economy. The private-property approach has increased in popularity in recent years. Tensions between government and property owners saw thirty-four states (as of 2006) voting on this issue and, as a consequence, producing the varied results shown in figure 7.1 (Jacobs 2007). While three states made no decision as of 2006 and thirteen failed to launch a change, the remaining either extended prohibition on development of private land or increased public protections.

THE CENTRAL PLANNING APPROACH

Particularly after World War II, and especially throughout Europe, another view became popular, wherein the role of a democratic government includes planning and regulating land systematically to protect the public good. Regulated planning is theoretically separated from taking private land with compensation and using it for public purposes. Following a long tradition of cultural appreciation of the intergenerational value of land, the Europeans took care in organizing land-use planning in well-constructed, integrated systems. Germany even put stewardship obligations in its constitution. In these jurisdictions, the historical assumption that a landowner could do anything that was not expressly forbidden by planning regulations changed into the principle that landowners could do only what was expressly allowed, everything else being forbidden.

Many countries emulate the European-style systems and incorporate planning systems in their laws, though implementation of the provisions remains a remote goal, especially where governance capacity is limited. The tension between owners and government planners is present even in communist countries and countries with highly centralized economies. China's 2007 constitutional amendment allowing private property is the best-known example. Lesser known is the Vietnamese reconstruction of its land laws to formalize land markets.

The tension between these two points of view is especially felt by nations seeking economic security. Private property is actively promoted by bilateral and multilateral international

Jacobs, H.M. 2007. Social Conflict over Property Rights. Land Lines. April. 14–19. Lincoln Institute of Land Policy. Cambridge, Mass.

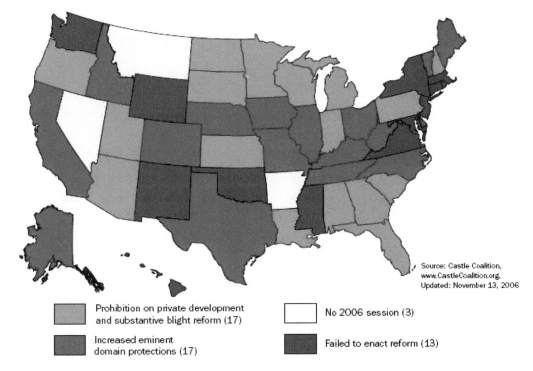

Source: Castle Coalition,
www.CastleCoalition.org.
Updated: November 13, 2006

Prohibition on private development and substantive blight reform (17)

No 2006 session (3)

Increased eminent domain protections (17)

Failed to enact reform (13)

Figure 7.1 The status of eminent domain legislation in the United States as of 2006 shows a shift toward protecting private-property rights.

development aid organizations, such as the U.S. Agency for International Development, the United Nations, World Bank, and nongovernment agencies. Africa, Asia, Latin America, and Central and Eastern Europe are all being encouraged that strong private property rights in land are good for economic growth (Jacobs 2007). The question, however, is how to balance owners' rights with the necessity and capacity of the government to regulate land use and development for the best interests of society. The answer is found in a country's land policy, which should set a reasonable balance between the role of landowners in managing their land and the role of government in providing services and regulating growth in support of sustainable development.

7.2 Planning control systems

Planning systems vary considerably in terms of scope, maturity and completeness, and the distance between expressed objectives and outcomes. The systems also vary in terms of the locus of power—e.g., centralization versus decentralization—and the relative roles of the public and private sector—e.g., the planning-led versus market-led approach (European Commission 1997). More generally, planning systems are influenced by the cultural and administrative development of the country or jurisdiction, the same way as for cadastral systems.

PLANNING APPROACHES

Approaches to spatial planning vary considerably throughout the world, reflecting historical and cultural developments as well as geographical and economic conditions. Across Europe, four major traditions of spatial planning can be identified (European Commission 1997):

♦ A **regional economic planning approach**, where spatial planning is used as a policy tool to pursue wide social and economic objectives, especially in relation to disparities in wealth, employment, and social conditions among different regions of the country. Central government inevitably plays a strong role. France is normally seen as associated with this approach.

♦ A **comprehensive integrated approach**, where spatial planning is conducted through a systematic and formal hierarchy of plans. These are organized in a system of framework control, where plans at lower levels must not contradict planning decisions at higher levels. Denmark and the Netherlands are associated with this approach. In the Nordic countries, local authorities play a dominant role, while in federal systems such as Germany's, the regional government plays a very important role.

- A **land-use management approach**, where planning is a more technical discipline in relation to the control of land use. The UK tradition of "town and country planning" is the main example of this tradition, where regulation aims to ensure that development and growth is sustainable.

- An **urbanism approach**, where the key focus is on the architectural flavor and urban design. This tradition is significant in the Mediterranean countries and is exercised through rigid zoning and land-use codes and a wide range of laws and regulations.

OPERATION OF PLANNING SYSTEMS

Another classification besides planning approaches can be made according to how European systems operate. Two characteristics can be identified:

1. the extent of discretion or flexibility in decision making to allow for development that is not in line with adopted planning regulations—that is, whether the adopted planning objectives and regulations are easily adapted to support changing priorities during actual development;

2. the degree of unauthorized development—e.g., as to whether there is a close, moderate, or distant relationship between the stated planning objectives and actual development.

By analyzing these two categories, European countries can be classified as follows (European Commission 1997):

- The United Kingdom has a discretionary system, yet there tends to be a close relationship between system objectives and actual development.

- Denmark, Finland, Ireland, and the Netherlands have a moderate degree of flexibility in decision making, and planning objectives and policies are close to the actual development that takes place.

- France, Germany, Luxembourg, and Sweden all have systems that have little flexibility in operation, and where development is generally in conformity with planning regulations.

- Belgium and Spain both have fairly committed systems while there is only a moderate relationship between objectives and reality.

◆ Finally, there is a group of countries, Greece, Italy, and Portugal, where systems are based on the principle of committed decisions in planning, but where in practice, there has been considerable discrepancy between planning objectives and reality.

Whatever the general overview, there are all kind of nuances that reflect the specific conditions and cultural traditions of individual countries.

LEGAL MEANS OF PLANNING CONTROL

The relationship between the public and private sectors is governed by the extent to which realization of spatial planning policy relies on public or private sources, and the extent to which development is predominantly plan-led or market-led.

The Danish system, for instance, is mainly plan-led and highly decentralized. The Ministry of the Environment establishes the overall framework in terms of policies, guidelines, and directives. Development possibilities are determined through general planning regulations at the local level (municipalities) and further detailed in legally binding local or neighborhood plans. Municipalities are also responsible for granting building permits, which serve as a final control in the system. Planning at the municipal level is comprehensive and includes determination of land policies, land-use planning, and land-use regulations in terms of urban or rural zoning. Municipal planning also establishes a regulatory framework for the content of more detailed and legally binding local/neighborhood plans that must be provided prior to any major development. The comprehensive municipal plan, as well as local or neighborhood plans, have to be submitted for public debate as well as public inspection and objections before final adoption. This facilitates public participation at all levels of the planning process. On the other hand, there is no opportunity for a public appeal, inquiry, or compensation regarding the contents of an adopted plan, even if local plans are binding on the community. Planning is considered politics, and the mechanics for public participation are regarded adequate to legitimize political decisions.

Planning regulations established by such planning systems are mainly restrictive. The system may ensure that undesirable development does not occur, but it does not guarantee that desirable development actually happens at the right place and the right time, because planning goals are mainly realized through private development as opposed to public. A development proposal out of line with the plan may be allowed, either through a minor departure from the plan or by changing the plan itself prior to implementation. This process includes public participation. Development opportunities are finally determined by the municipal council. On the other hand, development proposals that conform to adopted planning regulations (figure 7.2) are easily implemented without delay.

7.3 Urban land-use planning and regulations

Urban planning is an old tool used for designing new cities. Urban planning is still used today in the design of new cities or neighborhoods, but it is also used for planning and regulation of existing urban areas, urban regeneration, and more generally, improvement and protection of the urban environment.

The tools of urban planning and regulation may vary considerably from country to country, from very basic means of controlling urban development to very sophisticated systems of planning control covering social, economic, and environmental concerns.

URBAN PLANNING CONTROLS

Urban development in many countries accelerated between 1945 and the mid–1980s at a time of increasing affluence and mobility. Sunlight, fresh air, and green surroundings were given high priority when creating new urban areas of detached houses, blocks of apartments, and low-rise housing. The result was a huge suburban sprawl around cities and towns. Today, the urban areas in many European countries have virtually stopped growing, and the demographic trends show that the need for new dwellings is more or less nonexistent. Other countries have different demands, but the need for planning controls is similar.

Urban environments in developed countries are typically controlled by local councils through comprehensive municipal planning and binding local or neighborhood plans. Management of

Enemark, S. 1999, Denmark—the EU Compendium of spatial planning systems and policies. Brussels, used with permission.

Development proposal \ Political decision	Desirable	Undesirable
Conform to adopted planning regulations	Permission	Prohibition
Does not conform to planning regulations	Adjustment	Refusal

Figure 7.2 Decision options within legal planning control systems involve a political element.

local affairs should be seen in total. Municipal planning gives the council a procedural instrument well suited to linking sectors and coordinating overall political and economic activities.

Urban planning normally includes zoning for various land uses such as residential housing, retail, light or heavy industry, offices, public spaces such as parks, and so on. Detailed regulations are then imposed to determine development opportunities in terms of the minimum size of parcels, building density, heights of buildings, and so forth. These regulations may be further detailed in development plans that include schemes for new subdivisions with a detailed layout of a new residential neighborhood, for example.

Urban planning has a significant impact on the value of land by virtue of the determination of development opportunities. Allocation of land rights in terms of possible future land use is a major factor in relation to the land market, especially when the permitted use is changed from, say, agricultural land to an urban use such as residential housing. Such changes, or betterments, impose a major increase in land value that may be subject to taxation.

In areas where no planning regulations are in place (figure 7.3), some general land-use regulations may apply. These may be found in legislation such as a building act and may include regulations for the minimum size of parcels, maximum building density in residential areas, maximum building heights, and so on. General regulations for subdivision and housing developments are effective in controlling development in areas where detailed planning regulations do not apply.

BUILDING PERMIT CONTROL

Most planning regulations are mainly reactive in that they only regulate the possible future use and development of land. As mentioned previously, regulations can ensure that undesirable development does not occur, though they cannot guarantee that desirable development actually happens at the right place and time.

The control of actual development is normally exercised through the issuance of a building permit (or planning permission) prior to construction. The administrative process of issuing a building permit normally includes a check of the development proposal against adopted planning regulations, land-use restrictions, sectoral land-use provisions, and various other regulations such as building bylaws, including detailed regulations for safety and quality of construction.

Figure 7.3 Cairo, Egypt, is an example of megacities that develop mainly outside of formalized planning.

The system of building permit control should then act as a final check in the planning control system and ensure that any new development is consistent with adopted planning policies and land-use regulations and restrictions.

URBAN HERITAGE AND REGENERATION

In the 1960s and 1970s, the main focus of urban development in Western countries was on developing new settlement areas for residential purposes. Since the 1980s and 1990s, however, the focus has turned to urban renewal and restructuring, including the conservation and protection of valuable urban and building features. This process of urban regeneration also includes traffic and environmental considerations for the purpose of generating new life in old (historic) city centers. The process of urban regeneration is normally managed by local authority such as the municipal council by means of spatial planning and intensive public participation. Projects may be implemented partly by public investment in infrastructure, partly by urban renewal companies, and partly by private investment works.

URBAN CONSERVATION

Urban conservation is, to a large extent, taken care of by means of planning—for example, by providing legally binding local plans or bylaws for the protection and maintenance of historical urban quarters or city centers. In addition, urban renewal schemes may contribute to urban conservation. In many countries, permits are needed for demolition or alteration of existing buildings. The municipal council then may consider imposing a ban in order to provide a local plan for the protection of historical or architectural values.

INFORMAL URBAN DEVELOPMENT

Informal urban development may occur in various forms including squatting, where vacant state-owned or private land is occupied illegally and used for illegal housing; informal subdivisions and illegal construction work that do not comply with planning regulations such as zoning provisions; and illegal construction works or extensions of existing legal properties (Potsiou and Ionnidis 2006).

There is no simple solution to preventing or legalizing informal urban development, which is a function of the level of social and economic equity in society and the level of national economic wealth. Consistent land policies, good governance, and well-established institutions can promote integrated land-use control that obviates the need for informal settlements (see section 7.6, "Integrated land-use management"). Decentralization, comprehensive planning, and public participation are key.

Although some illegal development, such as in postconflict or postdisaster situations, may be difficult to stop, many other forms could be significantly reduced through government intervention that is supported by the public. Integrated land management can serve as a fundamental means to support sustainable development by preventing future informal development and legalizing the existing sector. The integration of land policies, land information, planning control, and land-use management should ensure that land-use decision making is based on relevant policies and supported by complete and up-to-date information on land use and rights in land. Land-use policy should also provide for establishing the relevant social and economic institutions that support legalizing the informal sector.

Control of land use will only be effective if it is administered locally through trusted local government services that are decentralized and sensitive to the local environment and community. Local planning officials must be empowered to effectively apply these policies and laws. However, a key element of effective decentralization is accountability, where local government is held accountable by both citizens and national government.

When dealing with informal urban development, it is important to remember that planning is politics, and the political decision-making process will only be legitimized if the public is truly engaged in the process. It is thus imperative that the planning process be transparent and inclusive and that citizens be encouraged to fully participate. This building of social capital will pay dividends as the public, through emerging "m-government" approaches based on mobile phone communications, comes to support the monitoring of urban development. This building of trust will take time, of course, as it requires considerable cultural and behavioral change on the part of all stakeholders.

Arguably, establishment of mature systems that are trusted by the public is also the key to preventing and legalizing informal urban development (Enemark and McLaren 2008). This goes for, at least, the developed part of the world. In developing countries, this approach must be supplemented by a range of measures that address the issues of poverty, health, education, economic growth, and tenure security.

However, when unauthorized informal development occurs, the planning system itself may offer only a partial explanation. Factors outside the formal planning system will often play a determining role in its operation and effectiveness. The historical relationship between citizens and government, attitudes toward land and property ownership, and implications of social and economic institutions in society will all play a role among a variety of other historical and cultural conditions (European Commission 1997).

INFORMAL SETTLEMENTS

The need for urban planning is a global problem. Today, there are about 1 billion slum dwellers in the world, while in 1990, there were about 715 million. More than 3.3 billion out of the world's population of 6.6 billion are now living in urban areas, with one-third of those living in slums. UN–HABITAT estimates that if current trends continue, the slum population will reach 1.4 billion by 2020 if no remedial action is taken. Current trends predict the number of urban dwellers will keep rising, reaching almost 5 billion in 2030 when 80 percent will live in developing countries. Over the next twenty-five years, the world's urban population is expected to grow at an annual rate of almost twice the growth rate of the world's total population (UN–HABITAT 2006a).

Focusing government policies and actions on informal settlements (figure 7.4) is essential, because one of every three of the world's city residents lives in inadequate housing with few or no basic services. Millennium Development Goal 7, Target 4 seeks to improve the lives of at least 100 million slum dwellers by 2020 (http://www.un.org/millenniumgoals/). City authorities tend to

Figure 7.4 The informal settlement of Kibera in Nairobi, Kenya, houses 1 million-plus people in an area of 150 ha, or about 0.6 sq. mi.

view most people living in slums as doing so illegally. Because of this, cities do not plan for or manage slums, and the people in them are overlooked and excluded. They receive none of the benefits of more affluent citizens, such as access to municipal water, roads, sanitation, and sewage. This attitude toward slum dwellers and management approaches that disregard them perpetuate the levels and scale of poverty, which impacts the city as a whole (UN–HABITAT 2004).

These issues have no clear solution. While approaches vary and systems differ in style and scale, in principle, planning problems in urban areas are axiomatic with lack of economic and governance development. The general state of many developing countries is characterized by an unequal distribution of land among inhabitants. Many poor inhabitants in these countries lack access to land or lack secure rights to the land they have settled on. Lack of tenure security is very often a central characteristic of informal settlements. Informal settlements are often neglected enclaves of settlements consisting of poor inhabitants living in distinctly poor conditions caused by bad housing and no access to basic services.

Provision of new infrastructure in existing informal settlements, redistribution of informal settlers, extension of services for those who build their own dwellings, and cooperatives undertaking development of affordable housing in combination with opportunities for business enterprise are all approaches to slum upgrading.

7.4 Rural planning and sectoral land-use regulations

The crisis of urban management is well known, but planning and development issues in rural areas are just as significant. Rural planning systems are complicated by separate systems of sectoral planning, which manage resources such as soil quality, landscape quality, raw materials, and water accessibility. In some systems, these interests are given priority in particular areas, with zoning reserved for agriculture, raw materials extraction, or special natural areas. Ideally, these sectoral controls should be integrated into the comprehensive spatial plans to form the basis for rural land-use administration. Many countries experience difficulties with sectoral land-use management. The basic mapping of natural resources, including groundwater, is frequently not available as a source of information to help balance regional-level plans with the administration of all sectoral land-use acts. Despite these overall difficulties, sectoral land-use management remains one of the world's principal means of planning national rural environments. These sectors all involve unique policies and applications.

RURAL ZONE DEVELOPMENT

A basic element of many mature planning systems is the division of a country into three zones: urban, recreational, and rural. In Denmark, for example, development is allowed in the urban and recreational zones in accordance with current planning regulations, while in rural zones, covering the majority of the country, developments or any changes of land use are prohibited or subject to special permission according to planning and zoning regulations. A typical exception is that construction necessary for commercial agriculture, forestry, and fishery operations often requires no rural zone permit. The rural zone development provisions are intended to prevent uncontrolled land development and installations in the countryside and to preserve valuable landscapes. Urban development can then only occur where land is transferred from a rural zone to an urban zone, which may be subject to a land-use tax, to be paid by the landowner.

NATURE PROTECTION AND MANAGEMENT

Many countries set aside land for nature and landscape protection in parks or reserves. Conservation systems can protect large areas or create protection zones along coastlines, around monuments of national interest, or protect landscapes and views. Typically, conservation regulations make it possible to set aside areas as nature reserves and determine how such areas shall be used. Standard provisions aim at maintaining aesthetic control by restricting advertising and ensuring that public structures in the countryside are located and designed so that the greatest possible consideration is given to scenic values and environmental interests. Location and design of roads, cables, and electric wires are also controlled.

Conservation has been an important instrument for nature protection and is mainly used to preserve areas of outstanding beauty or cultural value or to protect areas with valuable flora or fauna of specific national interest. Increasingly, conservation systems partner with individual landowners who are willing to enter into arrangements to preserve natural vegetation and landscapes to permanently protect conservation values. The major methods of providing protection involve state-run restoration and protection projects.

AGRICULTURAL POLICIES

Especially in European Union countries, legislation requires agricultural properties to be operated in accordance with agricultural and environmental considerations. The general protection of agricultural land can be abolished, however, if local planning deems that the land is to be used for other than agricultural purposes, especially when rural land is transferred to an urban zone for development purposes. These policies need to reflect changes in agricultural economics and allow for changes in the size of holdings that are necessary to maintain the rural population's ability to work the land.

Modern agricultural policy in developed nations now supports various forms of less intensive farming, which further reduces the total agricultural area while not reducing the food supply. The policy, which supports environmentally friendly land use in agriculture, includes support for reforestation (permitted for the total area of the holding), introduction of ecological farming methods, and environmentally friendly growing methods (without fertilizers and sprays), as well as permanent fallowing of agricultural land.

FORESTRY POLICIES

Sustainable forests require intensive management of multiple, and sometimes conflicting, policies. Production of wood chips, for instance, conflicts with provision of living spaces for wild fauna and flora.

Forestry policies often include protection and specific management regulations. In Denmark, for instance, twelve percent of the country's land must be used and operated as forests. The national forest policy implies that the Danish forest land is to be doubled within the next eighty to 100 years, which means that 5,000 ha (about 20 sq. mi.) are to be reforested per year.

In many countries, forestry policies take a multiuse approach that combines modern forestry production with protection of the environment and inclusion of recreational activities. This type of sustainable land use should apply whether the forestry land is state-owned or in the hands of private parties or companies.

NATURAL RESOURCE MANAGEMENT

Natural resources are a fundamental asset. Properly managed natural resources provide the foundation for maintaining and improving the quality of life of the world's population and can make invaluable contributions to sustainable growth.

Raw materials, such as metals, gravel, clay, and chalk, are finite resources. National policies, therefore, often aim to limit consumption to ensure a long-term supply of raw materials. Environmental considerations are often integrated with commercial activity and taken into account when permission for extraction is given according to the relevant legislation. Extraction should be based on an overall raw materials plan, which takes environmental and other interests into account.

ENVIRONMENTAL PROTECTION AND POLLUTION CONTROL

Environmental policies should emphasize that economic growth can be achieved simultaneously with improvements to the environment. Industries must be able to constructively and economically absorb environmental considerations within their enterprise. Policies may be based on the "polluter pays" principle, which is internationally recognized. Installations should be located at a site causing the least possible pollution and adopt measures to curb pollution to the greatest possible extent. These principles are the basis of recent global/national carbon-trading initiatives.

Environmental policies normally include provisions to prevent and control pollution of air, earth, and water, as well as provisions for noise and waste treatment. Requirements for use of the least-polluting technology should also be included. A statutory system of prior approval/ authorization should apply to the establishment of plants or activities that are considered potential sources of pollution. This approval should ensure that businesses or industry meet a number of environmental and technological standards so as to pollute soil, air, and water as little as possible. Environmental policies may also include provisions for wastewater treatment to be managed through the guidelines that safeguard the quality of watercourses.

Groundwater is becoming an increasingly important policy area and is now one of the major political subjects in most developed countries. The aim is normally to ensure sufficient

Figure 7.5 New Zealand has a pristine coastal environment to protect.

uncontaminated water resources to meet expected future needs. This may be achieved by using spatial planning and may include regulating future land use in the areas of special interest.

COASTAL ZONE MANAGEMENT

The land–sea interface is one of the most complex areas of land/marine management as it is home to an increasing number of activities, rights, and interests. The coastal zone is a gateway to ocean resources, a livelihood for local communities, a reserve for special flora and fauna, and an attractive area for leisure and tourism. Many nations are politically, economically, socially, and environmentally dependent on the coastal zone and so depend on proper management of this fragile environment to ensure sustainability and social justice.

The coastal zone is considered a vulnerable area and is often strongly regulated to ensure a balanced approach to development that includes all stakeholders. Land-use planning in coastal areas needs an integrated approach to accommodate interests in both the land and marine environment. Without strict control and regulations for a balanced development in the coastal areas, some pristine environments may disappear, as has happened in many regions in the world. A balanced development in coastal areas can only be achieved when all stakeholders and interests are taken into account. This will often necessitate an overall national policy for managing coastal zone interests (figure 7.5).

A special issue in coastal areas is achieving a balance among economic development, livelihoods and the quality of life in local communities, and protection of the natural environment. Conflicts may occur when the livelihood of the indigenous population and their access to coastal resources is overtaken by economic interests. These include tourism and leisure development, which do not necessarily benefit low-income people and the local community. In this extreme form, indigenous people are displaced from their original habitat and may need to relocate in informal settlements with limited basic services, unacceptable environmental conditions, and few or no work opportunities. Coastal management policies should ensure social equity in terms of access to coastal land and other coastal resources and be supported by pro-poor policy change and national poverty reduction strategies particular to marine areas (FIG 2008b).

7.5 Land consolidation and readjustment

Land consolidation adjusts the structure of agricultural holdings in rural areas to optimize conditions for agricultural production. In some regions, such as central Europe, the infrastructure in rural areas is inadequate, and the individual holdings may consist of many small parcels, which is inconvenient for effective agricultural production (figures 2.9 and 4.1). This structure may also be a result of inheritance where land is divided into small strips. A land consolidation scheme may then include a certain area where landowners allow their holdings to be restructured into larger and more convenient parcels of land that are more or less equivalent to the value and size of their original holdings. The process is normally undertaken by a lands department or by licensed surveyors. It is often initiated by individual landowners and is normally based on voluntary participation. The final result is agreed to by all parties, and the cadastre and land book are updated accordingly. The result of such a process in Denmark is shown in figure 7.6.

Land consolidation may also be used for adjusting the structure of agricultural or residential holdings to implement major infrastructure projects or nature management plans. Land consolidation is also used to facilitate urban renewal and downtown developments. In these cases, the process is normally referred to as land readjustment.

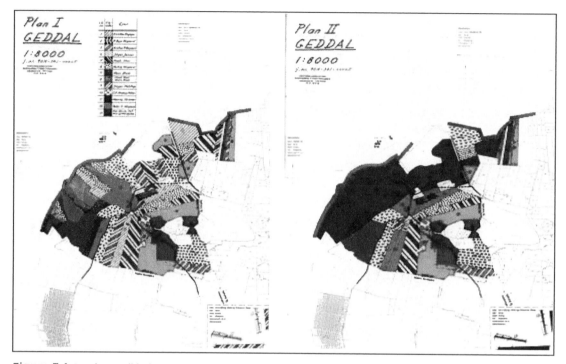

Figure 7.6 Land consolidation converts an area of scattered land parcels, left, into more homogeneous holdings, right, to optimize farming.

Land readjustment aims to repurpose the physical allocation of land into modern social and business uses. In Hanoi, Vietnam, for example, much of the old city is composed of small, very narrow allotments, causing high-rise development to get denser as the population increases (figure 7.7).

Other countries experience severe misalignment of land uses in rural areas and extensive conversion of rural to urban land. Land readjustment systems, which have been around for centuries, now form a common experience of land managers.

> *"The concept of land readjustment was used by President George Washington who formed an agreement in 1791 with landowners of the site where the city given his name was to be developed. A legal framework was first introduced in Frankfurt-am-Main in German in 1902. Different forms of land readjustment exist in many countries, including Germany, Japan, Taiwan, Republic of Korea, Western Australia (land pooling), India (plot reconstruction), and Indonesia. In Japan, about 30 percent of the urban land has been*

Figure 7.7 Small, narrow lots in Hanoi, Vietnam, have given way to dense urban development as the population has increased.

developed by land readjustment areas. ... In the Republic of Korea, 342 land readjustment projects have converted agricultural land into urban land." (UNESCAP 2007)

The UN Economic and Social Commission for Asia and the Pacific (UNESCAP) identifies the important prerequisites (2007) for successful implementation of readjustment:

◆ The scheme must be supported by national, regional, and municipal governments, with the national government providing regulations that ensure fairness in the system

◆ Land readjustment agency must be given powers to coordinate access to assistance from various government departments

◆ Land registration and cadastral systems need to be efficient

◆ The country must supply a sufficient number of skilled and highly dedicated professionals at the local level as well as objective and well-trained land valuers

◆ Processes must be based on public–private cooperation, the technique should be supported by the majority of landowners, and forced acquisition should be avoided

Other countries take a different approach. Many Australian states do not use it. Some countries, such as Thailand and Sweden, have regulations that make land consolidation difficult. Often, it is not used because of the inevitable political issues it raises and consequent government inertia.

7.6 Integrated land-use management

As the stresses on land grow, integration of land-use management is increasingly necessary to support sustainable development. Land policies, land-use control systems, and land information management must be integrated to ensure that existing and future land use are consistent with land policies as well as planning and sectoral regulations and that decisions are based on complete, up-to-date land information systems (figure 7.8).

Three key principles ensure successful system integration:

◆ Decentralization of planning responsibilities

  ◆ Creating local representative democracy responsible for local needs

  ◆ Combining responsibility for decision making with accountability in terms of economic, social, and environmental consequences

  ◆ Applying monitoring and enforcement procedures

◆ Comprehensive planning

  ◆ Combining aims and objectives, land-use structure planning, and land-use regulations in one planning document covering the entire jurisdiction

◆ Public participation

  ◆ Creating a broader awareness and understanding of the needs and benefits of planning regulations

  ◆ Enabling a dialog of government, citizens, and other stakeholders about management of the urban and rural environment

Integrated land-use management is based on land policies contained in the overall land laws, including cadastral and land registration legislation, and planning and building legislation. These laws identify the institutional principles and procedures for land and property registration, land-use planning, and land development. More specific land policies are established in sectoral land laws for agriculture, forestry, housing, natural resources, environmental protection, water supply, heritage, and so on. These laws establish institutional arrangements to achieve these objectives through permit procedures, information policies, dispute handling, and so forth. Sectoral programs collect relevant information for decision making within each area. These programs feed into the comprehensive spatial planning carried out at the national, state/regional, and local level.

Importantly, a mature system of comprehensive planning control needs to be based on appropriate and updated land-use data systems, especially the cadastral register, land book, property valuation register, building and dwelling register, etc. These registers need to be organized to form a network of integrated subsystems connected to the cadastral and topographic maps to form a national SDI for the natural and built environment.

In the land-use management system (i.e., the planning control system), the various sectoral interests should be balanced against the overall development objectives for a given location and thereby form the basis for regulation of future land use through planning permissions, building permits, and sectoral land-use permits according to the various land-use laws. These decisions are based on relevant land-use data, reflecting the spatial consequences for land and society. In principle, implementation that is consistent with adopted planning policies in support of sustainable development can then be ensured.

An integrated approach to land management depends on appropriate policies and structures of governance. Decentralization can be seen as the key to sustainable development. In many countries, the obvious local arena for land-use planning and decision making is the local

Enemark, S. 2004. Building Land Information Policies. Proceedings of Special Forum on Building Land Information Policies in the Americas, Aguascalientes, Mexico, 26–27 October 2004, used with permission.

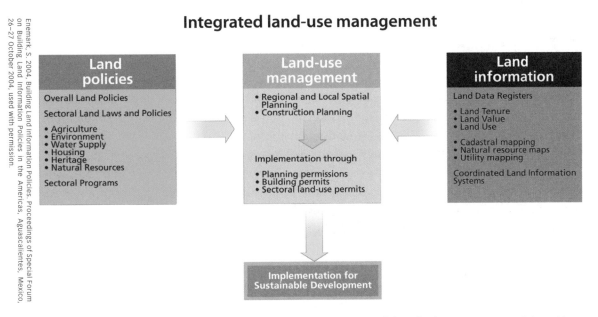

Figure 7.8 Land policies and land information must be integrated into land-use management to achieve sustainable development.

municipality. Whatever outcome may emerge from a decentralized system, the decisions must be assumed to be the right ones in relation to local needs. Decentralization thus institutionalizes the participation of those affected by local decisions. This argument is particularly valid in land-use decision making and administration. Land-use planning thus becomes an integrated part of local politics within the framework of policy making at the regional and national level. The purpose in administering tasks at the local level is to combine responsibility for decision making with accountability for the financial, social, and environmental consequences.

Integrated land-use management requires comprehensive planning that combines policies and land-use regulations covering the total jurisdiction into one planning document. This consolidated presentation of political aims and objectives as well as problems and preconditions should then justify the land-use plan and the more detailed land-use regulations. Public participation should be encouraged to create a broader awareness and understanding of the need for planning regulations and enable a dialog between government and citizens about the management of natural resources and the total urban and rural environment. Eventually, this dialog should legitimize local political decision making. In terms of informal urban or rural development, there is a need for a monitoring system—e.g., through continual updating of a large-scale topographic basemap and proper enforcement procedures to evaluate activities and trends in relation to overall land policies.

7.7 Land development

The term land development refers to the processes of implementing land-use planning or development proposals for building new urban neighborhoods and new physical infrastructure and managing the change of existing urban or rural land use through granting of planning permissions and land-use permits. Depending on the scale of the development project, the process may include a range of activities such as land acquisition, subdivision, legal assessment and planning consent, project design, construction works, and the distribution of development incentives and costs. The process also includes a range of actors such as landowners, developers, public authorities, building contractors, and financial institutions. The land development process is a multidisciplined activity.

Some development activities such as detailed design or actual construction work are not normally considered part of land administration. What it does cover, however, is the control of development proposals and change of land use in relation to adopted planning regulation and land-use laws. This also includes determination of property boundaries as the base location of construction works according to building regulations.

Figure 7.9 Dubai, United Arab Emirates, has seen an expanded period of rapid land development.

DEVELOPMENT CONTROL

Land development can be seen as the actual outcome of the planning process—the end result of implementing adopted land policy measures. Development control then means that public authorities should ensure that any development and construction activity is in line with adopted plans and regulations, thus contributing to a sustainable future. Such construction activity may be extensive (figure 7.9).

Almost all countries have systems in place to control the land development process. However, the efficiency of these systems varies considerably depending on the maturity of the institutional structures and the overall economic, judicial, and cultural conditions. The efficiency of LAS can be measured by the degree of unauthorized development—i.e., as to whether actual development is in line with the stated planning objectives.

The key means of controlling development is through building permits (or planning permission) and subdivision permits. The role of the building permit is basic. However, subdivision control is another important tool in the land development process as it regulates access to property. In the United States, for instance, subdivision regulations are a key means of governing the conversion of raw land into building sites. Locally adopted regulations normally include rules under which the developer cannot make improvements or divide and sell land until the planning commission has approved a plan for the proposed design of the subdivision. This is controlled against the standards set in adopted subdivision regulations. Regulations governing access to land may also include agreements that require longtime leaseholds to be approved by authorities before they are entered into the land book title register or registry of deeds.

In other countries, detailed subdivision regulations may be less common. Often, only more generalized rules are included in state laws — e.g., the minimum size of parcels. The important point, however, is that subdivisions should only be allowed when the purpose of the development is in line with adopted planning policies. In Denmark, for instance, the subdivision process, as undertaken by private licensed surveyors, must include documentation that the future use of the parcels complies with adopted planning regulations and relevant sectoral land-use laws.

Cadastral records and especially cadastral maps play a key role in facilitating land development control. The legal rights in land and the boundaries of existing properties represent the starting point of any development. Updating the cadastral records and maps is therefore essential for the ongoing process of land-use control.

THE DEVELOPMENT PROCESS AND THE ACTORS INVOLVED

In more general terms, the land development process involves converting undeveloped land into developed land, which directly affects the value of land. The development of land and its effect on land value can be divided into four phases.

Figure 7.10 Land value increases as a result of development.

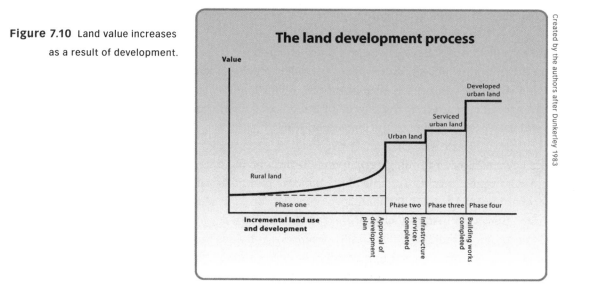

Created by the authors after Dunkerley 1983

In most countries, land value or property prices are determined by the market. Figure 7.10 shows that when expected development opportunities for undeveloped land such as agricultural and forestry areas are expected (phase one), then the value of the land tends to be affected accordingly. Once the land is approved for development purposes—i.e., through adoption of a detailed land-use plan (phase two)—the land value will reflect this new land–use opportunity. In some countries, this increase in value is subject to taxation, since the added value is created by societal development and not by actions of the landowner. Phase three appears once the individual parcel within the detailed plan is slated for construction, and service fees are paid for subdivisions and the installation of roads, water, and sewerage systems. Phase four appears once the land is fully developed. The final value of the land and the individual properties will, of course, vary depending on the extent, usage, and quality of design and construction. This final value is eventually determined by market forces (supply and demand) and may, in some cases, be lower than the actual costs of development.

The property development process may be organized in different ways depending on the role of the developer. This may be the landowner, a professional developer, or a public authority such as the municipality. For a specific development project, the process may include a whole range of activities and procedures, which are typically concept design, site appraisal, and a feasibility study, including the land acquisition and development option, detailed design and evaluation, approval of the project from planning and building authorities, contracting and construction, and, finally, marketing, management, and disposal of property (Ratcliffe and Stubbs 1996).

For a developer to assess the full potential of a property, it may be necessary to (1) determine the best possible uses to which a piece of land or property can be put in the future, with regard to the planning consent (with possible conditions) likely to be granted; (2) estimate the market value of the land when put to this use; (3) consider the time that will elapse before the land can be so used; and (4) estimate the costs of carrying out the works required to put the land to the proposed use together with such items as legal costs, the agent's commission on sales and purchase, and the cost of financing the project (Britton, Davies, and Johnson 1980).

Typical actors involved in the development process include the following (developed from Cadman and Austin-Crowe 1993):

◆ **Landowners**, whether a private party or a legal person and whether public or private, play an important role since they hold the legal rights for any development or change of land use.

◆ **Developers**, such as private-sector development companies, may act as an entrepreneur taking the risk to produce a development project for a profit, or they may act more as a project manager controlling and coordinating the project throughout its various phases.

◆ **Financial institutions**, such as banks, insurance companies, and investment funds, play an important role in lending capital for financing developing projects on the basis of relevant risk analysis.

◆ **Planning/building authorities** should ensure that the development proposal is in line with adopted planning policies and regulations and thereby prevent "undesirable development." They also act as facilitators to ensure that "desirable development" actually appears at the right place and the right time. These roles may include negotiations with the landowners or developers to achieve optimal results.

◆ **Building contractors** undertake specialized activities within the construction process based on a contract with the landowner or developer setting the terms for delivery, quality and risk management, and payment.

◆ **Professional advisers** may include a whole range of professionals to support and advise the landowner or developer on specific issues. These professionals include lawyers, architects, engineers, surveyors, accountants, etc.

◆ **Third parties** may play an important role in the development process in the form of objectors who may delay the whole process through appeals and public inquiries. Objectors may be neighbors (often claiming "not in my backyard") or more specialized nongovernmental organizations (NGOs) defending specific interests such as heritage and nature protection.

Any development process is unique in terms of scope, processes, and actors. Yet the process almost always includes the balance of economic interests against the overall aim and objectives of the relevant land policies and regulations to ensure a project meets the defining goal of sustainable development.

LAND ACQUISITION AND FINANCIAL INCENTIVES

Countries and regions vary widely in how they go about implementing adopted land-use policies through actual development. It may be a predominantly public-sector approach, or it may be led by the private sector. In any case, there are a number of mechanisms to ensure that plans and policies are carried out. A key incentive, of course, is to provide local infrastructure and public services in terms of health and educational facilities, thereby encouraging individuals and private companies to locate in the area in accordance with the goals of the adopted land-use plan.

The main types of land policy instruments used to achieve plan objectives include

- **Land acquisition by agreement**, where public authorities such as the municipality acquire land through private agreement with the landowners to achieve its development objectives. Land acquisition by agreement (or by buying a development option) is also used by professional developers.

- **Land banking**, where municipalities in particular may build up large areas of publicly owned land and thereby control the supply of land for development in certain areas. Such strategic purchases place the municipality in a key position for controlling future development through phased disposal of serviced land.

- **Expropriation or compulsory purchase**, a well-known means in most countries that enables any tier of government to purchase land in the public interest against full compensation of the market value. The public interest may be public roads, parks, and service facilities such as schools and health care. Public interest may, however, also be considered the reason to implement an adopted detailed plan. In many countries, however, expropriation is seen as a time-consuming and politically sensitive process, and is therefore used as the last resort.

- **Preemption rights**, which require, in principle, that landowners offer their property for sale to the municipality first and normally at market value. This means can be used in different forms to ensure that the public interests in a certain area can be achieved.

- **Financial incentives**, which may include subsidies to encourage specific development at a certain time and place. Incentives could lower land prices, provide property tax abatement over a number of years, or lower the cost of development loans. In some countries, however, public authorities are not entitled to offer economic incentives. In Demark, for instance, the activities of public authorities are limited under the general principles of equality and objectivity; and those activities must not interfere with the general conditions of market forces or benefit individual persons or companies.

There is a range of other means to be considered such as public–private partnerships, which potentially can be very useful in implementing larger regeneration schemes — e.g., when converting old industrial areas, so-called brownfields, into modern urban use, often with multiple land uses. Another strategy includes promoting and marketing mechanisms for branding specific developments.

Figure 7.11 Dubai, United Arab Emirates, is a prime example of large-scale land development in a coastal environment.

INFRASTRUCTURE DEVELOPMENT

In most countries, local or regional authorities and utility companies are responsible for providing and maintaining local infrastructure in terms of roads, water supply, sewage systems, communication networks, and the like. In some cases, however, the developer may undertake some of these responsibilities as part of implementing a major project based on a special agreement with authorities. The costs of these infrastructure facilities are normally paid by the end users through fees to be calculated according to local bylaws. Major infrastructure facilities such as highways, bridges, and electricity transmission lines are normally undertaken by state authorities or state-authorized entities.

The design and implementation of local infrastructure is often carried out as an integrated part of the development process—e.g., roads and sewer systems for a major subdivision of a new urban neighborhood. It may then be a condition of development that the road and sewage networks are completed by the developer and the fees are paid, often via development fees.

URBAN DEVELOPMENT

Urban development is a generic term that covers a wide range of activities from implementing new full-scale urban areas or towns (figure 7.11) to simply building a new dwelling or an extension of an existing one. It can include building new urban neighborhoods, urban water- or harborfront facilities, a commercial center, business complex, or industrial plant. Or it can be as simple as adding on new apartments. In principle, any change of land use in urban areas can be considered urban development. The development process will vary according to the scale of the development.

A key issue is the ability to control urban development at all levels and ensure that it is functional, sustainable, and in line with adopted planning policies. The issue will be especially relevant in terms of controlling development in the future megacities of the world (figure 7.12).

As the global population increases, the issue becomes still more acute. According to UN–HABITAT, the year 2007 was when the globe became urban. More people around the world are now living in cities than in rural areas, while in 1950, it was less than 30 percent. Also in 1950, there was only one megacity (New York) with more than 10 million inhabitants. Today, there are more than twenty megacities, with some of them holding more than 20 million inhabitants. How can we deal with the social, economic, and environmental consequences of this mass development—the resulting climate change, social and legal insecurity, environmental pollution, infrastructure chaos, and extreme poverty? Managing megacities is likely to be an overwhelming challenge throughout the next century.

Another problem of urban development is "urban sprawl," found in most major cities throughout the world, particularly in the United States. The result is often huge developments, which suffer from a lack of identity. A range of social and environmental problems are connected to this kind of

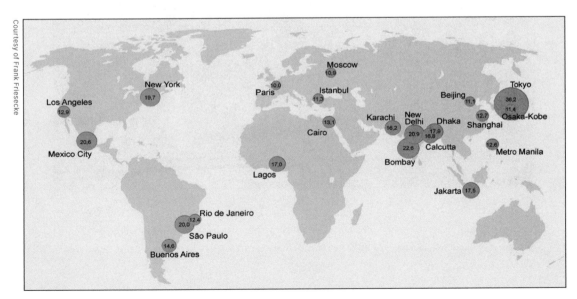

Courtesy of Frank Friesecke

Figure 7.12 Projected spatial distribution of the world's megacities in 2015.

Ørestad, as framed in red below at left, is an example of building a new urban neighborhood only five minutes from the center of Copenhagen. Located near the fixed road and rail link between Denmark and Sweden and only a few minutes from Copenhagen Airport, the area is ideal for a secondary center to Copenhagen City. Ørestad is intended to be a town in the classic sense with various spatially integrated functions closely linked to a new light-rail system and the overall road and rail system to Sweden. Ørestad is divided into smaller districts along the light rail like pearls on a string. Each district has its own characteristics, but all have mixed uses. The northern area is the university district, while the southern area comprises Ørestad City with shopping and leisure facilities. Ørestad, at 3.1 square kilometers, will be developed over a period of twenty to thirty years. When finished, 50,000–60,000 people will work in Ørestad, mainly in the finance, services, IT, health, and research and development sectors, while around 20,000 people are expected to live in the new Ørestad area.

Courtesy of Niels Ostergard

Figure 7.13 Left: The new urban development of Ørestad is located close to the heart of Copenhagen, Denmark. Right: The motorway link that runs west to east through the southern part of Ørestad toward the fixed road and rail link between Denmark and Sweden connects to the Copenhagen Airport in the top of the picture.

"endless" development. Solutions are found in more comprehensive planning that encompasses a variety of urban facilities and social activities. The term "smart growth" was introduced in the United States to describe urban planning as a means of combating sprawl and its problems (Frumkin 2002).

A more traditional example of urban development (figure 7.13) includes a new metropolitan area of about three square kilometers, or about one square mile, close to the city center of Copenhagen, Denmark.

RURAL DEVELOPMENT

While the focus in urban areas is on economic development, the focus in rural areas is more linked to industries such as agriculture, forestry, and mining as well as overall protection of the natural environment. Rural areas often harbor a range of competing interests. Modern and efficient production methods often compete with leisure and preservation interests. In many countries, rural development is strictly controlled to ensure a sustainable environment. In Denmark, for instance, no development is allowed in rural areas without special permission, except for development that's connected to agriculture, forestry, or fishery.

Conflicting interests in rural areas, including the coastal zone, call for comprehensive planning that combines commercial interests with environmental, leisure, and conservation objectives. Comprehensive planning should act as a basis for controlling land use and development in rural areas. It should also include the perspective of rural-urban linkages that preserve the rural-urban connection. Adopting this kind of planning policy in rural areas should ensure sustainable living while preventing unnecessary migration out of rural areas.

INTEGRATED LAND DEVELOPMENT AND LAND USE

Land development should be seen as an integrated part of LAS, especially in terms of assuring that actual development is sustainable and in line with adopted land-use policies. The new land-use pattern established as a result of the development is then recorded. Maps and registers are updated to form the basis for future planning and administration. This dynamic interaction is ongoing and should be carefully understood and managed as a total system.

While this book focuses primarily on aspects of tenure and spatial enablement, incorporating and understanding issues of land use and land development are central to ensuring the land management paradigm delivers sustainable development. As a result, this brief introduction to managing the use of land is considered essential in understanding and implementing the paradigm.

Chapter 8

Marine administration

8.1 The need to improve marine administration

8.2 Challenges in building marine administration systems

8.3 Existing marine administration

8.4 The marine cadastre concept

8.5 Marine registers

8.6 Developing a marine SDI

8.7 Using the land management paradigm to meet marine needs

8

8.1 The need to improve marine administration

Oceans cover almost two-thirds of the Earth's surface. They regulate weather patterns and sustain a huge variety of plant and animal life (United Nations 2003). The diversity of marine environments requires effective economic, social, and environmental management that is just as comprehensive as land management. Coastal zones are now the most important area in terms of human population growth and sustainable development, and their significance will only increase.

> *"The coastal zones of nations are complex, and involve finely balanced ecosystems within a narrow band of land and sea. They provide homes for millions of people. Coral reefs are home to more than a million species. Coastal zones are economically, politically, and socially critical to many nations. Coasts are used by millions of people for recreation. Major transport hubs are situated in or near the coastal zone where ports and harbors are vital to commerce and trade.*

"This narrow band of sea and land occupies only 20 percent of the world's land area. Half the world's population, some 3 billion people, lives within 200 km. of the coast, and it is estimated that by 2025, this figure may double. Our cities use some 75 percent of the world's resources and discharge similar amounts of waste." (Greenland and Van der Molen 2006, 3)

Simply, the interests of a nation do not stop at the land–sea interface. They continue into the marine environment. Therefore, the responsibilities and opportunities of governments to provide infrastructure for land and resource management extend to marine areas. This has brought with it an increased need to more effectively and efficiently manage marine resources to meet the economic, environmental, and social goals of sustainable development.

8.2 Challenges in building marine administration systems

While land administration systems traditionally stop at the coastline, many countries apply land-based tenures, measurement and identification systems, and registration systems as initial solutions to marine management problems. Indeed, some countries even extend existing organizational structures beyond their traditional terrestrial boundaries. Indonesia, for instance, uses the national land agency to manage many close-to-shore marine uses. New South Wales in Australia uses planning processes to initiate leases for oyster production. Alongside these extrapolations of systems from the land to marine environment, the unique marine activities of fishing, navigation, aquaculture farming, and many others, including pollution cleanup, tend to be separately managed. As a result, nations tend to produce diverse, silo-based, and generally unsatisfactory marine management and, consequently, insufficient, uncoordinated marine information.

Concerted international efforts over the past decade have sought to introduce coherence and capacity in marine management by identifying the unique features of a marine administrative system, such as a lack of markers and changing boundaries, and integrating marine and land management wherever possible, particularly in coastal zones. Modern LAS theory illustrates these efforts. The theory demands a unified management approach, particularly in applying the land management paradigm. That means including coastal and marine zones within the "land" concept. To be successful in this broad context, a marine system must be subsumed into a national approach to management of land, coastal, and marine environments. This then will yield sustainable development.

The ultimate aim is to replace management by silo agencies with integrated management that institutes systems designed to meet the management imperatives delivered through a cadastral

PCGIAP 2004, International workshop on marine administration—the spatial dimension, Working Group 3, used with permission.

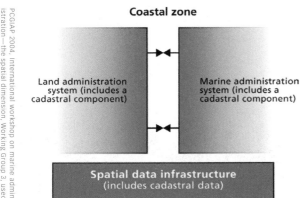

Coastal zone

Land administration system (includes a cadastral component)

Marine administration system (includes a cadastral component)

Spatial data infrastructure (includes cadastral data)

Figure 8.1 Successful marine administration demands seamless integration of both marine and land management.

system (figure 8.1). Thus, three major components of LAS—the cadastre, the register, and the SDI—now find their place in international discussions of marine management. Though each needs to be uniquely configured for the complexities of marine areas, best practice suggests they all should be extrapolations of their land-based cousins, not separate organizations and agencies.

The coastal zone, the overriding feature of the marine environment, is of vital importance, forming the glue that joins land and marine areas together. It is the centerpiece of the marine management system of every nation that enjoys sea boundaries, and it is ever changing. In short, if a nation fails to manage its coastal zones effectively, neither its land management nor its marine management will work. This is especially true for nations formed by archipelagoes or whose coastlines are extensive comparative to their land mass. Thus, states with extensive coastlines (Vietnam, Mozambique, Canada, Chile, Australia, and Costa Rica, shown in figure 8.2), island states (such as New Zealand and Madagascar), and many archipelagic nations (Indonesia, the Philippines, and Japan) need specially designed LAS that incorporate the marine environment. As with land, the cadastral component forms the fundamental information layer that supports successful integrated management. The international trend in modern marine management is to build a holistic approach to jurisdictional management and simultaneously create systems that improve regional management as well.

8.3 Existing marine administration

The International Hydrographic Organization (IHO), an intergovernmental consultative and technical organization established in 1921 to support safety in navigation and protection of the marine environment, currently represents more than seventy-seven member nations and other

Figure 8.2

Costa Rica, a country
defined by its coastal zone,
needs LAS that incorporate
the marine environment.

international organizations seeking to improve marine management. It faces a difficult task. The existing management framework is problematic, because the oceans are broken up into various national and international jurisdictions depending on the distance from a country's coastal baseline (figure 8.3). These jurisdictions are governed by a complex web of local, state, and national legislation, international conventions, and maritime practices. This type of framework creates complex interactions among overlapping and sometimes competing rights, restrictions, and responsibilities (RRRs) across various activities in the marine environment and coastal zones.

Added to this, the United Nations Convention on the Law of the Sea (UNCLOS) has, since 1994, provided the overarching international arrangement governing use of the oceans for those of the 192 UN members who have accepted it. The UNCLOS zones of maritime governance do not cover the rivers, streams, deltas, mangroves, underground water systems, and so on that allow fresh water transfer into marine areas. These and other inland waters are totally within a nation's jurisdiction, and no foreign vessel has any right of passage. After a country sets its coastal baseline (usually the low-water mark), the UN-recognized zones broadly establish a country's capacity to regulate activities as follows:

◆ **Coastal waters**, those not regulated by UNCLOS, operate in some federated states
   to distribute regulatory opportunities between state and national governments.

◆ **Territorial waters** allow a nation to set laws and use any resource. Within
   any particular nation, state and local governments often vie for jurisdictional
   opportunities.

◆ **The contiguous zone** (between 12 and 24 nautical miles from the baseline) allows the nation to enforce laws on resource access and to regulate smuggling and illegal entry.

◆ **The exclusive economic zone** extends to a distance of 200 nautical miles and gives the nation the right to use all natural resources, especially fish, and oil and gas, and to lay pipes and cables.

◆ **The continental shelf** may extend beyond 200, and up to 350, nautical miles depending on the extent of the shelf and allows nations to use mineral and oil resources under the sea floor.

◆ **The high seas** are not managed by any nation, but through a loose framework of international conventions—for example, a regime to control mineral resource exploitation in deep seabed areas outside national jurisdiction depends on a nation's engagement in the International Seabed Authority. UNCLOS protects the opportunity for scientific research on the high seas. A right of access through transit states to and from the sea, without taxation, is given to landlocked states.

Typically, a large number of stakeholders has management rights, interests, or obligations in coastal zones. The players' opportunities and responsibilities overlap and sometimes conflict or compete. Australia, for example, has one of the most geographically coherent coastlines of any nation, yet it has created more than 300 pieces of legislation relating to marine management (Binns 2004). A theoretical and practical framework to permit consistent marine management, to holistically identify spatially defined rights and responsibilities, and to incorporate those that cannot be defined spatially awaits development. By contrast, rights to and responsibilities

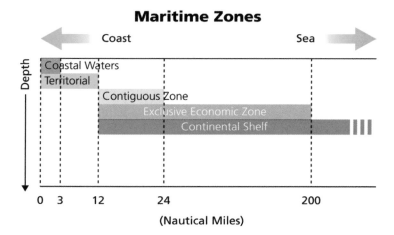

Figure 8.3 The maritime zones sanctioned in the UNCLOS framework allow countries certain regulatory authority.

for land, especially individual property rights, are historically and politically mature and are generally easy to define and physically realize.

The first-generation attempts to build a framework for holistic and seamless marine management rely on extending the familiar LAS components of cadastres, registers, and SDIs.

The **marine cadastre** adapts and extends the land cadastre to account for time-referenced marine interests, a lack of markers, use of GPS-based information, overlying interests, and shifting boundaries (for example, tidal boundaries and ever-changing coastlines).

In contrast to the land environment, **marine registers** do not regularly operate in areas where separate marine-use planning schemes apply. Marine registers therefore principally focus on such activities as managing marine oil and gas exploration rather than intense trading of interests in these rights through large-scale markets. While the immediate contrasts with land registries are large, the basic functions of the registration systems remain consistent: They define the what, who, when, where, and how particulars related to production opportunities.

The **marine SDI** demands interoperable data from seafloor bathymetry to water temperatures, from fishing zones for deep sea through coastal waters to the terrestrial sphere. Interoperability of land-generated and marine-generated information is essential.

8.4 The marine cadastre concept

The concept of the marine cadastre evolved to bring coherence to the various approaches. The design of the marine cadastre was influenced by the environmental movement and its effect on politics and society; by emerging technologies for realization and visualization of marine information and boundaries; and by the need to deliver regional, rather than merely national, marine management.

The term "cadastre" is unfamiliar to marine management, and arguably inappropriate to the marine environment, but according to Cindy Fowler and Eric Treml (2001), "Many (and some may argue all) of the cadastral components such as adjudication, survey, and owner rights have a parallel condition in the ocean." National marine cadastres are being built because of increasing awareness of the importance of spatial data to management of the marine environment and the need for a structured and consistent approach to the definition, maintenance, and management of offshore legal and administrative boundaries. Moreover, cadastral systems operating in

Figure 8.4 A floating fishing village in Vietnam is an example of informal arrangements.

marine environments are proving to be effective in many countries, including Canada, the United States, New Zealand, and the Netherlands (Nichols, Monahan, and Sutherland 2000; Fowler and Treml 2001; D. Grant 1999; Barry, Elema, and Molen 2003), with the major research initiatives in the United States, Canada, and Australia.

In the land administration theory underpinning these developments, the marine cadastre contributes to the offshore SDI by incorporating marine administrative processes and institutions for better marine management. In essence, the marine cadastre provides the basic means for delineating, managing, and administering legally definable offshore boundaries and generates essential information about related marine activities in an orderly and comprehensive way. Formal international maritime boundaries, internal maritime boundaries, and administrative and jurisdictional boundaries defining marine protected areas, commercial aquaculture farming areas, restricted fishing zones, and other areas where operational restrictions apply can be identified along with informal arrangements like floating villages (figures 8.4 and 8.5).

Given its stage of development, a marine cadastre has many definitions. Bill Robertson, George Benwell, and Chris Hoogsteden (1999) describe the marine cadastre as

> *"A system to enable the boundaries of maritime rights and interests to be recorded, spatially managed, and physically defined in relationship to the boundaries of other neighboring or underlying rights and interests."*

Figure 8.5 Aquaculture farming is an important part of Vietnam's marine environment.

Sue Nichols and others (2000) highlight the value of information, introducing concepts of ownership and the need to record rights and responsibilities in addition to the recording of boundaries.

> *"A marine cadastre is a marine information system, encompassing both the nature and spatial extent of the interests and property rights, with respect to ownership and various rights and responsibilities in the marine jurisdiction."*

These ideas were the starting point for development of an Australian marine cadastral concept. The diagram in figure 8.6 demonstrates the need to develop the marine cadastre in the context of the terrestrial environment. The majority of maritime activities occur in and around the coastal zone. This zone straddles both land and sea and is the public access point to the marine environment. Urban and industrial development and other land-based activities are also sources of pollution in the marine environment. The linking of the marine and terrestrial cadastres will enable a more seamless integration of spatial data at the land–sea interface, facilitating integrated and effective coastal zone management, including pollution regulation and control.

The diagram also shows the range of stakeholders and related activities that occur within seas and oceans. Comprehensive capacity to represent the diversity of interests in the marine environment remain the greatest challenge facing designers of a workable marine cadastre. These interests range from tourism and recreational activities—principally fishing, boating, diving, and swimming—to protection of marine ecosystems and the disposal of waste, such as ammunition and chemical dumps (table 8.1).

As with a land cadastre, the focus is on the administrative and legal boundaries that govern when and where these activities occur. The rights, restrictions, and responsibilities that go along with the boundaries must also be recorded. For example, marine protected areas have defined boundaries for the purpose of excluding or restricting the opportunities and rights of marine stakeholders. The restrictions are only effective if this information is attached to the boundaries available to stakeholders and the public. In essence, the marine cadastre provides the means for delineating legally definable offshore boundaries for the purpose of managing and administering activities in the defined areas.

The tangible outcome of the marine cadastre is the ability of users and stakeholders to "describe, visualize, and realize" spatial information in the marine environment (Todd 2001). Ideally, the cadastre describes the location and spatial extent of RRRs in the marine environment, including management boundaries, coastal planning guidelines, ocean parcel boundaries, and legally defined areas. These spatial extents should then be visualized through the continual updating of accurate

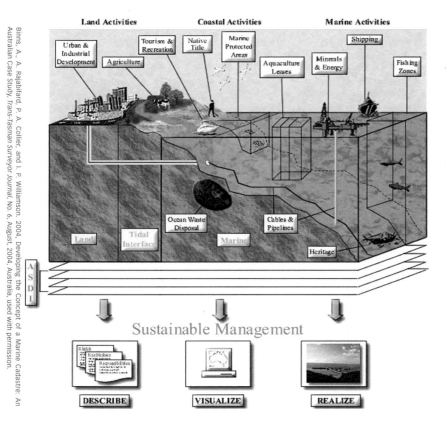

Binns, A. A. Rajabifard, P. A. Collier, and I. P. Williamson. 2004, Developing the Concept of a Marine Cadastre: An Australian Case Study, *Trans-Tasman Surveyor Journal*, No. 6, August, 2004, Australia, used with permission.

Figure 8.6

The marine cadastre concept incorporates both land and sea data to enable sustainable management of the coastal environment.

TABLE 8.1 – RANGE OF ACTIVITIES IN THE MARINE ENVIRONMENT	
ACTIVITY	**INCLUDES**
Tourism and recreation	Diving • Boating • Fishing • Swimming
Marine protected areas	Marine national parks • Marine sanctuaries
Shipping	Commercial shipping • Freight haulage • Local transport
Cables and pipelines	Oil and gas pipelines • Telecommunications • Electricity cables
Human occupation	Housing over water • Houseboats • Permanent mooring of boats
Aquaculture leases	Mussel farms • Abalone farms • Spat gathering areas • Oyster farms
Minerals and energy	Mineral exploration • Oil and gas exploration • Resource extraction
Native title	Nonexclusive access to the sea and seabed.
Ocean waste disposal	Ammunition dumps • Chemical dumps • Jarosite dumps • Scuttled vessels • Land-based sources
Heritage	Shipwrecks • Indigenous artifacts

digital spatial data in a maintenance environment. The three-dimensional nature of the marine environment and time dimensions of interests should also be visualized. This ability to describe and visualize maritime boundaries will enable users to realize them in practice. This in turn will aid in managing and creating new fisheries or aquaculture leases, policing marine protected areas, conducting exploration and mining, and laying offshore cables and pipelines. This then leads to an integrated and practical approach to a jurisdiction's management of its maritime extent.

8.5 Marine registers

Most management systems demand both cadastral and spatial capacity, as well as naming and identification capacity. Developed countries use registers of marine interests to assist administration of significant marine activities, including offshore oil and gas exploration and extraction, transportation, mining, fishing, and aquaculture.

The necessity to manage resources, including activities in the marine environment, highlights the need for an administrative registration framework that provides accountability by measuring work and extraction activities and monitoring compliance.

However, in the marine environment, both the spatial identification and information systems can be much more flexible. Indeed, marine interests may not have a geographic location at all; opportunities may be tied to simply holding a license or being the owner or operator of a licensed fishing vessel. Interests may be held in common, or by groups of owners who share a history or commercial arrangement.

Typically within each nation or jurisdiction, existing systems are separately administered, prioritized, and legislatively or administratively authorized. Within each system, opportunities are allowed to a person or organization for specific purposes. Most registration systems manage the primary relationship between the owner or holder of the opportunity and the granter (the agency, government, or authorizing manager). Trading, subsequent to the grant, is usually a derivative function of the registration system, though such systems are well developed in offshore gas and petroleum industries as well as inshore aquaculture leases such as oyster farms. Even so, these systems remain less developed in other resources and applications.

Marine registers need to serve a variety of purposes. They should be part of a policy-based national system and provide interactive opportunities to accomplish broader management goals — such as

- Monitoring pollution and oil spills, identifying cause, and supervising cleanup
- Monitoring shore-to-sea pollution, warm water, silt, and nitrogen
- Protecting natural features
- Monitoring shipwrecks and historical assets
- Maintaining shipping channels
- Protecting marine parks
- Updating hydrographic information
- Managing shipping risks, including piracy
- Managing shore degradation
- Protecting pipelines and cables
- Monitoring weather patterns and tides
- Monitoring bird, fish, and mammal populations

To achieve this kind of multipurpose role, resource registers linked to marine cadastres need to be more comprehensive than traditionally organized land registries. Table 8.2 identifies the kinds of standards needed for a modern and adaptable marine asset register. Overarching issues such as property acquisition and compensation, risk management, forgery and fraud, and relationship to corporate and other business registers are just some of the essential details for legal infrastructure that a marine register supports.

8.6 Developing a marine SDI

In response to the need for integrated land information, an SDI was developed to create a secure environment that enables users to easily access and retrieve complete and consistent spatial datasets. A similar facility is needed for marine management. A marine cadastre can delineate, manage, and administer legally definable offshore boundaries. It can manage ownership and work activities. Nevertheless, the marine environment requires an overarching spatial information platform that facilitates coordinated use and administration of these tools.

Most current SDI initiatives direct their attention landward with limited consideration of marine and coastal SDIs. Yet there is a growing and urgent need to create a marine SDI to facilitate marine administration. This SDI should deliver a seamless model that creates a spatially enabled land–sea interface and bridges the gap between the terrestrial and marine environments. Ideally, this would result in harmonized and universal access, sharing, and integration of coastal, marine, and terrestrial spatial datasets across regions and disciplines.

Management of the various RRRs in modern systems is ideally achieved through the cadastre and integrated registers, with an SDI as a tool to coordinate access to spatial data across a jurisdiction, and to strengthen and support management. SDIs enable a uniform approach for maximum integration and security of data, effective resource use, and development of comprehensive information systems. Most countries separate LAS from their emerging marine administration system, impeding management of the coastal zone. Replacement of two separate systems by a seamless platform would allow robust administration of both coastal and offshore resources and assure maximum return on investments in spatial data and management systems.

The idea of a seamless administration system that covers both the marine and terrestrial environments is generally accepted and noncontroversial. A synchronized SDI is an essential implementation strategy that allows integrated spatial management of interoperable data from both environments. The marine cadastre delivers the fundamental datasets that are especially

TABLE 8.2 – MARINE RESOURCE TITLING STANDARDS
(AFTER WALLACE AND WILLIAMSON 2005)

PROPERTY ASPECTS

Resource	Clear identification of the resource involved.
Property nature	Statement that the title is property and what type—e.g., for taxation, trading, foreign investment, or inheritance purposes. Identification of relationship between this property and other competing property.
Proprietary extent	Definition of proprietary characteristics: exclusivity, transferability, divisibility, inheritability, and any limitations on these, particularly whether owners are personally required to harvest the resource. (If they are, the interest is not proprietary in character.)
Application process	Statement of criteria: who will make the judgment or payment, who will have issuing authority, who has entitlement to apply, limits on applications (e.g., to existing users or vessels of a particular size), and processes of prioritization of competing applications.
Access	Statement of the nature of access: exclusive or shared. Also, a statement of access limitations vis-à-vis other marine activities.
Use aspects	Defined opportunities to use, to consume, to "waste."
Fees and royalties	Statement of the fees, royalties, and payments required and the means of adjusting them periodically.
Transfer	Statement of terms (if any) upon which the title can be transferred.

THIRD-PARTY PROPERTY ASPECTS

Priorities	Priority system based on "first to register" or some other simple system.
Layering	Statement of how the activity relates to others vis-à-vis recreational fishing, navigation, emissions controls, and so on.
Overriding claims	Statement of overriding claims—native titles, recreational fishing, prior rights (if any).
Security	Statement of whether the title can be mortgaged and opportunities for lender to gain possession, sell, or foreclose, and priorities among serial lenders.

Continued on next page

Continued from previous page

TABLE 8.2 – MARINE RESOURCE TITLING STANDARDS	
TITLE ASPECTS	
Name	Lease, concession, or license; preferably a name the public understands.
Pro forma requirements	Provision for a standard "lease" in plain language, available in digital and Web format; easily accessible generic print versions, preferably established by subordinate legislation or other authoritative public source rather than "in-house" arrangements.
Grant	Authoritative description of the grant process, particularly identifying when property exists. Ideally, registration is the only determinant.
Time period	Specification of beginning and end dates.
Renewal	Specification of arrangements, including whether renewal is available after expiration of the title.
Conditions	Statement of all the special conditions applying to the title (which may be included by reference rather than quoted in full).
Trading mechanisms	Guide to standard transactions: negotiated, resource banks, auction, return to government for reissue to new owner.
Trading information	Provision for public disclosure of information relating to trades and prices.
Equity constraints on transfer	Limits on transfers, limits on individual transferable quotas (ITQ), transfers across vessel size (Colby 2000), and other limitations.
Management of resource	Retirement of access policy if resource is shrinking or depleted (fish, oil, gas) or if management considerations require moratorium or suspension of access, including opportunities to claim compensation (if any).
Forfeiture	Statement of situations in which forfeiture is available.
Termination	Statement of situations in which title is ended other than forfeiture.
Link use with resource	Program for linking opportunities for use with available supply and methods of assuring adequate supply for the public good.
Bond or security	Statement of bond, security, or guarantee given to ensure compliance.
Work plan/activity	Clear statement of the activity provided.

Continued on facing page

Continued from previous page

TABLE 8.2 – MARINE RESOURCE TITLING STANDARDS	
INFORMATION ASPECTS	
Capture	Description of information collection, verification, archiving, and access functions.
Sharing	Specification of interoperability, metadata standards, access systems, privacy, commercial in confidence.
Multipurpose uses	Specification of availability of information about permitted marine activity for better management of other activities and for regional (not merely jurisdictional) management.
COMPLIANCE ASPECTS	
Work plan/activity	Clear statement of the activity required.
Reports	Specification of the nature of information collection, including form and timing.
Insurance	Statement of types and amount of insurance required and information to be provided.
Cessation of title	Statement of condition of site on relinquishment and forfeiture of bond or security.
Enforcers	Identification of the agencies and officers able to enforce.
Entry of enforcers	Identification of opportunities for officers to access site.
Enforcement officer powers	Inspection of site, books and records, taking statements, and so on.
Enforcement incentives	Fines, penalties, reduction of next year's permit, revocation.

vital to coastal zone management. The functionality of a cadastre in supporting the SDI is now recognized after a protracted debate about how to use and adapt land-based tools to service marine needs. In modern theory, the cadastral component and the SDI are fundamental to the way marine information is developed and shared, and ultimately for competent marine administration.

The need for a marine cadastre was recognized at the Permanent Committee on GIS Infrastructure for Asia and the Pacific (PCGIAP) Workshop for Administering the Marine

THE IMPORTANCE OF MARINE BOUNDARIES

The state of Victoria, Australia, was involved in bringing a court action against a New South Wales (NSW) licensed abalone diver in 2008, because the Victorian Parks Authority claimed he was diving for and collecting abalone in the ocean twenty-one meters across the NSW-Victoria state maritime border (figure 8.7) from the end of his diving air hose. This brought the diver within the boundaries of a Victorian state marine park where abalone fishing is prohibited, even though the boat he was operating from was determined to be in NSW waters. This location is one of the remotest parts of Australia with access only by sea or helicopter. This emphasizes the growing importance of maritime boundaries near the coast that, in general, can only be determined by satellite positioning.

Courtesy of R. Webb, Land Surveyor, Eden, NSW, Australia. In conjunction with Surveyor General, Victoria.

Figure 8.7 A marker in a remote coastal area shows the boundary between New South Wales and Victoria, Australia.

Environment held in Malaysia in 2004 and endorsed by the United Nations through a resolution passed at the 17th United Nations Regional Cartographic Conference for Asia and the Pacific (UNRCC-AP) meeting in Bangkok in 2006. A resolution that aimed to define the spatial dimension of the marine environment was passed, defining the terms marine cadastre and marine SDI within the context of marine administration (see figure 8.1). The marine cadastre was seen as a management tool that spatially describes, visualizes, and realizes formally and informally defined boundaries along with their associated RRRs in the marine environment. This tool, in turn, is central to the SDI, facilitating the use of interoperable spatial information relevant to the sustainable development of marine environments (figure 8.8).

The UN meeting recommended that countries with an extensive marine jurisdiction and administrative responsibilities be encouraged to develop a marine administration component as part of a seamless SDI covering both land and marine jurisdictions to ensure a continuum across the coastal zone (UNRCC-AP 2006).

The importance of the spatial dimension in administering marine environments was recognized by FIG Commissions 4 and 7 (FIG 2006).

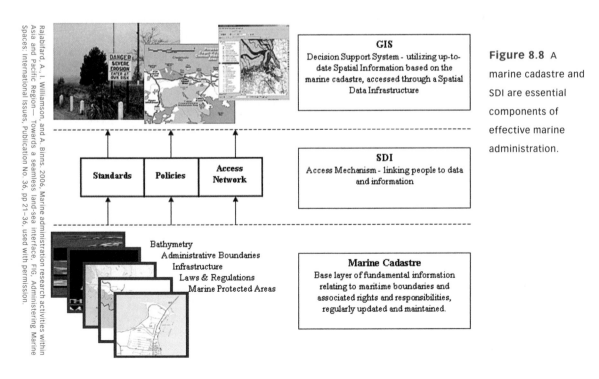

Rajabifard, A., I. Williamson, and A. Binns. 2006. Marine administration research activities within Asia and Pacific Region— Towards a seamless land-sea interface, FIG. Administering Marine Spaces: International Issues, Publication No. 36, pp 21–36, used with permission.

Figure 8.8 A marine cadastre and SDI are essential components of effective marine administration.

GIS
Decision Support System - utilizing up-to-date Spatial Information based on the marine cadastre, accessed through a Spatial Data Infrastructure

Standards   Policies   Access Network

SDI
Access Mechanism - linking people to data and information

Bathymetry
Administrative Boundaries
Infrastructure
Laws & Regulations
Marine Protected Areas

Marine Cadastre
Base layer of fundamental information relating to maritime boundaries and associated rights and responsibilities, regularly updated and maintained.

International organizations, including the IHO, are getting increasingly involved. IHO is working on a strategy to implement a marine SDI to better manage global marine activities. At the IHO International Workshop on Marine SDI, held February 12–13, 2007, in Havana, Cuba, IHO discussed the role of a marine SDI and the requirements and strategies to facilitate its development.

8.7 Using the land management paradigm to meet marine needs

This summary of the past decade of activities in marine management is drawn from the land administration perspective. From this vantage point, the activities of marine scientists, biologists, climate theorists, and many other experts are acknowledged as indispensable to future policy making and management of our fragile marine environments. The monitoring of impacts of climate change and its far-ranging consequences in tropical, polar, and temperate regions requires concerted and refined capture and analysis of information. This information is crucial to building a robust system of marine administration capable of providing government with a framework to deliver sound marine management policies. In turn, this administrative infrastructure improves the chain of information and facilitates the measurement and evaluation of management processes. Sound administration is vital to effective implementation of policies that stem from scientific research for the sake of our oceans and the marine environment. And land administration tools, developed to better suit marine needs, are the first stage of a comprehensive marine management system.

The treatment of land, coastal zones, and marine areas in a unified management and information system relies on the land management paradigm. Yet achieving an integrated approach will require a change in attitude by the organizations and institutions that nations use. The focus of LAS on "land" as merely the dry area (sometimes including fresh water) is inadequate for modern needs, especially when coastal zones are under increasing pressures. The management of marine resources requires regional and international cooperation, even more so than for land. The new marine-focused tools of cadastres, SDIs, and registers will play a crucial role in building and strengthening these processes.

Chapter 9
SDIs and technology

9.1 Why do land administration systems need an SDI?

9.2 Introducing the SDI

9.3 Integrating information about the natural and built environments

9.4 Making ICT choices

9.5 Land administration and cadastral data modeling

9.6 Maintaining momentum

9

9.1 Why do land administration systems need an SDI?

Design of land administration systems, either to improve an existing system or develop a new one, can benefit from improvements in technology. Making the right decisions about the use of technology is obviously important. We can no longer build support systems that freeze out opportunities to manage land holistically. Nor can we approach technology as if it is just about the use of computers. In fact, technology is about the way institutions work and operate. Modern land administration needs to make the most of new technologies.

History helps explain the nature of modern technology choices in the context of LAS. From the 1980s, LAS institutions generally relied on digitizing their internal systems, typically by install-ing databases and using computer-aided design (CAD) systems to assist surveying. The arrival of GIS used in the earth sciences made little difference to LAS design. The land administration components of registries, cadastres, valuation systems, and planning systems were generally regarded as silo institutions, so their technological support systems were similarly isolated and stand alone. GIS functions meanwhile became the playground of mapping agencies and the people who used spatial data, especially for environmental management. The two key professional groups, surveyors and cartographers, did not have much in common, nor did their technical support systems. The arrival of the Internet meant that both the land administration and GIS practitioners took separate paths into Web-enabled environments. These arrangements became institutionalized, and, in most countries, that separation continues.

These old models of land administration and mapping are inadequate to solve the demands of a modern sustainable society. A broader view of what is required was needed to explain the inadequacies and identify a way forward in terms of LAS design. Thus, the land management paradigm was identified to guide decision makers through the complicated processes of building modern systems and justifying their decisions and expenditures according to one ultimate aim: delivery of sustainable development.

Most countries approach bridge building between the silo agencies and their respective information and technical systems by adopting a spatial data infrastructure (SDI) strategy. The SDI is more significant than most people realize. As can be seen from figure 9.1, the organizational structures for land management must take the ever changing local cultural and judicial settings and institutional arrangements into account to support implementation of land policy and good governance. Within each individual country, the land management activities needed to support sustainable development may be described by the three components of land policy, land information infrastructure, and land administration functions. In this regard, the SDI

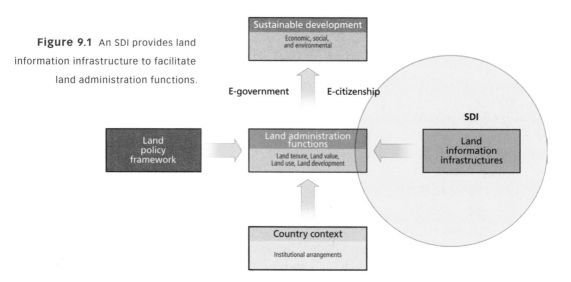

Figure 9.1 An SDI provides land information infrastructure to facilitate land administration functions.

plays a central role in facilitating a country's land information infrastructure. Increasingly, large-scale "people relevant" data derived from LAS drives the development of SDIs.

The designers of SDIs realize the need for infrastructure that can facilitate sharing and integrate data while guaranteeing the delivery of both information and services. Integration inevitably improves the information available to decision makers and helps them make sound decisions about sustainable development since it requires the integration of data from disparate data sources. Most of the key information needed by land policy makers, businesses, and society in general is parcel-related cadastral information about the built environment that is generated through land administration. This data needs to be integrated with other forms of data if sustainable development is to be achieved. Thus, integration also streamlines the processes and services needed for overall land management, more than just environmental management, by describing the total impact of people on land. The butterfly diagram in chapter 5 (figure 5.6) highlights the opportunities created by an effective SDI (figure 9.2).

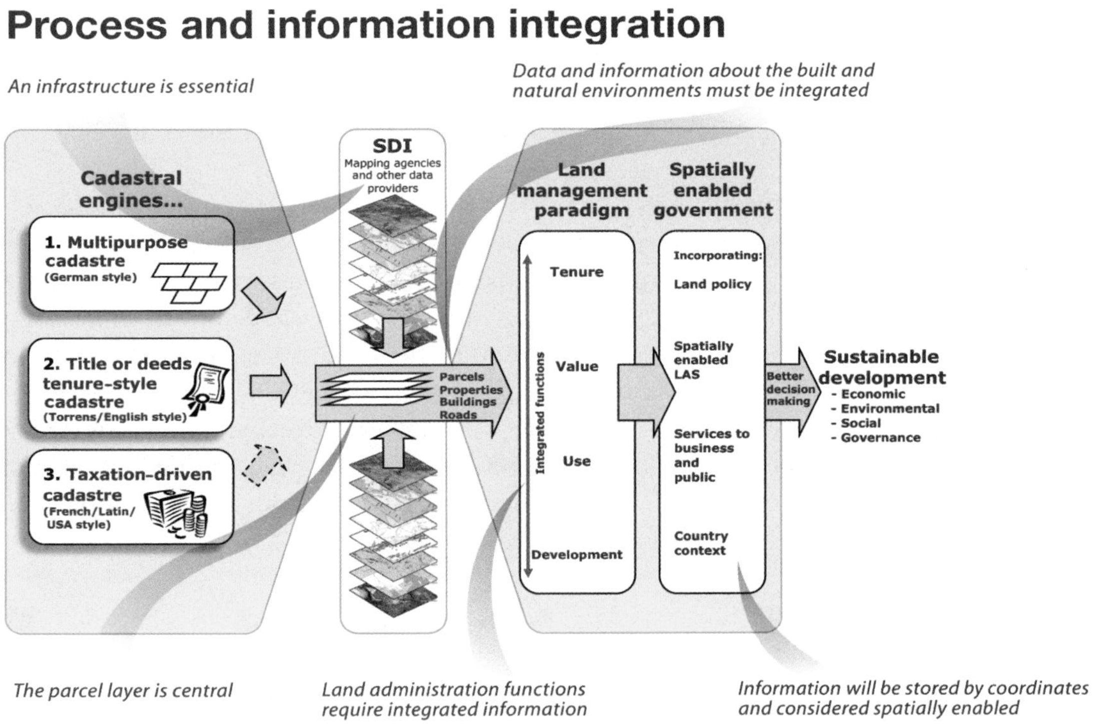

Process and information integration

Figure 9.2 The butterfly diagram shows how an SDI is essential for integrating land information and the cadastre to spatially enable government and lead to sustainable development.

Figure 9.2 introduces a new capacity for SDIs that evolved around the year 2000—spatial enablement through widespread dissemination of spatial information over the Internet. Examples of the power of spatial knowledge lie in systems such as Google Maps, Google Earth, and Microsoft Bing Maps for Enterprise. These, and the many other spatial systems competing in a growing world market, show that location can be used as a sorting tool to organize not only information, but business processes. The emerging world of spatial enablement and information must be accommodated in modern LAS design.

Future land administration will rely on the SDI as an enabling platform to facilitate essential functions and opportunities. Having said that, the potential of an SDI can only be realized if it has a strong cadastral component that institutionalizes the land administration paradigm. Within this context, access to complete and up-to-date information about the built and natural environments is essential for managing processes associated with the four land administration functions.

In this emerging modern context, professional tools and systems, particularly the cadastre and the SDI, continue to evolve. Most countries began by implementing SDI tools at the national, state, and local level without sufficient consideration of the central role of the cadastre. Today, much worldwide SDI activity is still at this stage, because designers focus on national mapping initiatives rather than concentrating on coordinating spatial information at all levels. However, this is changing.

Now, highly developed SDIs increasingly focus on large-scale, people-relevant data (land-parcel-based data or built environmental data) that is essential for land administration and policy implementation. New institutional and policy arrangements are being created by countries to aggregate large-scale spatial datasets (cadastre, road networks, street addresses, and political boundaries) and integrate them with small-scale, national, natural resource, and topographic datasets. As a result, the historic roles of traditional national mapping agencies and land registries are especially challenged by the evolution of the SDI concept and the need to share spatial information throughout government, not merely in those agencies that use GIS technology. Without a strong cadastral component, an SDI cannot support the land management paradigm, and governments cannot capitalize on the opportunities offered by the new spatial technologies.

The emerging vision for SDIs is an enabling platform that links services across jurisdictions, organizations, and disciplines. This cross-jurisdictional approach aims to provide users with access to and use of information related to both the built and natural environments in real time—something that nonintegrated silo organizations cannot deliver (Gore 1998). This information is then used to enhance decision making and in turn supports the achievement of the economic, environmental, social, and governance objectives of sustainable development.

9.2 Introducing the SDI

SDI CONCEPTS AND HIERARCHY

Countries that seek to improve their land management capacity by implementing the land management paradigm require an SDI. In this context, SDIs facilitate the sharing and integration of multisource datasets with data specifically related to land administration, particularly the cadastre. It is an important key to spatial enablement, or the usability of spatial information, especially for information generated by land administration processes.

Descriptions of the components and operation of SDIs and their integration into the spatial data community are readily available. SDIs encompass the policy, access networks and data-handling facilities (based on available technologies), standards, and human resources necessary for the effective collection, management, access, delivery, and utilization of spatial data for a specific jurisdiction or community. The complex relationships among the technological, institutional, organizational, human, and economic processes need to be reflected in SDI design. So does its operation as an intermediate mechanism that facilitates the transfer of information for the public good across jurisdictions.

More significantly, the role of SDIs in society must be defined, so that an SDI initiative is accepted by the public and aligned with spatial industry objectives. SDIs are, by necessity, more effective than the sum of their individual components. An SDI is not a "database." It is an infrastructure that provides a policy framework, access technologies, and standards that link people to information. In particular, the SDI is the key to the spatial enablement of modern land administration. Once location or place is used to organize government information, government processes can be reengineered to deliver better policy outcomes. In more developed nations, the infrastructure involves an integrated, multilevel hierarchy of interconnected SDIs based on partnerships at the corporate, local, state/provincial, national, regional (multinational), and global level. An effective SDI can save resources, time, and effort for users who need to acquire new datasets by eliminating the duplication and expenses associated with the generation and maintenance of disparate data and then integrating that data with other datasets. To date, the SDI has effectively met user needs up to a point. However, satisfaction of user needs in a dynamic and fast growing user market now requires a collaborative environment, such as a Virtual Jurisdiction, in which spatial information providers from various backgrounds can work together with current technologies. Rapid advancement of information and communications technology alone cannot meet these differing needs. Achievement of this vision requires the interaction of various agencies using integrated datasets in order to serve the public interest.

SDIs can be expensive. However, having an SDI can be justified if it leads to effective economic, social, and environmental decision making (Rajabifard 2002). SDIs now have the potential to determine the ways spatial data is used throughout an organization, a state or province, a nation, different regions, and the world at large. Inefficiencies occur if a coherent SDI is not in place, and opportunities to use geographic information to solve problems are lost (SDI Cookbook 2000). By reducing duplication and facilitating integration and development of new and innovative business applications, SDIs can produce significant human and resource savings and returns.

The design and implementation of SDIs involve technology, design of institutions, creation of legislative and regulatory frameworks, and acquisition of new types of skills (Remkes 2000). Balancing these elements facilitates both the intra- and interjurisdictional dynamics of spatial data sharing (Feeney and Williamson 2000; Rajabifard, Feeney, and Williamson 2002a). New relationships and partnerships among different levels of government and between public- and private-sector entities must be created to achieve this objective. This is an important design element, particularly when different organizations are brought together to share and integrate data to serve land administration processes. These partnerships require organizations to assume responsibilities that differ from those of the past (Tosta 1997). An SDI has to ensure consistency of content at least within its own jurisdiction. All cooperating agencies need access to accurate and consistent spatial databases capable of informing local and interjurisdictional decisions. Provision for nonparticipating members to join is also essential.

The components of an SDI can be categorized in different ways, depending on their role within the framework (Rajabifard, Feeney, and Williamson 2002a). The important ways people use data provides one possible categorization; a second consists of the main technological components—the access networks, policy, and standards. Both types are needed. They are both dynamic, reflecting changes in communities (people) and in data. An integrated SDI cannot be composed of spatial data, value-added services, and end users alone. Otherwise, it cannot evolve to meet the continual advances in technology and the evolution of rights, restrictions, and responsibilities. Interoperability, policy, and networks must be integrated into the SDI as well.

Early discussion of the SDI concept focused on nations as an entity. Now, more attention is given to understanding the SDI hierarchy, which is made up of interconnected SDIs at the various levels as illustrated in figure 9.3 (Rajabifard, Escobar, and Williamson 2000). In general, the various levels are a function of scale. Local government and state-level SDIs manage large- and medium-scale data, leaving national SDIs to manage medium- to small-scale data, with regional and global SDIs adopting a small scale for their activities. The improved understanding of the SDI hierarchy has challenged different jurisdictions to improve the relationships among the different levels and to coordinate spatial data initiatives.

Most successful SDIs are built through mutually beneficial partnerships that build inter- and intrajurisdictional relationships within the hierarchy. These partnerships adopt a focused approach to SDI development, creating business consortiums to develop specific data products or services for strategic users. Thus, early identification of the human and community issues involved in these partnerships is essential.

Other kinds of relationships also exist within the hierarchy and need to be understood so that the SDI can deliver benefits to any jurisdictional level. In addition to the vertical relationships between different jurisdictional levels, complex horizontal relationships within each political or administrative level need to be analyzed. The vertical and horizontal relationships within an SDI hierarchy are very complex because of their dynamic inter- and intrajurisdictional nature (Rajabifard, Feeney, and Williamson 2002b). Users of an SDI thus need to understand all the relationships involved in the dynamic partnerships it supports.

THE CHANGING ROLES OF THE SDI

When the SDI concept was originally conceived, it was regarded as a mechanism to facilitate access to and sharing of spatial data hosted in distributed GIS formats. This initial concept has now evolved. In its place, a new business model offers chains of Web services through distributed GIS frameworks. This new SDI model is a "virtual jurisdiction" or "virtual enterprise," promoting partnerships among public and private spatial information organizations, widening data and services, and increasing complexity beyond the capacity of individual partners.

Rajabifard, A., F. Escobar, and I. P. Williamson, 2000, Hierarchical Spatial Reasoning Applied to Spatial Data Infrastructures, Cartography Journal, Vol. 29, No. 2, Australia, used with permission.

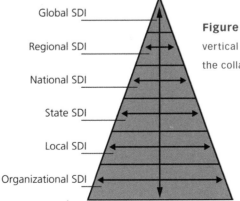

Figure 9.3 The SDI hierarchy has both horizontal and vertical relationships among its jurisdictional levels expanding the collaborative use of information.

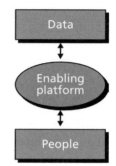

Figure 9.4 An effective SDI connects people to data, allowing easy access to and sharing of information.

To apply this new model, countries need an integrated platform to support the linkage of services across participating organizations. The virtual jurisdictions, and the opportunities offered by ICT and the Internet, are essential to building this environment. These initiatives were also driven by other demands. Many spatial information organizations were forced to work in a more tightly coupled mode to deliver more complex products or services beyond their internal capacity. Increasing an organization's share of the spatial information market and strengthening the role of spatial data in e-government services also supported development of these new SDI models. The roles of subnational governments and the private sector in SDI development also changed in response to demands for more emphasis on delivering sustainable development. The advantages of the new business models lie in their more holistic and technologically advanced support mechanisms for land administration and land policy implementation in general. This relationship to policy supporting sustainable development will continue to have a significant influence on future SDIs.

THE SDI AS AN ENABLING PLATFORM

Use of spatial data and spatial information in any field or discipline, particularly land administration, requires an SDI to link data producers, providers, and value adders to data users. The SDI provides ready access to spatial information to support decision making at different scales for multiple purposes. Initially, the infrastructure links data users and providers on the basis of the common goal of data sharing. Potentially, it enables sharing of business goals, strategies, processes, operations, and value-added products—making a virtual jurisdiction. The spatial capability of government, the private sector, and the general community will be enhanced by development of an SDI as an enabling platform (figure 9.4).

Land management, emergency management, natural-resource management, water-rights trading, and animal, pest, and disease control are all fields that require precise spatial information in real time about real world objects along with the ability to develop and implement cross-jurisdictional and interagency solutions. In response, the SDI will be the main gateway to discover, access, and communicate spatial data and information about a jurisdiction. The infrastructure will permit sharing of business goals, strategies, processes, operations, and value-added products, as well as data. All types of participating organizations (including government, businesses, communities, and academia) can gain access to a wider share of the information market. Organizations can provide access to their own spatial data and services and, in return, gain access to the next generation of new and complex services. These services would be structured and managed so that third parties would see them as a single enterprise. The benefits from this more effective data sharing and use of technology should facilitate improved decision making.

The creation of an enabling platform allows easier access to and use of spatial data not only for government and the wider community, but in particular, the spatial information industry. If barriers are minimized, users can pursue their core business objectives with greater efficiency and effectiveness. Reduction of information costs encourages industries to invest in the capacity to generate and deliver a wider range of spatial information products and services to expanding markets. The design of an integration platform requires development of a set of concepts and principles that facilitate interoperability.

9.3 Integrating information about the natural and built environments

The ability to observe and monitor changes in both the natural and built environments is essential to planning ahead and delivering sustainable development. Access to such data is therefore crucial. So is integration of the datasets. In formal terms, this involves integration of cadastral (built) and topographic (natural) spatial data to support sustainable development (Rajabifard and Williamson 2004), as illustrated in figure 9.5.

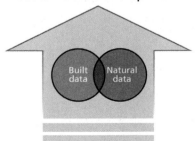

Figure 9.5 Integration of datasets for both the natural and built environments facilitates sustainable development.

Cadastral and topographic datasets are the most important spatial datasets in any country. These datasets provide the foundation for modern market economies (Groot and McLaughlin 2000).

Cadastral datasets are mostly the accumulation of individual property boundary surveys undertaken by land surveyors. By its very nature, cadastral data is large scale and very different from topographic data, which is produced at medium to small scales over large regions using a range of different techniques. Countries generally develop two separate foundation datasets for unrelated purposes, and most continue to manage them separately. These separate institutional and data arrangements impede delivery of sustainable development, especially because of unjustifiable duplication and increased costs for data collection and maintenance. These datasets should be organized by the same overarching philosophy and data model to achieve multipurpose data integration, both vertically and horizontally (Ryttersgaard 2001).

A national SDI aims to integrate multisource spatial datasets. It is a difficult task. Many reports highlight the heterogeneity and inconsistency of these initiatives and activities and attempt to address these impediments by documenting the technical inconsistencies (Fonseca 2005; Baker and Young 2005; Jones and Taylor 2004; Hakimpour 2003). Technical inconsistencies tend to arise from nontechnical aspects and fragmentation of the social, institutional, legal, and political arrangements affecting individual data custodians and organizations (Mohammadi et al. 2006). Moreover, the single most encountered impediment to data sharing is militarization of mapping information so that it cannot be used by cadastral agencies, resource managers, or land administrators. For developing countries, military control of images, aerial photos, and satellite data is a serious impediment to economic growth, rational land management, and overall sustainable development.

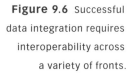

Figure 9.6 Successful data integration requires interoperability across a variety of fronts.

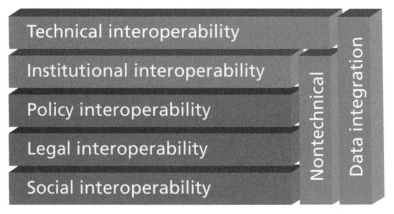

SPATIAL DATA INTEGRATION CHALLENGES

Each of the technical and nontechnical inconsistencies and challenges impeding data integration needs to be identified and resolved. In most countries, each dataset is managed by a data custodian that follows unique strategies and policies for data creation, coordination, sharing, and usage. Thus, most of the integration steps are not technical. Data integration involves much more than the geometrical and topological matching of data and ensuring that feature attributes correspond (Usery, Finn, and Starbuck 2005). It also requires addressing all nontechnical legal, policy, institutional, and social factors that affect interoperability (figure 9.6). These integration issues need to be framed in the context of the SDI model and the history and priorities of the jurisdiction. Each part of an SDI, including the cadastral layer, needs to be based on national organizational, economic, social, and other priorities.

Different kinds of issues are associated with effective data integration in developed nations (Mohammadi et al. 2006). Technical interoperability is perhaps the most straightforward and has received the most attention. Nontechnical issues, by contrast, remain open to resolution. Any country that seeks to develop an SDI must address all the issues (figure 9.7).

Technical issues in the SDI framework include computational heterogeneity (standards and interoperability), maintenance of vertical topology, semantic heterogeneity, reference system and scale consistency, data quality, existence and quality of metadata, format consistency, consistency in data models, and heterogeneity in attribution. Institutional issues include collaboration among stakeholders, business models and associated funding models, users' awareness of data, and, finally, data management approaches. Policy issues include policy drivers, national priorities,

Mohammadi, H., A. Rajabifard, A. Binns, and I. P. Williamson. 2006. The Development of a Framework and Associated Tools for the integration of Multi-Sourced Spatial Dataset. 17th UNRCC–AP, Bangkok, Thailand, September 18–22, 2006, used with permission.

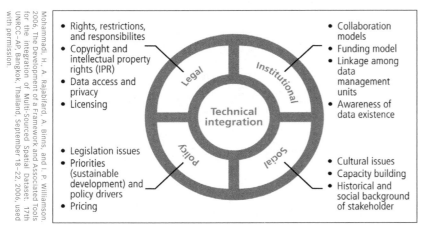

Figure 9.7 A vast array of nontechnical issues comes into play in the integration of technical data.

pricing, and institutional structures. Cultural differences, capacity building, and the social background of spatial data stakeholders are also obvious social issues. Legal issues include

- ◆ Rights, restrictions, and responsibilities (RRRs)
- ◆ Copyright and intellectual property rights (IPRs)
- ◆ Data access and privacy
- ◆ Licensing

Efforts to establish an SDI will fail unless a coordinated approach is used to address all the issues and inconsistencies associated with multisource data integration, summarized in table 9.1.

To create an environment in which different datasets can be integrated across applications, the infrastructure should provide a suite of tools and guidelines, including standards, policies, and collaboration requirements.

DATA INTEGRATION AND SDIS

One of the most important tasks of an SDI is effective data integration. This is accomplished by providing all technical and nontechnical options and requirements, including a network, standards, and policy tools (Rajabifard and Williamson 2001) as shown in figure 9.8.

TABLE 9.1 – INTEGRATION ISSUES			
TECHNICAL ISSUES	**NONTECHNICAL ISSUES**		
INSTITUTIONAL ISSUES	**POLICY ISSUES**	**LEGAL ISSUES**	**SOCIAL ISSUES**
Computational heterogeneity (standards and interoperability)	Existence of supporting legislation	Definition of rights, restrictions, and responsibilities	Cultural issues
Maintenance of vertical topology	Consistency in policy drivers and priorities (sustainable development)	Consistency in copyright and intellectual property rights approaches	Weakness of capacity-building activities
Semantic heterogeneity			Different backgrounds of stakeholders
Reference system and scale consistency	Pricing		
Data quality consistency		Different data access and privacy policies	
Existence and quality of metadata			
Format consistency			
Consistency in data models			
Attribution heterogeneity			
Utilization of consistent collaboration models			
Funding model differences			
Awareness of data integration			

SDI design must resolve data issues. Technical issues related to data standardization and provision of access channels can be addressed by appropriate standards and compliances. Nontechnical issues and interaction between people and data can be achieved through the policy component.

On the technical end, a lack of vertical topology will hinder the capacity to analyze datasets. Consistent data models permit effective analysis. The quality of data, including accuracy, coverage, completeness, and logical consistency, is important both for integration of data and to avoid mixing high-quality and low-quality data. Good metadata improves users' ability to integrate data. Lack of a reference system and format heterogeneity hamper efficient data integration.

Characteristics of the network that affect its suitability for data integration must also be considered along with Web services issues.

Institutional arrangements, legal and social issues, policy considerations, and collaboration and funding models can all facilitate data integration. Capacity building, cultural considerations, and engagement of stakeholders in the design of an SDI help remove social barriers to data integration. Outlining the intellectual property and access paths that affect source data is important for successful final integration. The political priorities of jurisdictions and how data coordination and integration that are reflected in legislation form policy directly affects data integration. Pricing and licensing arrangements play a key role as does raising awareness about the benefits of merging related data.

Social impediments to integration of data have received little attention from SDI designers. These are the most challenging, because social arrangements are typically complex and intangible. The impediments need to be resolved through medium- to long-term processes, not avoided or considered only in short-term mandates.

Rajabifard, A. and I. P Williamson 2001, Spatial Data Infrastructures: Concept, SDI Heirarchy and Future directions, GEOMATICS '80 Conference. Tehran, Iran, used with permission.

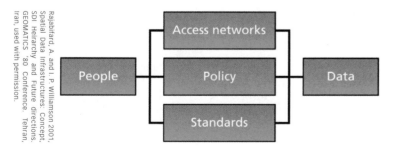

Figure 9.8 The SDI model incorporates standards, policy, and access networks to connect people and data.

9.4 Making ICT choices

ICT IN LAND ADMINISTRATION

Agencies engaged in land administration in market economies rely heavily on technology. They share this reliance with other government agencies and services that increasingly promote administrative efficiencies through Web-enabled systems and e-government. Use of information and communications technologies (ICT) for land administration, however, involves special considerations.

The previous concentration on government institutions will be widened by engagement of utilities, the spatial sciences, and other businesses in the construction of land information products. The transitions are shown in figure 9.9.

Initiatives undertaken by land administration organizations to deliver information, and sometimes even services, to the public over the Internet and to facilitate interorganizational workflows are common. Analysis of these experiences can help determine good practices, and effective and innovative ways to reengineer existing services. As part of the evolution of ICT in land administration, the first attempt to build a comprehensive interactive framework involved introduction of e-land administration for a truly integrated digital land information system. In general, e-land administration means utilization of ICT capabilities to deliver land administration functions and services online.

Australia is a regional leader in e-land administration, providing ten online land information services. Its jurisdictions have also initiated projects for electronic conveyancing and electronic lodgment relying on Internet-based systems to process settlement and lodgment of land dealings online (Kalantari et al. 2005).

In New Zealand, the Landonline program began in 1996, following amalgamation of the two government departments responsible for cadastral survey and land title registration (Grant 2004).

Figure 9.9 IT in land administration has evolved from the manual systems of the 1970s through computerization and Web-enabled land administration in the 1980s and 1990s to interoperability and e-land administration over the past few years that will lead to spatially enabled government (i-land) in the near future.

The program relies on a fully digital cadastre incorporating the various records, plans, and images in an intelligent data form, and transformation of institutional knowledge and expertise into business rules, to produce an integrated information system. The information system automates data flows and processes and integrates record and business rules.

In Great Britain, the Land Registry proposes a fully electronic conveyancing system for England and Wales. This would include e-lodgment of applications, e-certificates, and deeds and electronic settlement of payment due on completion. The database for this system combines information from the Land Registry with other information relevant to users, especially buyers and sellers who will be able to launch single, comprehensive searches of property. The Land Registry's role will be to provide an electronic system linking conveyancers to their respective systems and to the Land Registry's database (Beardsall 2004).

In the Netherlands, all deeds since 1999 have been scanned as a first step in utilizing ICT in land administration processes. New deeds are now immediately scanned on receipt, and proof of receipt is automatically generated in real time. The digital signature is an essential part of this process (Louwman 2004; Stolk 2004). Their emphasis on improved workflow management increases efficiency, especially the timeliness of information about land administration processes (Louwman 2004).

The cadastral administrations of all German states are currently developing the official cadastral system ALKIS, which will integrate cadastral data of the older Automated Real Estate Register (ALB) and Automated Real Estate Map (ALK). In addition, the data model of ALKIS will be identical to the updated Authoritative Topographic and Cartographic Information System (ATKIS). The challenge of this project is to reach interoperability among different systems within towns and counties, because in most cases, different GIS models are installed for different applications (Bruggemann 2004).

The Polish government is working on two major stages of e-land administration, including building a technological framework and modernizing the organizational, institutional, and legal framework. Its aim is to gain an integrated electronic platform, a new land book, and improvement of the fiscal cadastre. It is also introducing an e-conveyancing system to its traditional notary services (Sambura 2004).

The Austrian CYBERDOC is the electronic document archive of civil-law notaries. The documents are scanned as they are generated (on the client computer), given attributed key words, and sorted permanently and unalterably in electronic form on the archive server (Brunner 2004).

CARIS Land Information Network (CARIS LIN), the land information system in the New Brunswick province in Canada, features one centralized, authoritative database. The database is distributed through a provincial intranet. CARIS LIN supports a semiautomated land transaction business and workflow. It facilitates the transfer of a name-based system to a parcel-based system and online title conversion. External information portals for searching and reporting and online user access to the land information system are other highlighted features of this e-land administration initiative. These features lead to a "virtual office" allowing users of real property information to "serve themselves" (Ogilvie and Mulholland 2004).

The variety of approaches to using state-of-the-art ICT in LAS makes it difficult for other nations to learn from these leading-edge scenarios. While ICT is heavily utilized by land administration organizations to improve service delivery, satisfy customers, and reduce operating costs, full realization of its benefits remains elusive. Comprehensive use of ICT in land administration awaits an effective land information system, which, in turn, depends on the SDI. The difficulties are compounded when the data is coupled with complicated technologies and bureaucratic management. Achievement of a spatially enabled society is dependent on a new vision that could be termed "i-land" (information about land).

ICT DEVELOPMENT PHASES IN LAS

A model that shows the transformation of LAS into e-land administration can help administrators and policymakers understand the phases of ICT that are involved.

The transformation from current LAS to e-land administration involves four phases (figure 9.10) as highlighted by M. Kalantari (2008).

Kalantari, M. 2005, e-Land administration. Doctoral Confirmation Report. The University of Melbourne, Australia, used with permission.

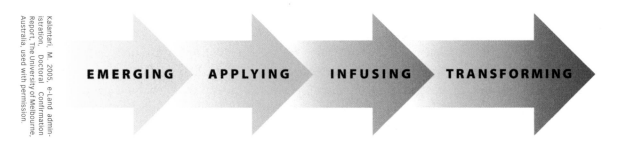

EMERGING   APPLYING   INFUSING   TRANSFORMING

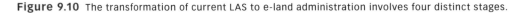

Figure 9.10 The transformation of current LAS to e-land administration involves four distinct stages.

The phases are as follows:

◆ **The emerging phase:** In this initial phase, ICT enablement is just beginning. The model of ICT-enabled processes has been designed but is not operational. Process partners are just starting to explore the possibilities and consequences of using ICT, but they are still firmly grounded in traditional practice. The conversion plan reflects an increase in basic skills and an awareness of the uses of ICT.

◆ **The applying phase:** In this next phase, process partners use ICT for tasks already carried out in LAS. Traditional processes largely dominate, but they are ICT enabled, and the use of ICT with various partners is increased. This phase assists movement to the next phase if so desired.

◆ **The infusing phase:** The next stage embeds ICT across the processes. Process partners change their productivity and professional practices by exploring new ways to deliver services. Traditional approaches no longer dominate.

◆ **The transforming phase:** LAS designers who use ICT to rethink, renew, and streamline processes are at the transforming phase. ICT becomes an integral, though invisible, part of productivity and professional practices.

ICT OPTIONS FOR LAS

Within the land management paradigm, LAS perform the processes associated with land tenure, value, use, and development. LAS are therefore the primary collector, recorder, and disseminator of essential land information. In this context, ICT stands for available technical options for determining and recording the data mentioned and the technical options for its dissemination. ICT can further assist LAS by providing infrastructure for effective coordination and communication between data management and data dissemination. This infrastructure involves a series of technical options discussed as follows that include data management tools, data dissemination tools, and enterprise facilitator tools for coordination and communication. Here, "communication" is more about connectivity and the exchange of information than the communication infrastructure itself.

Data management tools: These tools facilitate and manage the development of land information to support the land management paradigm. They provide the capacity for data modeling, data capture, database systems, data cataloging, and data conversion as a means of keeping land information standard, making it deliverable across multiple servers for access and sharing (Kalantari et al. 2005).

Data modeling tools: These tools specify a database and describe what sort of data will be held and how it will be organized. The most common alternative approaches to data modeling involve the entity relationship (E-R) and the Unified Modeling Language (UML) (Simsion and Witt 2005).

The E-R approach dominated the development of spatial databases until the late 1990s, when object-oriented analysis and design emerged and the UML approach gained in popularity. UML is a richer language that provides a set of graphical notations with significant benefits to both system designers and database designers. UML can therefore be used not only for spatial databases but also to describe the business processes of land administration and the relationship between subsystems and external entities (Van Oosterom et al. 2004).

Data capture tools: Technology for measuring distances and angles has steadily improved. Modern instruments, like "Total Stations" used in boundary surveying, measure angles to within 5 seconds of arc and distances of 1,000 meters to a precision of better than 5 millimeters. Besides that, precise GPS can also locate points to centimeter accuracy in real time. Digital cameras that take aerial images can automatically include GPS coordinates.

Ground survey techniques have been extensively used for cadastral mapping associated with surveying. The photogrammetric methods used extensively in other mapping processes are much less popular. Under suitable conditions, however, photogrammetry can produce maps and measurements that are as accurate as those obtainable by standard ground methods. The use of this alternative depends on the method of producing cadastral maps and on whether a sporadic or systematic approach to boundary demarcation and identification is adapted. Currently, the most common method of building digital cadastral databases is by digitizing boundaries from two-dimensional (2D) hard-copy cadastral maps. There are many systems in use to improve the accuracy of this type of data, including "rubber sheeting" or adjusting to control from GPS or photogrammetric sources (Elfick, Hodson, and Wilkinson 2005; also see sec. 12.3, "Professional tools").

Database system tools: Databases are traditionally used to handle large volumes of data and to deliver the logical consistency and integrity that is essential to successful handling of spatial data. Integration of spatial data such as land parcels with nonspatial information, including ownership, value, and use, in one database, called a geodatabase, has significantly improved, especially through the efforts of the Open Geospatial Consortium Inc. (OGC 2003).

To date, mainstream database systems have implemented spatial data types and spatial operators more or less according to the specifications of OGC (Zlatanova and Stoter 2006). However, the LAS context raises difficult issues related to the spatial dimensions of registered objects and associated interests. The first is the incorporation of the attributed interest into a spatial dimension. This involves differences among spatial characteristics of RRRs. The object might be a polygon or a 3D object, a line, or a point. Take an easement over a land parcel, for example. The right can be represented by a line with associated attributes or as a polygon. The next challenge is the relationship among legal land object layers and how they can be connected in a spatial database, which is the subject of ongoing research (Kalantari et al. 2006).

Data catalog tools: A data catalog describes and provides links to available data just like a card catalog organizes library books. In particular, a data catalog can organize land information distributed in subsystems held in in-house databases within LAS.

A data catalog is usually accompanied by metadata, or data about data. Metadata elements and schema are used by data producers to characterize data. Metadata facilitates data discovery, retrieval, and reuse. Users rely on land administration metadata to better access and use the data for various applications. The OGC (2003) has developed a standard conceptual metadata schema, intended to be used by information systems, program planners, and developers of spatial information systems like cadastral databases.

Data conversion tools: For LAS to be spatially enabled, data must be available in various formats to accommodate the diversity of spatial databases. Format requirements can be met in two ways: special-purpose translators or the use of a common format like Geography Markup Language (GML) or LandXML. GML is an XML language written in XML format for the modeling, transport, and storage of geographic information. The key concepts GML uses to model the world are drawn from the OpenGIS abstract specification and the ISO 19100 series. GML provides a variety of objects for describing geography, including feature classes, coordinate reference systems, geometry, topology, time, units of measure, and generalized values (ISO and OGC 2004).

LandXML is a new international standard for a digital interface with surveyor's software. The LandXML schema facilitates the exchange of data created during the land planning, civil engineering, and land survey processes. Land development professionals can use LandXML to make the data they create more readily accessible and available to anyone involved with a project (`www.landxml.org`). GML provides coordinate (projection, geographic, and geocentric) systems and simple-feature geometry models. Transforming LandXML design data to GML provides a way to propagate complex land design geometries into GIS databases.

Data dissemination: The butterfly diagram (figure 9.2) shows that dissemination of land information is one of the most important aspects of modern land administration. The processes of information dissemination involve complexities arising from the diversity of organizations, clients, and users and the variety of highly specialized processes. Dissemination may include the order, packaging, and delivery, offline or online, of the data (SDI Cookbook 2004).

Meanwhile, evolution in Internet and WWW technologies offers a variety of tools for data access and sharing that are increasingly attractive and popular. Sharing tools facilitates the development of Web-based access to land information in a seamless and integrated view. These tools provide interoperable sharing techniques based on international standards. Technically, land information dissemination is driven by GIS services supported by interoperability and Web services and distributed computing technology such as Grid computing, peer-to-peer (P2P), and Agent (Yang and Tao 2006) described as follows.

Web services tools: The Web is an immensely scalable information space filled with interconnected resources. A service is an application that exposes its functionality through an application programming interface (API). A Web service is therefore defined as an application with a Web API. Web services rely on service-oriented architecture (SOA) that defines a set of patterns to connect a client to a server. The standard technologies for implementing SOA are Web service description language (WSDL); universal description, discovery, and integration (UDDI); and simple object access protocol (SOAP). Web services support heterogeneous communication, because they all use the same data format, XML. Web services communicate by sending XML messages (Manes 2003).

The OGC proposes a series of specifications for GIS services (OGC 2005) that include the Web Map Service (WMS), Web Feature Service (WFS), and Web Coverage Service (WCS). Common aspects include operation request and response contents; parameters included in operation requests and responses; and encoding of operation requests and responses. OGC specifications and standards are encompassed by GIS service vendors like ESRI (ESRI 2006).

DISTRIBUTED COMPUTING TECHNOLOGY TOOLS

The processes used within the land management paradigm need to be managed with appropriate land information infrastructures, ideally providing complete, integrated, and up-to-date information about both the natural and built environments. These processes involve many computations, which can be managed by using distributed computing technologies, including agent computing, P2P computing, grid computing, and so on.

Agent computing: An agent is a computer system suited to an environment that is capable of autonomous action in order to meet its design objectives. There are several characteristics for an agent—some of them ideal and far from reality. But some characteristics such as mobility, communication ability, reactivity, and inferential capability can enhance GIS service applications in various fields (Kalantari 2003). Agent technology can be used for registry search, service discovery and integration, parallel administration, and parallel service evaluation. The combination of agent computing and GIS services can therefore enhance the performance of complex land administration processes to facilitate spatially enabled LAS.

P2P computing: The Internet as originally conceived in the late 1960s was a peer-to-peer system. P2P computing is contrasted with newer client/server architecture, which now dominates. Now, the basic strategy is to use P2P-networked computers to serve as both clients and servers simultaneously. P2P computing provides an infrastructure for sharing the widely untapped computing power within in-house computers in LAS (Oram 2001). Communication and coordination among all peers remain issues, however. Agent technology, especially mobile agent systems, is considered a useful alternative to address these problems (Kalantari 2004).

Grid computing: Grids are persistent environments that enable software applications to integrate instruments and display computational and information resources that are managed by diverse organizations in widespread locations (Foste and Kesselman 1999). They bring together geographically and organizationally dispersed computational resources and human collaborators to provide advanced distributed high-performance computing to users (Foste and Kesselman 2004).

Core grid technology is developed for general sharing of computational resources and is not especially designed for geospatial data and land information. To meet this need, a geospatial grid has to be able to deal with the complexity and diversity of geospatial data and large volumes of land information.

ENTERPRISE FACILITATORS

Land administration functions and processes can be facilitated by many tools to serve end users. These tools include many enterprise facilities such as electronic banking, digital signature, and electronic documents.

Electronic banking: Electronic banking, known as electronic fund transfer (EFT), uses computer and electronic technology as a substitute for checks and other paper transactions. EFTs are initiated through devices like cards or codes that let users, or people authorized by users, access accounts. Many financial institutions use ATM or debit cards and Personal Identification Numbers (PINs) for this purpose. Some use other forms of debit cards, such as those that require, at the most, a signature or scan. Electronic banking (e-banking) is a foundation for building an electronic conveyancing system (e-conveyancing). E-conveyancing and online settlements are now considered essential for driving efficiency in finance and property transactions.

Digital signature: A digital signature is an electronic signature that can be used to authenticate the identity of the sender of a message or the signer of a document, and possibly to ensure that the original content of the message or document is unchanged. Digital signatures are easily transportable, cannot be imitated by someone else, and can be automatically time-stamped. The ability to ensure that the original signed message arrived unchanged prevents the sender repudiating it later. Digital signatures remain a serious issue for management of significant transactions and important documents such as property transfers. They are another foundation for e-banking and e-conveyancing systems.

Electronic documents: Electronic documents contain information expressed in an electronic-digital form with properties allowing their authenticity to be verified. They should be accompanied by identification of the natural or legal persons who send them, or on whose behalf they are sent, with the exception of people who act as intermediaries. Electronic documents also identify the natural or legal persons to whom they are addressed. The validity of the documents is assessed against an electronic documents circulation system, which is a collection of processes used to check completeness and validity.

E-LAND ADMINISTRATION

Digital technologies adopted early in land administration activities were often intended to enhance the performance of specific government programs or activities. Various land administration agencies established computer systems for record keeping, financial and personnel management, printing, and other internal operations. However, these systems were developed in isolation within agencies and generally stood alone. The trend was to develop computer systems that were independent from and not interoperational with other systems. By contrast, e-land administration efforts are becoming interinstitutional, based on partnerships among agencies, and between government and the private sector.

Meanwhile, the confluence of microcomputing and the information and telecommunication technologies that began in the mid-1980s with the arrival of the graphic-based Web in 1994 brought the potential for e-land administration (Aldrich, Bertot, and McClure 2002). This led to a definition of e-land administration as the transformation of land administration through the use of ICT.

This definition covers two perspectives. One is the sum of all electronic communications among land administration agencies, the private sector, and citizens. The other perspective is the sum of electronically provided products and services created to satisfy mandatory land administration regulations (Greunz, Schopp, and Haes 2001).

A remaining question is whether e-land administration should focus on what citizens want or conversely on what agencies want. This question raises another perspective. There is a need to evaluate e-land administration to assess the degree to which anticipated agency and citizen outcomes are being met and the level of synchronization of outcomes anticipated by the agency and users (Aldrich, Bertot, and McClure 2002).

The five phases in development and implementation of an e-land administration system (e-LAS) are shown in figure 9.11 (Kalantari 2008). The first phase is Internet-based land administration. This includes delivering organizational information to customers over the public Internet and through private internal networks to agencies' own staff via the intranet. Most governments in developed countries have managed to go this far.

The second phase is transacting with customers over the Internet. This requires an organization to offer products and services to its customers over the Internet.

The third phase is integrating Internet applications with transactional e-land administration by connecting internal enterprise applications and transactional e-LAS.

The fourth phase is external integration with partners and suppliers through connecting internally integrated applications to the enterprise applications of external partners.

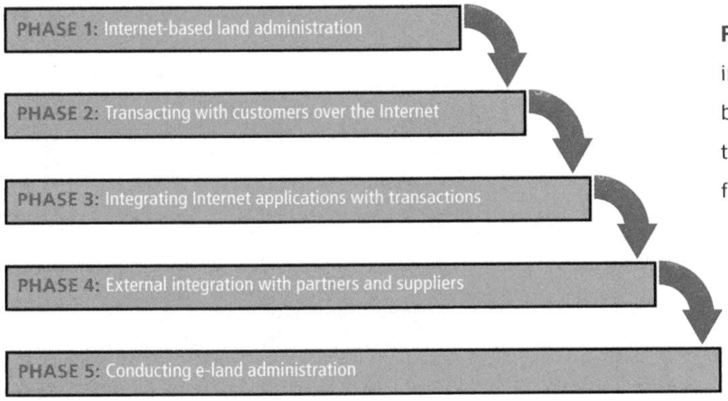

PHASE 1: Internet-based land administration

PHASE 2: Transacting with customers over the Internet

PHASE 3: Integrating Internet applications with transactions

PHASE 4: External integration with partners and suppliers

PHASE 5: Conducting e-land administration

Figure 9.11 The five phases of implementing e-land administration begin with providing information over the Internet and culminate in full-circuit e-land administration.

The final phase is undertaking real-time monitoring and understanding of e-land administration services (Kalantari 2008).

CHALLENGES OF IMPLEMENTING E-LAND ADMINISTRATION

Implementation of e-land administration is not easy (Jaeger and Thompsom 2003). The major challenge is to prevent conflicting goals by coordinating local, state, and national e-land administration initiatives. Conflicts of interests and goals among the agencies performing related tasks make implementation problematic, even if the digital processes are used at various levels of government to improve service quality and reduce costs.

Given the main goal of e-land administration to improve services to citizens, a citizen-focused approach needs to be continuously emphasized. Because e-land administration is a modern way of serving people with high-technology infrastructure, a well-educated audience is essential. Importantly, e-land administration should be as simple as possible to encourage even people without a technical background to get involved. Additionally, building human-resource capacity must be taken into account.

Protection of personal privacy in a virtual environment is still one of the most challenging tasks in designing an electronic service. Financial settlements, the digital signature in e-conveyancing, and other person-specific information and transactions require special protection. Appropriate security control must also be implemented in an e-plan lodgment process.

Elimination of as many paper-based processes as possible is another challenge for e-land administration. Reengineering data models to fully digitize the process offers a good solution. Also, the technical infrastructure, including special exchange language for cadastral databases and special infrastructure for delivery of spatial information, must be robustly maintained. Finally, sustainable e-land administration processes depend on development of methods and performance indicators to assess the services delivered and standards used.

EVALUATING E-LAS

Governments in developed countries worldwide are recognizing the importance of delivering services that put citizens at the forefront and provide a single interface to access all (or a range of) government services. These trends influence the adoption of ICT in land administration and the design of SDIs. Existing approaches, however, indicate a lack of harmonization and integration and tend to eliminate the opportunity for a single-portal e-government service for land administration functions. The first step in improving these existing approaches is an assessment of their capacity and effectiveness.

Monitoring of e-LAS requires the measurement of several characteristics. Quality of service, for instance, involves performance, functionality, user requirements, and popularity (also see sec. 13.3, "Evaluating and monitoring land administration systems").

Performance: The two major criteria for measuring performance are throughput and response time. Throughput is a server-oriented measurement that identifies the amount of work done in a unit of time. Response time is the amount of user-perceived time between sending a request and receiving the response (Peng and Tsou 2003). An interaction for retrieving a Web page is needed in order to measure performance. Any interaction can be broken into four stages, and the combined time for each stage represents the total time of the transaction. D, or DNS (domain name services), is the resolution time, which is the time it takes for the Internet system to connect when users utilize normal English descriptions to communicate with servers that must be addressed by an IP address. T, or TCP, is the connection time, which is the time it takes for the connection process to the server. The FirstByte Time of F represents the time the browser waits between the request and receipt of the first byte of data from the Web server responding to that request. The content time is directly related to the size of the downloaded file, or H.

Recognition of the time it takes for each stage helps to identify delays and problems within various stages. The network, server, and client machine can all be factors. It is also important to note that the complexity of functionality influences the response time in delivery of a service. For example, the processing time for finding parcel information in a database is different from the time it takes to zoom in on or pan a similar dataset.

Nevertheless, for some services, the files are large, but the response time is short, meaning they comprise proper network settings. For that reason, overall performance is dependent on the combination of service, network, and client machine, not an individual component alone. More specifically, overall system performance depends on bottlenecks caused by the slowest component. Therefore, the first step to enhance system performance is to identify the weakest link.

User interactivity and functionality support: The amount of functionality a service provides depends on usability. Three factors — technical, general, and cadastral — influence the functionality of e-LAS.

To use data provided by e-LAS services, online applications can be either lightweight download, JavaScript for a more dynamic system, or specified in metadata or data catalog support.

The benchmarks for the download rate of the Web site are the number of objects, number of requests, and number of scripts on the page. Using fewer images on the site or reusing the same image can promote system performance, as an image should easily fit into one TCP/IP (Transmission Control Protocol/Internet Protocol) packet. Most services have additional external HTML files, which load another page or object and slow the total display time of the page. However, most browsers can multithread several HTML files on the page. Web site pages for many land administration Web services have a moderate amount of images. But in some cases, they take advantage of caching, using fewer images on the service, or reusing the same image on multiple pages.

Minimization of requests improves the performance of an e-land administration service. This relies heavily on the client side of transactions. Using JavaScript, clients' requests over the server can be sorted and arranged before they are submitted to the online service, effectively reducing the number of interactions between the user and the online land information service (Green and Bosomair 2001). Although multiple scripts on the page will increase the time of download, they assist the interaction and decrease network load during the process.

Important general functionality issues in e-land administration services include displaying wide regions on a small screen; supporting various methods of zooming and panning (more than 10 percent of server requests is spent on these tasks); producing different views with the scale change; producing different cartographic displays for a special object in different scales; and allowing users to express ad hoc queries and other orders and to receive information from the system (European Commission 2002).

Accessibility to information on property RRRs, descriptions of their extent, support for land transfers, provision of evidence of ownership, information for property taxation, monitoring of land markets, and support for land market and land-use planning are important cadastral functionalities in the context of land administration that can be supported by online services.

The development of the cadastre is dynamic and has been driven more by institutions and technology than by users. However, models for services, especially on the user side of the business (owners, buyers, and lenders), should focus on the needs of users rather than the historical development of cadastral systems.

Analyzing user requirements: Understanding users' abilities and goals can positively influence the entire design, development, and customization of the service European Commission 2002). Various kinds of users of land information services require unique facilities. Information specialists usually need raw data and functionality to produce information from data. Services for this group should be large, accessible, flexible, and linked to other packages and services.

For decision makers, the service should provide proper and optimized decision-making models. Availability of strategic data is also crucial, although services should be compact, small, and manageable and provide interfaces to other similar services that assist policy makers.

Services for general users should be close to real life, possibly to solve their day-to-day location-related problems, and data provided must be meaningful. For example, providing topographic data may not make sense, while street information and addresses and sales price histories of properties are very useful to the public. Small and efficient services will attract and satisfy these users, who need an intuitive interface for their requests. Services need to address the needs of nonexpert users and interested citizens by providing simple user interfaces and relevant information.

Popularity: There are several criteria for measuring the popular appeal of e-land administration services, including findability, number of return visits, length of time on the Web site, and so on. The number of Web references to a site and number of visitors can be used to measure the popularity of the Web site within an Internet network. Measurement services include Link-Popularity.com (`http://www.linkpopularity.com`). Being linked to a popular Web site can dramatically increase traffic to a specific Web site. Assessment results show that popularity is related to the number of citations over the Web. If the popularity of a service is high, then the number of references to the site by other organizations is also high.

INTEROPERABILITY IN LAND ADMINISTRATION

Similar to other systems, LAS comprise many components and tools. In order to deliver a consistent and robust result and services, these components and tools need to be interoperable. This is especially true of the information produced by LAS processes. Interoperability in information systems is the ability of different types of computers, networks, operating systems, and applications to work together effectively, without prior communication, in order to exchange information in a useful and meaningful way (Inproteo 2005). Interoperability assumes the capability to communicate, execute programs, or transfer data among various functional units in a manner that requires the user to have little or no knowledge of the unique characteristics of those units (Rawat 2003).

In the domain of spatial information, interoperability was originally about cooperation among different organizations—specifically, the compatibility of an information system to run, manipulate, exchange, and share data related to spatial information on, above, and below the Earth's surface. Any kind of application can then serve society over computer networks (Rawat 2003). The idea was then extended to businesses and organizations, in addition to public administration, to improve collaboration and productivity in general, increase flexibility, enhance service efficiency, and add to productivity while at the same time reducing costs.

The complexity of LAS raises nontechnical interoperability issues. Semantics, legal, and intercommunity issues need to be addressed to achieve interoperable e-LAS. Once established, an interoperability framework in e-land administration facilitates cost-effective linking of LAS processes—sharing resources, finding data, and serving the public. Thus, effective e-land administration is at the core of sustainable development.

THE INTEROPERABILITY FRAMEWORK OF E-LAND ADMINISTRATION

An essential feature of a successful SDI is the interoperability of information. The SDI shares reliance on interoperability with e-land administration. Interoperability covers four aspects: semantic, legal, intercommunity, and technical as illustrated in figure 9.12 (Kalantari 2008).

Semantic interoperability: The concepts of land and land administration may be viewed from different perspectives. The ordinary citizen and physical planner may think of land as the actual space in which people live and work. The lawyer may think of the assets of real property rights, while the economist and accountant may see economic commodities. In some contexts, nationhood and cultural heritage are included (United Nations 2004). Whatever the perspective, the information infrastructure used in land administration must match terminology to optimize land management capacity. Lack of semantic interoperability and heterogeneity

Figure 9.12 E-land administration is interoperable on four levels.

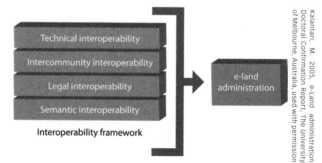

Kalantari, M. 2005, e-Land administration, Doctoral Confirmation Report, The University of Melbourne, Australia, used with permission.

occurs where there is a disagreement about the meaning, interpretation, or intended use of the same or related data in various domains (Tuladhar et al. 2005). Moreover, concepts and semantics need to be aligned even in a specific domain, such as the cadastral domain. Different but related domains also need to be harmonized, including registry, taxation, and planning information. A single standard might not be possible, but a core standard based on common concepts should be achievable; common concepts must be created to allow "talking across boundaries" (Lemmen et al. 2005). Semantic interoperability represents agreed terminology and interpretation of concepts such as a unique definition among all land administration organizations of the third dimension and what it consists of.

Legal interoperability: Land administration organizations have internal process and workflow management solutions; however, effective administration across the related organizations needs guidelines and policies. For example, a framework of land and property laws is needed to ensure the optimum use of space and to enable the land market to operate efficiently and effectively (United Nations 2004). The framework facilitates legal interoperability among organizations. A uniform description of the cadastral domain is needed for cost-efficient construction of data transfer and data interchange systems among different parts of the system (Paasch 2004).

From an international perspective, property registration infrastructure remains mainly regional or local, while banking infrastructure is global. The real estate market can, at least for a subset of people, become global as well (Roux 2004). The global land market needs internationally accepted policies. Legal interoperability will generate directives, rules, parameters, and instructions for managing the business workflow—i.e., using information and incorporating communication among businesses.

Intercommunity interoperability: Intercommunity interoperability involves the coordination and alignment of business processes and information architecture that span people, private partnerships, and the public sector. Intercommunity interoperability leads to LAS that are built for the whole sector, so that users should not have to turn to a number of systems to get the whole picture (Ljunggren 2004).

A recent World Bank comparative study of LAS realized the lack of national interoperability in various areas (World Bank 2003b). For example, the existence of multiple agencies with overlapping land administration roles and responsibilities, each supported by empowering legislation, is a critical issue in some countries in Asia. A similar issue for almost every Latin American country involves separation of the property registry from the cadastre at the information and institutional levels. Coordination is also a critical issue in African countries, where major problems surround the flow of spatial information for land administration purposes within government, between departments at the national level, between national and lower-level tiers of government, and between government and the private sector and users.

Intercommunity interoperability raises the issue of building a unique portal to perform various tasks and applications in land administration. A single portal for intercommunity interoperability with a simple user interface that hides complex logic and operations is the optimum result for land administration and real estate management (Roux 2004).

Technical interoperability: Many types of heterogeneity arise because of differences in technical systems supporting land administration—for example, differences in databases, data modeling, hardware systems, software, and communications systems.

The differences in database management systems (DBMS) largely come from data models, which have direct influence on data structure, constraints, and query languages (Radwan et al. 2005). Moreover, in order to satisfy market needs, the data must be reliable and timely for all users. In order to minimize data duplication, data-sharing partnerships among data producers are coordinated so that there are fewer conflicts in data standards (Tuladhar et al. 2005). Technical interoperability issues also arise when Web services are built—for example, for cadastral information. The services need to operate with any kind of platform, regardless of programming language, operating system, or computer type (Hecht 2004).

Technical interoperability is maintained by continued involvement in the development of standard communications; construction of data exchange, modeling, and storage as well as access portals; and interoperable Web services equipped with user-friendly interfaces. A toolbox developed to achieve interoperability in land administration is described as follows.

INTEROPERABILITY TOOLBOX FOR E-LAND ADMINISTRATION

E-land administration needs to process all kinds of land information and related data, including new datasets created by a busy land market. Examples include establishment of a parking register, water-trading register, natural-resource register, or aboriginal-heritage register, as well as the familiar information generated by land market processes in a land registry. Broad e-LAS of this kind require a range of tools to deliver interoperability. The SDI can facilitate its delivery (Rajabifard, Binns, and Williamson 2006).

The semantic, legal, and intercommunity aspects of interoperability lie more at the administrative and political level. They involve the arrangement of data sharing and processes among the land administration subsystems. Interoperable e-land administration is realized through technical interoperability tools.

A technical interoperability toolbox works at all levels by providing tools for managing data, including modeling, capturing, converting, and so on. The toolbox also provides tools to adapt the organizational structure of LAS to a digital and electronic format. Access and sharing tools are needed to facilitate data and information exchange throughout the subsystems of land administration. The toolbox not only provides accessible data in a proper electronic architecture, it also supplies appropriate models and functionalities that aid decision making. The technical interoperability toolbox includes four types of tools, illustrated in figure 9.13 (Kalantari 2008).

Data management tools: These tools facilitate and manage the development or enhancement of land information from multiple, distributed sources. Cadastral data that is stored for use in local databases, for instance, can often be used in external applications once it is published. Data management tools facilitate data description, data modeling, data capture, database design, data cataloging, and data conversion and migration as a means of standardizing the way cadastral information is held and delivered across multiple servers.

Enterprise architecture design tools: These tools facilitate and support development of plug-and-play enterprise systems and architectures using a Web-based foundation.

Land Administration System

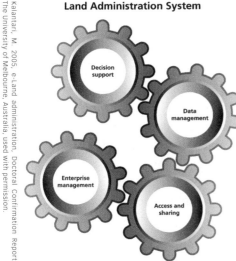

Figure 9.13 The land administration interoperability toolbox includes four types of tools that allow easier sharing of information between the public and private sectors.

Applications are based on compositions of services discovered and launched dynamically at runtime (OGC 2003). Service (application) integration looks to be the innovation of the next generation of e-business. One approach involves interoperability with external software through the use of Web service standards (Hecht 2004).

Access and sharing tools: These tools facilitate the development of Web-based access in a seamless and integrated view. They provide interoperable sharing techniques based on international specifications—for instance, OGC (2003) and ISO International Standards. Access may involve the order, packaging, and delivery, offline or online, of the data (SDI Cookbook 2004). Once the data management and sharing techniques have located and evaluated the cadastral data of interest, the Web services manage access.

Exploitation tools: These tools allow consumers to do what they want with the data for their own purposes. Decision support and exploitation tools, especially in the land-use and land development functions of land administration, facilitate decision support applications that draw on multiple, distributed cadastral data resources.

9.5 Land administration and cadastral data modeling

Data modeling looms in importance as a method of sharing information among agencies involved in land administration. A subset of data modeling is cadastral data modeling. A database is specified by a data model that describes the sort of data held and how it is organized. Data modeling is a design activity, like architecture. There is no single, correct answer for any particular database. Furthermore, data modeling processes must be flexible enough to accommodate a variety of different solutions. The processes need to be creative and allow choices. Data modeling is like a "prescription" that should be distinguished from data analysis, which is like a "description."

Data modeling is important in terms of leverage. Even a small change to the data model may have a major impact on the system. For example, a cadastral database with spatial identifiers should provide topology between layers, while a cadastral database with nonspatial identifiers does not necessarily require topology. In addition, the program design to use the data depends heavily on the data model. A well-designed data model can make programming simpler and cheaper, since poor data organization is often expensive to fix. Data modeling is a powerful tool for expressing and communicating business specifications. It can take users more directly to the heart of what their business needs.

Three approaches can be considered data modeling. The first approach is function driven and focuses on the function specified by the system. The second approach is data driven and emphasizes developing the data model before performing detailed functions. The third method is prototyping, which involves a learn-by-error approach. A prototype is built based on the data and delivers iterative functionality through the processes of show, modify, and show again. Evaluation of the data model should consider completeness, nonredundancy, enforcement of business rules, data reusability, stability and flexibility, simplicity, and communication effectiveness.

CADASTRAL DATA MODELING AND DATA MANAGEMENT

Studies show that data management of LAS is one of the most expensive and costly parts of any system and absorbs between 50 percent and 75 percent of related total costs concerned with a LAS computer environment. Data costs include such items as data modeling, database design, data capture, data exchange (Roux 2004), and data cataloging.

Cadastral data must be able to be updated and kept current (Meyer 2004). Real-time updating of digital cadastre databases (DCDB) is particularly important. Although recent advances in data capture technology make this prospect easy, typically these initiatives are made in isolation, and no common ground is formulated for the handling of cadastral and related data. Consequently, the datasets cannot be easily integrated and shared, because they lack consistency. Further, no effective measures or supporting digital tools exist for direct data access and propagation of updates between them in order to keep datasets up-to-date and in sync (Radwan et al. 2005). The processes of capturing built boundary data provide an example of this problem. To gain maximum benefit from existing data, the building process should not only extract data from the documents and build the boundary network, but it should also analyze the data and provide a measure of the reliability and accuracy of the computed coordinates. This opens the way to coordinates being used more widely, especially as the primary means for surveyors to convey instructions on how to locate the physical boundaries of a property (Elfick, Hodson, and Wilkinson 2005). Effective data management in land administration is possible if efficient and cost-effective methods of capturing cadastral data, including spatial and nonspatial data, are realized in the cadastral data model.

The cadastral database should join attribute data with spatial data and present both in an integrated portal, because attributes are as important as spatial information for decision support (Meyer 2004). However, current integrated portals do not necessarily allow attribute data and spatial data to be joined. They enable users to access various distinct databases using a unique portal. After about 2000, systems architecture design changed in response to the growing need to deliver simultaneous access to datasets that were developed within various divisions of a large organization, edging a step closer to joint accessibility of the data. These datasets increasingly have to be accessed at an integrated level (Vckouski 1998). Similarly, in cadastral data modeling, new systems architecture must facilitate access to cadastral databases, whether spatial or nonspatial.

Data must be standardized so that information can be shared across jurisdictional boundaries (Meyer 2004). Therefore, cadastral data needs to have its own exchange language to better communicate among various organizations. Because of the nature of cadastral data, the language must have a spatial component to permit exchange and migration of the data.

Metadata provides linkages to more detailed information that can be obtained from data producers (Meyer 2004). The catalog provides consistent descriptions about the cadastral data. The objective of the cadastral data catalog is to develop a description of each object class, including a definition, a list of allowable attributes, and so on (Astke, Mulholland, and Nyarady 2004). A cadastral data model that includes a data catalog facilitates data publication across a network.

Figure 9.14 illustrates the role of modeling in data management. It formulates the method of capturing spatial and nonspatial cadastral data. It is the basis of database design. The modeling component allows the data catalog to fit metadata in the proper position, whether it is separate from or integrated with other data. Modeling also introduces standards for the exchange and conversion of data among the various services of different organizations.

CADASTRAL DATA MODELING AND COORDINATION AMONG SUBSYSTEMS

An effective cadastral data model must describe what is fundamental to a business, not simply what appears as data. Entities should concentrate on areas of significance to the business. Existing

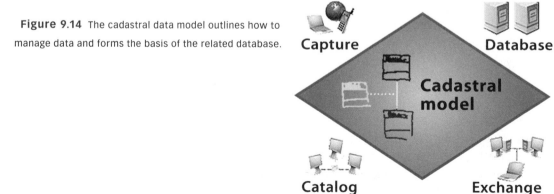

Figure 9.14 The cadastral data model outlines how to manage data and forms the basis of the related database.

Capture

Database

Cadastral model

Catalog

Exchange

Kalantari, M. 2005, e-Land administration, Doctoral Confirmation Report, The University of Melbourne, Australia, used with permission.

cadastral data models include the subject, the object, and the rights associated with them. Most of the efforts to build a data model follow the classic concept of the cadastral domain within land administration based on historical arrangements made for land registration, surveying, building, and maintaining the cadastre (Wallace and Williamson 2004). Increasingly, efforts are being directed toward incorporating flexible and informal land arrangements within the model, particularly social tenures. This flexible approach is part of the land administration response to meet Millennium Development Goals of security of tenure for the millions of people who hold land through informal arrangements (also see the UN–HABITAT Global Land Tool Network at `www.gltn.net`).

In general, utilization of ICT in land administration now focuses on electronic submission and processing of development applications, e-conveyancing, the digital lodgment of survey plans, online access to survey plan information, and digital processing of title transactions as a means of updating the database. Modern land administration needs to incorporate the requirements of all the processes of all subsystems within the cadastral data model (figure 9.15). An expanded cadastral data model that realizes both land taxation and land registry requirements, for instance, can facilitate the processes within an e-conveyancing system.

Kalantari, M., A. Rajabifard, I. P. Williamson, and J. Wallace. 2006. The new vision on cadastral data model, Proceeding FIGXXIII Congress, Shaping the Change, Munich, 2006, used with permission.

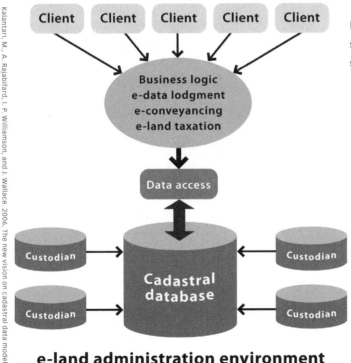

Figure 9.15 A cadastral data model should incorporate information from all subsystems to benefit users.

For example, the e-conveyancing system should be developed in conjunction with the land taxation and land registry subsystems to ensure that all land transfer requirements are met in one simple process. The tax systems rely on properties, not parcels, and use property identifiers that link the title, local government, and tax systems. These systems are concerned with property price and land use. The descriptions of vacant land, residential property, industrial property, rural property, and commercial property are crucial to many taxation regimes. Only some of that information can be accessed from land registries.

Local governments independently gather data layers, such as those for dog exercise parks and sites, walking trails, recreation and horse riding clubs, as well as open spaces within local government boundaries. This sort of information is associated with land parcel and property layers that are not found in the DCDB of any country or state.

An expanded cadastral data model that accommodates both large-scale and local land information can facilitate dataflow among subsystems. It allows easy plug-and-play access between local land information and the cadastral database.

In modern land administration, cadastral data modeling is a basic step toward efficient service delivery, because data is defined in the context of business processes. Every single process in the land administration subsystems should directly influence the core cadastral model. The modeling process should recognize business processes to mirror them in the core cadastral model. Two fundamental changes in the cadastral data model are needed to meet the challenges of technology and to tie land administration processes to land management policy (Kalantari et al. 2008). One involves changing the building block of land administration from physical land

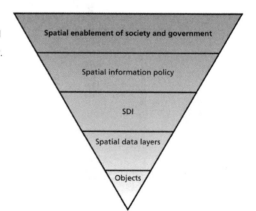

Figure 9.16 A coherent SDI leads to the spatial enablement of government and, by extension, society.

Spatial enablement of society and government

Spatial information policy

SDI

Spatial data layers

Objects

parcels to legal property objects. This allows a wider range of RRRs to be incorporated in the cadastral fabric. The second is to use the spatial referencing system as the legal property object identifier, making the spatial reference the center of the cadastral information system. The latter change promotes interoperability and simplicity in data exchange processes, particularly when it comes to upgrading and updating cadastral databases.

9.6 Maintaining momentum

Two areas of land administration, capacity building and technology, must be approached with flexibility and imagination. They are interdependent. An overview of what is happening in those countries able to afford the latest systems can inform strategic decisions in developing countries. A well-organized and robust computer support system can compensate for the lack of human and institutional resources, provided the design of the system is sensitive to detail and local conditions. Many of the successes in the use of technology, especially in the availability and utility of satellite images, offer nations opportunities to build much more useful and multipurpose solutions to their land administration needs. Even the poorest nations can generally start with images and basic maps. These can be used to collect vital land information and share it with planners, owners, and farmers.

Not counting the legacy technologies that hold back change in developed nations, new opportunities for building and improving LAS through clever technology can be achieved by most nations, including the poorest. In this context, the importance of LAS to supporting concepts like spatially enabled government and society needs to be considered. The creation of economic wealth, social stability, and environmental protection can be achieved through the development of products and services based on spatial information collected by all levels of government. These objectives can be facilitated through the development of a spatially enabled government and society (figure 9.16), where location and spatial information are regarded as common goods made available to citizens and businesses. This flow of information then encourages creativity and product development.

Improving the ability to "find, see, and describe" is only the beginning of spatial enablement. The design of the land information system, therefore, needs to be sufficiently comprehensive to take these objectives into account and manage them through an SDI as described further in section 14.3, "LAS to support spatially enabled society." Spatial enablement is ultimately a transformational technology to assist efficient organization of government and its administrative systems to benefit society.

Chapter 10
Worldwide land administration activities

10.1 Land administration projects

10.2 Recent land administration and cadastral activities

10.3 The Worldwide Cadastral Template Project

10.4 Improving capacity to make global comparisons

10

10.1 Land administration projects

PROJECTS AIMED AT IMPROVING PEOPLE'S LIVES

Land administration theory and practice underpin worldwide attempts to improve the living standards of millions of people who reside in slums, conflict- and disaster-affected areas, transitional agricultural areas, designated forests and national parks, and many other situations where uncertainties about the future are evident. Institutionalization of land administration systems capable of both reflecting and improving existing people-to-land relationships is the typical focus of many international aid and antipoverty initiatives, generally called land administration projects (LAPs). Many similar initiatives involve agricultural aid or provision of services in peri-urban areas and slums, rather than projects focused on land, but these peripheral projects also depend on adequate LAS for sustainability, and many projects include LAS components. Building land administration capacity is also an essential part of national economic transition and maturity of governance capacity for countries seeking upward mobility in their international status.

These projects and related activities have produced an enormous body of literature and critical evaluation. The project papers themselves are, in some cases, available through the aid agencies and international agencies. However, access to project documents in many countries continues to be problematic, with consequential limitations on the growth of the literature base and critical evaluation. While limitations on access are sometimes justified, restrictions imposed by some countries' development assistance agencies are unfortunate. The major UN agencies, including the

TABLE 10.1 – TYPES OF WORLDWIDE LAND ADMINISTRATION AND RELATED PROJECTS	
Recognition of indigenous titles	New laws and title systems protecting indigenous people (Fitzpatrick 2005)
Redesign of native title	Native title, in countries where formal recognition is successful, is undergoing reconstruction to drive the communal titles to individual ownership—for example, in some American Indian tribal lands and Northern Territory (Australia) Aboriginal land.
Disaster response	Major disasters, such as Cyclone Tracy in Darwin, Australia, in 1974; the tsunami in Aceh, Indonesia, in 2005; flooding in Mozambique in 2000; earthquakes in Bam, Iran, in 2003 and Kashmir, Pakistan, in 2005; and Hurricane Katrina in New Orleans, Louisiana, in 2005, required LAS support and reconstruction.
Postconflict management	Large-scale, post–WWII reconstruction delivered huge success in re-creating LAS in Axis and Nazi-occupied countries. More recently, reorganization followed the breakup of Yugoslavia and emergence of minor nations (Fitzpatrick 2006).
Reconstruction of government postrevolution	The plight of the African continent, by contrast, shows how lack of governance, and the consequent inability to manage land, undermine economic growth (Chauveau et al. 2006).
Postcolonial nationalism	Timor-Leste is the most recent in a long list of countries seeking to reassert local land policies and management systems after independence.
Centralist government reconstruction	Vietnam, China, and Laos are moving into more market-driven agriculture and land management.
Urban slum management and reconstruction	The plight of millions living in urban slums is the focus of major LAP activity and rethinking about how to deliver secure tenures and basic services of sanitation and water in the absence of administrative infrastructure.

Continued on facing page

World Bank, apply policies of public access, mostly free Internet access, to thousands of documents in well-organized digital libraries and sell major publications reporting on world development and LAPs for a reasonable price, as do many of the major foreign aid agencies. Table 10.1 provides a highly abbreviated list of the many kinds of LAPs from these sources, but it is far from exhaustive. Projects involve a number of fronts, including legal reform, establishment of institutions, cadastral reform, and even the physical relationships of people and land. The choice of a central focus also varies remarkably, depending on the source agency, national government policy, funding arrangements, and circumstances. Thus, one of the problems for researchers is simply identifying which projects among all the aid and development activity are relevant to land administration.

Continued from previous page

TABLE 10.1 – TYPES OF WORLDWIDE LAND ADMINISTRATION AND RELATED PROJECTS	
Delivery of security of tenure	UN–HABITAT runs multiple programs for bringing services to the urban poor and upgrading tenures. Food and Agriculture Organization (FAO) focus is on similar strategies for the rural poor to deliver food security and alleviate poverty.
Titling for the poor	The most well-known proponent, Hernando de Soto (2000) and the Committee for Legal Empowerment of the Poor, seek "passporting" of land, or identifying the rights and interests of the poor in land, as a means of documenting capital in the hands of the poor.
Management of mass relocation	Human-rights-based tenure systems, work-based tenures, occupancy tenures, and possessory tenures are all under construction, with the UN–HABITAT's GLTN being the managing agent. These tenure initiatives are focused on the most vulnerable type of people-to-land relationships, basically where the relationship is mere occupancy. They do not use the formal tools that underpin thriving land markets, but scale down the tools or adapt them substantially to provide minimum security.
Large-scale national or systematic projects	The most famous is the successful Thailand Land Titling Project. By contrast, the Philippines project provides the greatest challenges (Burns 2006; Bruce et al. 2006). The largest EU-funded project, in Greece, also presents challenges.
Latin American land reform projects	A long history of projects sought to address unequal land distribution, large agricultural laborer populations, insecure small holder property rights, and conflicts with indigenous people (Bledsoe 2006).
Central and East European countries projects	Accession to the European Union provides the major incentive for commodification of land through property rights and LAS (Bogaerts, Williamson, and Fendel 2002; Dale and Baldwin 2000).
Resource administration	Another branch of the project world seeks to provide sustainable access to forests, marine areas, and minerals.

LAPs range in cost from hundreds of thousands to billions of dollars. They can involve small grants to deliver tenure security in carefully chosen micromanagement areas, such as secure vegetable plots for the urban poor in Indonesia, or national programs for infrastructure to support secure tenures. Large-scale and pilot titling programs are run in most of the former Soviet Republic and former communist countries of Central and Eastern Europe. Southeast Asia has programs in Indonesia, Laos, Cambodia, Vietnam, Sri Lanka, Bangladesh, and elsewhere. Latin America has long been the focus for large-scale projects. African aid is also frequently framed around large-scale projects for delivery of tenure security. However, this generalized account of project types leaves out many of the worldwide LAP activities whose recurring theme is to organize how people relate to land to improve their daily lives. No single way of achieving this goal applies to all situations. Nor is there a precise demarcation between projects focused on land reform and land administration.

ANALYSIS OF LAND ADMINISTRATION PROJECTS

Despite the size and coverage of the body of existing literature (recent contributions include Lindsay 2002; Bruce et al. 2006; Chauveau et al. 2006; Burns 2007), further analysis is necessary to provide a picture of the world's achievements and failures in LAPs. Major research is under way to fill these gaps and to better understand the way that administration of land can deliver sustainable development. Comprehensive, well-designed projects stand a much better chance of success than technically focused, "one size fits all" approaches. Broadly designed projects address the risk of antipoor bias by careful selection of language that is comfortable for intended beneficiaries, consideration of the amount of fees charged for technical and administrative services, accessibility of services, recording of local kinds of land rights in addition to standard kinds of rights, and effective accounting, transparency, and oversight mechanisms (Kanji et al. 2005).

The idea behind the land management paradigm is that paying attention to policy in project design will improve LAPs. Especially since 2000, the analytical literature has revolutionized the understanding of tenures, the institution of property, and the relationships among sale and rental land markets, labor markets, product markets, and credit markets. The reasons for project failure have also been expanded.

> *"However strong a land right may be in terms of substantive law, it is severely weakened in practice if there is not a functioning institutional and legal apparatus that allows it to be exercised and enforced. In this respect, many of the land administration systems around the world are seriously flawed, in one or more of the following ways:*

◆ *Procedures are often cumbersome, complicated, and difficult to understand and use. Key institutions, such as registries and courts, are frequently located far from many users. Factors such as these contribute to systems that are expensive and inaccessible, and that people seek to avoid, for example, through informal, off-the-record transactions.*

◆ *Some systems are 'overdesigned,' by including technical standards and requirements for precision that exceed the needs of users, are difficult to implement in practice, and, again, add to costs.*

◆ *Some approaches to land administration functions may be poorly adapted to specific social contexts, such as adjudication and recording techniques that fail to reflect various secondary or derived rights in land.*

◆ *Unclear allocation of authority between different agencies within the system creates confusion, overlap, duplication of effort, and conflict." (Lindsay 2002, sec. 3.1.4)*

Many land administration agencies suffer from insufficient financing and human capacity, with the result that cadastral maps are unfinished, transactions go unrecorded, information is unreliable, and disputes are unresolved. Other reasons compound the problems, most of which relate to corruption or lack of transparency and failure to build in the minor functions of LAS, including organization of easements, usufructs (in civil-law systems, a kind of servitude over property of another, typically for a limited duration), organization of transactions, and social transitions. Perversely, failures also occur when land rights are effective, so that the poor who were the intended beneficiaries sell out and move to city-fringe slums. A land administration system that is designed to reflect the land management paradigm aims to improve project design by ensuring that the inherent complexities of people-to-land relationships are recognized at the earliest stage and by delivery of self-sustaining systems to accommodate the essential four functions of land management—tenure, value, use, and development.

A continuing theme of LAP activities involves the differences between rural and urban land and the different approaches needed for each (Dalrymple 2005). Security of tenure is a universal requirement for stability in land access and use, but delivery mechanisms vary in rural, peri-urban, and urban areas. This dichotomy between urban and rural land is reflected in LAS approaches throughout the globe—in the administration of the United Nations, where UN–HABITAT manages urban land and FAO manages rural land through national governments, and in land practices used from place to place. The distinction is also embedded in the different characteristics of urban and rural poverty. For the rural poor, secure access to land

and water and fair employment practices in agriculture are the primary concerns, whereas in urban areas, labor opportunities and housing issues predominate.

The mechanisms used to deliver security in urban areas tend to reflect the higher value of land and the need to provide services, especially sanitation and water. But in rural sectors around the world, the physical and administrative infrastructure is typically inadequate.

THE EXPENSE OF LAPS

While it's hard to generalize about LAPs, a common approach is to assess the scope of a land reform project, then build in infrastructure to deliver sustainable land administration capacity, focusing first on expensive urban land, then the rural areas. Within these generalizations, specific approaches vary around the world. Nevertheless, the cost factor is a universal challenge. Even relatively small countries, both in geographic size and size of population, need high levels of funding to build the infrastructure LAS need. Irrespective of a country's wealth, these projects are expensive and take from fifteen to twenty years to deliver national coverage. For very poor countries, like Honduras (2004–08 LAP, $38.9 million), Nicaragua (2003–08 LAP, $38.5 million), and Panama (2001–09 national LAP, $58.57 million), the projects represent a significant national commitment. The populations of Honduras, Nicaragua, and Panama are 7.1 million, 5.6 million, and 3.15 million, respectively.

What compels countries to pay for these programs directly out of their own pockets and by way of loans repayable over time is the confidence that these LAPs will deliver effective land markets, security of tenure, and poverty alleviation. Thus, relying on the success of land markets in the West, even the poorest countries seek similar infrastructure and institutions to develop and manage their own assets in land. According to the worldwide pattern of land administration, if land registries sustain registration of secondary and derivative transactions (those coming after the land registration program is initiated), they will have an income stream capable of sustaining the organization. But the cadastre, the core of successful LAS, will not automatically generate such a revenue stream, and funding the cadastre with no immediate source of income is universally problematic.

The basic reason for designing projects that combine the cadastre in conjunction with land registry functions is the pragmatic need to identify both the process and the funding streams that can deliver the most expensive, least understood, but most essential part of a national land administration system. This initial pragmatism in project design has the happy result of delivering an institutional and organizational structure that is capable of supporting implementation of the paradigm in countries that use it.

Although it is expensive, an intelligently designed LAP that focuses on the cadastre can effectively arm a nation for its pursuit of sustainable development. Thus, building a cadastre is justified, not because it supports markets necessarily, but because the land management paradigm approach makes it the basic tool for managing land and resources. The cadastre then supports government policy making in a much wider area of activities than mere land administration. The question is not whether a country should build schools rather than a cadastre. If phrased in this way, only one answer is possible: Schools come first. The question rather is whether a country should build a cadastral tool that focuses on all aspects of government: health, education, environmental management, demographics, services, and so on. In this way, the cadastre upholds sustainable development, since it has the unique capacity to support all government activities and policy implementation at once.

10.2 Recent land administration and cadastral activities

COLLECTING INFORMATION

Evaluation of land administration and cadastral systems is ongoing in developed and developing countries with an eye toward identifying improvements and addressing future needs. Implementing and reengineering aspects of LAS, particularly the cadastre; comparing systems; and identifying best practices within nations of similar socioeconomic standing are activities occurring worldwide (Steudler, Rajabifard, and Williamson 2004). Initially, many countries failed to recognize that institutional and managerial issues were more critical than technical aspects of LAPs (Onsrud 1999). Efforts to improve coordination of cadastral projects, especially in the last decade, led to international appreciation of the path to improvement.

The collection of information about national systems increased interest in land administration and cadastral systems. Comparison and evaluation of the systems led to identification of best practices (Steudler et al. 1997; Steudler, Rajabifard, and Williamson 2004). These initiatives were mainly carried out by the International Federation of Surveyors (FIG) Commission 7, the Permanent Committee on GIS Infrastructure for Asia and the Pacific (PCGIAP), and UNECE, among others.

UNECE was a catalyst in broadening the focus from cadastral systems to land administration during the 1990s through initiatives summarized as follows (Steudler, Williamson, and Rajabifard 2003). Most of the questionnaires and results are available at `http://www.unece.org/env/hs/wpla/`.

- FIG–Commission 7 in 1995: Questionnaire on characteristics, privatization, fees, strengths and weaknesses, reforms, and trends of cadastral systems (thirty-one country replies)

- FIG–Commission 7 in 1997: Questionnaire on characteristics, privatization, fees, strengths and weaknesses, reforms, and trends (fifty-four country replies) (Steudler et al. 1997)

- Meeting of Officials on Land Administration (MOLA) in 1999: UNECE Documentation of Land Administration in Europe (carried out by Austria)

- MOLA in 1999: Study of legislation relating to cadastre and land administration in UNECE member states. Compilation of legislation in UNECE member states relating to cadastre and land administration

- FIG–Commission 7 in 2001: Standardized Country Report: Statistical Indications and Basic Characteristics (thirteen country replies)

- FIG–Commission 7 in 2002: Benchmarking Cadastral Systems (Steudler and Kaufmann, eds. 2002)

- EUROGI in 2002: Questionnaire on cadastres in preparation for the EUROGI presentation at the First Cadastral Congress in the European Union (Granada, May 15–17)

- Working Party on Land Administration (WPLA) in 2002: Inventory of restrictions of ownership, leasing, transfer, and financing of land and real properties in UNECE member countries (thirty country replies, carried out by Russia)

- WPLA in 2003: Survey on the restrictions of public access to information about land administration, landownership, land transfer, and mortgaging (carried out by Slovakia)

- WPLA in 2003: The use of public-private partnerships (PPPs) in the development of land administration systems (carried out by the United Kingdom)

- WPLA in 2005: Inventory of Land Administration Systems, fourth edition (fifty jurisdictions in forty-two countries, including Canada, carried out by the UK). Its forerunners of the previous three editions remain available.

- PCGIAP–FIG in 2003: Worldwide Cadastral Template Project

Each study used a different methodology or approach for the assessment and comparison of cadastral systems, which consisted mainly of one of the following:

- Inventory method
- Benchmarking approach
- Case study method
- Cadastral template approach

A World Bank initiative, "Comparative study of land administration systems" (Burns 2006), used a case study approach to examine LAS worldwide, starting in 2003. This study provided a basis for a more informed assessment of land administration initiatives by reviewing the characteristics, accessibility, costs, and sustainability of different land titling and registration options based on information compiled in participating countries.

However, the many studies and reports dealing with land administration since 1995 give little attention to the basic issues facing cadastres or to the role of cadastres in national spatial data infrastructure (SDI). The integration of cadastral information into the SDI generally was, however, one of the major issues facing LAS designers. More organized information is needed about the variety of systems, the need to integrate new technologies, and opportunities to test the experiences of others in local contexts. A discussion of one of the major studies will aid understanding of the many complex issues associated with comparative analysis of land administration and cadastral systems. A good example is the PCGIAP–FIG cadastral project that used a template approach to describe cadastral systems worldwide. The availability of information has the secondary, but more important, consequence of encouraging independent comparison and evaluation of land administration systems and approaches.

10.3 The Worldwide Cadastral Template Project

OVERVIEW

The PCGIAP–FIG cadastral template project aims to improve the comparative analytical capacity in cadastral design within national land administration, governance, and professional and educational considerations but on a global scale. Its attractions are its ability to highlight the diversity and complexity of cadastral and land administration activities and its construction of a typology for collecting this variable information so that it can be used both by one-time users and by those interested in advanced comparative research. The information is a major resource in itself. Most importantly, the project facilitates analyses that contribute to international understanding of how each nation approaches land administration design. By sharing the results in a variety of languages at **www.cadastraltemplate.org**, the project hopes to attract an ever-increasing number of participating countries.

This comparative evaluation effort has collected cultural and technical descriptions of national cadastral systems since 2003. The project depended on collaboration between the PCGIAP Working Group 3 (Cadastre) and FIG–Commission 7 (Cadastre and Land Management) to create a cadastral template broad enough to include the variety of cadastral systems used throughout the world and to collect information that can be compared and analyzed. Project design reflected the experiences of FIG–Commission 7, particularly the work of Daniel Steudler and others (1997).

This project was designed to enable policy makers to monitor cadastral developments and changes in response to improvements in organizational capacity, technology, and spatial information. Also, it helps countries determine which questions to ask when reforming their systems. Best-practice techniques and ideas for reengineering have improved the ability of nations to build LAS, ranging from those embarking on new systems to countries with mature systems. The project developed benchmarks and contributed to an improved understanding of the complex relationships among cadastral, LAS, and national SDI initiatives. The template's far-ranging importance was demonstrated by its support by the fifty-six PCGIAP Asia–Pacific member countries together with FIG and other regional organizations such as the UNECE Working Party on Land Administration, United Nations Economic Commission for Africa (UNECA), Committee on Development Information (CODI), and the Permanent Committee of Geospatial Data Infrastructure for the Americas (PCIDEA). Additional financial support was made

> **The Permanent Committee on GIS Infrastructure for Asia and the Pacific (PCGIAP) was established following Resolution 16 of the 13th United Nations Regional Cartographic Conference for Asia and the Pacific (UNRCC-AP) in Beijing, China, in 1994.**
>
> **PCGIAP aims to "maximize the economic, social, and environmental benefits of geographic information in accordance with Agenda 21 by providing a forum for nations from Asia and the Pacific."**
>
> **PCGIAP objectives are pursued by four Working Groups: Regional Geodesy, Fundamental Data, Cadastre (now Spatially Enabled Government), and Institutional Strengthening (PCGIAP 2000).**

available by the Australian government through the Department of Education, Science, and Training. The project was established under a UN mandate by Resolution 4 of the 16th UN Regional Cartographic Conference for Asia and the Pacific (UNRCC-AP) in Okinawa, Japan, in July 2003. The development and building of the template involved extensive work by the international spatial information community. Specific country contributions are the work of policy makers and government officials in member nations.

The template (figure 10.1) is available in English, Spanish, and Portuguese. By 2006, thirty-nine nations from Africa, the Americas, the Pacific, Asia, Europe, and the Middle East had completed the template. The reason for many countries, such as the United States, not participating was the difficulty in identifying an appropriate person or organization to complete the questionnaire or, for some countries, lack of involvement in the activities of the FIG. All the data is integrated into the Web site, on both a country-by-country basis and a data field format to enable multiple comparisons. Statistical data is also presented in graphical charts.

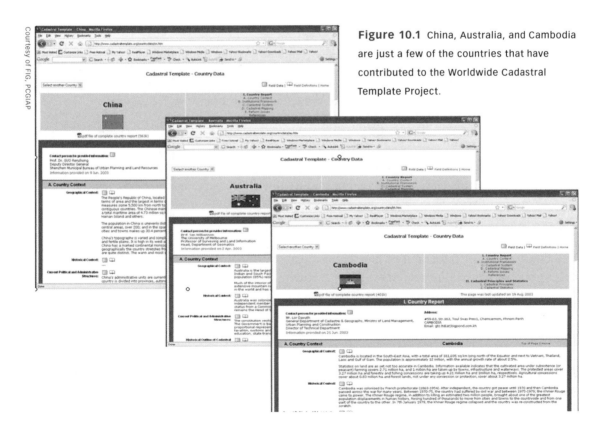

Courtesy of FIG, PCGIAP

Figure 10.1 China, Australia, and Cambodia are just a few of the countries that have contributed to the Worldwide Cadastral Template Project.

DESIGN, STRUCTURE, AND OUTCOMES OF THE PCGIAP–FIG CADASTRAL PROJECT

The objective of the project is to identify the basic social, conceptual, and institutional context of a country's cadastral system and how this context relates to building SDI. A comprehensive framework for comparing and evaluating LAS could be built from the results, identifying best-practices lessons and initiating a methodology for comprehensive evaluation of a land administration system. In later developments, the relationship between cadastral and topographic mapping in the establishment and maintenance of the SDI of each member nation could be explored, particularly the justification and associated conceptual, institutional, and technical issues of integrating the two broad datasets.

Basic principles for the design of the template were its suitability for the member nations of the PCGIAP, as well as of FIG–Commission 7 member nations (mainly European with some African, South American, and Asian representatives), ease of completion, simple structure, and capacity to reflect the main issues of cadastral systems. Brevity was essential, as was question design, to ensure consistent responses.

To reflect the variety of problems facing individual nations, the template had to

- Get an indication of the order of magnitude of the basic tasks in a cadastral system — i.e., how many parcels there are to survey and register

- Get an indication of the magnitude and problems involved in the informal occupation of land

- Reflect the role of the cadastre in land administration and related SDI activities and monitor the completeness, comprehensiveness, use, and usefulness of spatial cadastral data

- Reflect existing capacity and future needs (Rajabifard et al. 2007)

The two sections of the template consist of a country report and a short questionnaire. The descriptive report of the national cadastral system covers country context; the institutional framework; cadastral system (purpose, types, and content of the system); cadastral mapping (example of a cadastral map and role of cadastral layers in the SDI); and reform issues (cadastral issues and current initiatives).

The questionnaire identifies the basic cadastral principles of a country and statistical information, including penetration of registration in urban and rural areas and the approximate number of professionals participating in the cadastral system, as well as an indication of the efficiency of the system.

The cadastral project provides a flexible template for the collection of information and allows comparisons. The first statistical and descriptive data analysis used thirty-four returns to build a worldwide comparison of cadastral systems as well as identify best practices and opportunities for improvement of national cadastral systems (Rajabifard et al. 2007). Since then, an additional five countries, including some from the Middle East, have been added. The structure of this first comparative analysis can be reused and the process repeated to include new participants as they join the project.

Principles and associated indicators, shown in table 10.2, were used in the analysis to develop a list of performance indicators to assess the operations of the cadastral systems of each country. Averages were used in the indicators when attempting to compare LAS within each country while weighted averages were used when attempting to gain an overall picture of an indicator from a particular perspective (e.g., population).

TABLE 10.2 – PRINCIPLES AND ASSOCIATED INDICATORS	
Principles	**Indicators**
STATISTICAL ANALYSIS	
Cadastral principles	Indicator 1: Registration systems
Cadastral statistics: Population and parcels	Indicator 2: Parcels vs. population Indicator 3: Strata units Indicator 4: Percentage of parcels registered
Cadastral statistics: Professionals	Indicator 5: Surveyors and lawyers Indicator 6: Surveyors vs. lawyers
DESCRIPTIVE ANALYSIS	
Educational and professional bodies	Indicator 7: Education and professional bodies
Cadastral reform issues and current SDI initiatives	Indicator 8: Cadastral reform issues and current initiatives

REGISTRATION SYSTEMS (INDICATOR 1)

Table 10.3 shows the mixture of registration systems and registration methods. The most common form was compulsory registration, used in half of the participating countries, irrespective of whether the system was deeds or title based (figure 10.2). Too few countries used a mixture of deeds and title systems for any correlation to be drawn. The major difference between the deeds and title systems is the registration of instruments (chain of deeds), as opposed to registration of title usually guaranteed by the government. Generally, the differences are no longer as significant as they once were because of the computerization of indexes, though they remain important in developing systems because of the cost and capacity issues involved in upgrading.

Table 10.4 shows the percentage mixture of registration systems and establishment approaches. Countries using title registration were more successful in recording all their properties (35 percent) than those relying on a deeds system (13 percent). This might seem to show the comparative effectiveness of a titles system compared to a deeds system. The figures, however, do not take into account the prevalence of other mechanisms supporting land administration. The matrix also shows the reliance on a systematic approach to registration in deeds systems, with 71 percent of deeds systems in countries without universal coverage utilizing this

TABLE 10.3 – MATRIX OF REGISTRATION SYSTEM VS. REGISTRATION METHOD (%)				
% of countries	Compulsory	Optional	Both	Total
Deeds	14.7	5.9	2.9	23.5
Title	50.0	14.7	2.9	67.6
Mixed	2.9	5.9	0.0	8.8
Total	**67.6**	**26.5**	**5.9**	**100**

Figures do not add up to 100 because of rounding.

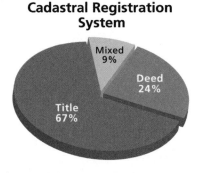

Cadastral Registration System

Figure 10.2 Chart shows the percentage of cadastral registration systems for countries participating in the template project that are based on deeds or titles or have mixed systems.

approach. Countries with mixed registration systems tended to use both systematic and sporadic approaches to establish cadastral records (figure 10.3).

PARCELS AND STRATA UNITS VS. POPULATION (INDICATORS 2 AND 3)

The populations of the thirty-four countries in the template project range from 80,000 in Kiribati to 1.2 billion in the People's Republic of China. The average population is 97.8 million. However, when the two highest (China and India) and two lowest (Kiribati and Brunei) values are excluded, the average population lowers to 36.6 million.

The distribution of the population between urban and rural areas is also significant from historical, land tenure, and cultural perspectives. Urbanization in the jurisdictions ranges from under 20 percent in Cambodia, a nation with high agricultural activity, to 100 percent in the Macao Special Administrative Region (in China), a small metropolitan area of less than 30 square kilometers. The urbanization average value is 63 percent (average of all the countries' urbanization percentages). However, when weighted according to population, this value is lowered to 55.2 percent.

TABLE 10.4 – MATRIX OF REGISTRATION SYSTEM VS. ESTABLISHMENT APPROACH (%)						
%	Systematic	Sporadic	Both	All properties recorded	Other	Total
Deeds	14.7	2.9	2.9	2.9	0	23.5
Title	23.5	5.9	14.7	23.5	0	67.6
Mixed	0.0	0.0	8.8	0.0	0	8.8
Total	38.2	8.8	26.5	26.5	0	100

Figures do not add up to 100 because of rounding.

Cadastral System Establishment Approach

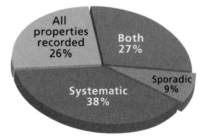

Figure 10.3 The greatest number of countries use the systematic approach to establish their cadastral systems, though some use both systematic and sporadic approaches.

The number of parcels per million people in each jurisdiction is important, because it reflects the tenure, history, laws, and politics of each country. While the number of parcels in each jurisdiction varies dramatically, it is primarily because of population, as illustrated in table 10.5.

TABLE 10.5 – LAND PARCELS AND POPULATION AND NUMBER OF STRATA TITLES PER MILLION				
COUNTRY	NO. OF LAND PARCELS [MILLION]	--> LAND PARCELS PER 1 MILLION POPULATION	NO. OF STRATA TITLES	--> STRATA TITLES PER 1 MILLION POPULATION
Australia	10.2	531,300	750,000	39,100
Belgium	9.4	940,000	1 million	100,000
Brunei	0.06	172,700	0	0
Cambodia	7	583,300	n/a	
China	246.5	205,400	188 million	156,700
Czech Republic	21.6	2,099,300	1,033,484	100,300
Denmark	2.5	471,700	200,000	37,700
Fiji	0.09	118,600	271	300
Germany	61.5	745,500	14 million	169,700
Hong Kong	0.3	44,800	2 million	298,500
Hungary	7.3	722,800	2 million	198,000
India	210	205,500	41 million	39,900
Indonesia	84.5	361,100	3,000	13
Iran	50	763,400	5 million	76,300
Japan	200	1,575,500	0	0
Jordan	0.86	172,000	320,000	64,000

Continued on facing page

Continued from previous page

TABLE 10.5 – LAND PARCELS AND POPULATION AND NUMBER OF STRATA TITLES PER MILLION				
COUNTRY	**NO. OF LAND PARCELS [MILLION]**	**--> LAND PARCELS PER 1 MILLION POPULATION**	**NO. OF STRATA TITLES**	**--> STRATA TITLES PER 1 MILLION POPULATION**
Kiribati	0.3	3,529,400	0	0
Korea (Rep. of)	35.8	756,500	6,497,308	137,400
Lithuania	> 2.0	578,000	> 4 million	1,156,100
Macao	0.01	23,100	n/a	
Malaysia	7.2	288,900	260,000	10,400
Mexico	30.7	314,800	no answer	no answer
Namibia	0.15	83,300	7,000	38,900
Nepal	24	1,025,600	5,000	200
Netherlands	7.5	466,800	900,000	56,000
New Zealand	2.3	575,000	120,000	30,000
Philippines	50	714,300	189,572	2,700
South Africa	18.0	401,800	1 million	22,300
Sri Lanka	8.5	441,800	10,000	500
Sweden	8	888,900	0	0
Switzerland	4.0	550,900	200,000	27,500
Turkey	35	516,200	10 million	147,500
Uzbekistan	8	307,700	1 million	38,500

The larger countries such as China and India, with huge numbers of people and parcels, face major problems in design and implementation of any national approach, although each has common historical, cultural, legal, and social dimensions. The smaller jurisdictions have more options since they tend to have more consistent land tenure relationships. The average number of parcels per million people (by averaging the figures for each country) is just under 630,000, or 1.6 persons per parcel (this is a key indicator since it reflects issues in the cadastral system for each country and is a good basis for comparing cadastral issues across countries). When this is weighted by population, the weighted mean is just over 350,000 parcels per million, or 2.85 persons per parcel.

The country with one of the lowest ratios of people to parcels is Kiribati (0.27 people per parcel), which has issues dating back to the establishment of its cadastral system that are still causing problems today. The main problem is its unique approach to how it divides land among inhabitants.

As shown by its cadastral map (figure 10.4), the land was simply divided into strips, with buildings, roads, and sports fields running haphazardly through parcels. The Czech Republic also has a low level of people to parcels (0.48 people per parcel); its approach of registering buildings and gardens as two separate parcels, and then combining these as a "property," contributes to this figure. These issues are specific to Kiribati and the Czech Republic, but they illustrate the problems involved in defining the terms "parcel" and "property" (Steudler, Williamson, et al., "The Cadastral Template," 2004).

The extremely low ratio of parcels per population in jurisdictions such as Macao and Hong Kong is predominantly because of their high population densities. The high densities indicate the existence of condominiums and apartments in these jurisdictions. Given that the parcel dataset does not include strata or condominium titles, the ratios are misleading. Hong Kong has only 300,000 land parcels, but about 2 million strata or condominium units. Including these units in this ratio changes the number of parcels per million people in Hong Kong to just over 340,000, which is close to the weighted mean. The number of strata and condominium units for Macao was not supplied.

Fiji also has quite a low ratio (112,500 parcels per million), which can be largely attributed to native lands administered by the Native Land Trust Board (NLTB). Because there are many residents of these large native lands, the number of parcels per person is greatly diminished.

Namibia's ratio is also quite low; moreover, strata units are uncommon. Namibia has only 7,000 strata units. This statistic is most likely because of its high level of informal (illegal) occupation of land—38 percent in urban areas and 70 percent in rural areas (see indicator 4 below). As stated in the country report (Owolabi 2003), all land that is not "otherwise lawfully owned" belongs to the state. The poor economic state of the country and the culture of its people (for example, extended families living together) would also have an impact on this result.

With regard to the strata unit, according to the analysis and as shown in table 10.5 and figure 10.5, India and China have an extremely large number of strata units, reflecting large populations. At the other end of the scale, Brunei, Sweden, Japan, and Kiribati have no strata titles. However, as discussed in the country report (Österberg 2003), Sweden has recently embarked on a project to administer the incorporation of strata titles within its Real Property Register. Within the Japanese country report (Fukuzaki 2003), there is no reference to strata titling. However, the country may use different methods to classify strata titles. Brunei and Kiribati appear to have no initiatives to promote their use.

The noticeable outlier in this dataset is Lithuania. This is because strata titles are issued to buildings, engineering utilities, and premises. The other extreme value is Hong Kong, where the high population density results in a very high ratio of strata parcels per million population. Indonesia has quite a low number of strata parcels per million, only 12.8. There is no indication in its country report (Nasoetion 2003) as to why there are only 3,000 strata titles included in its 84.5 million-parcel cadastral register. The complexity of its strata system and cost of units in a

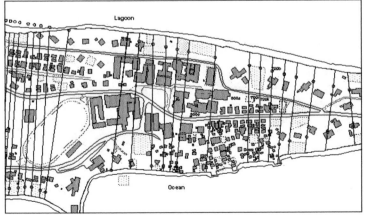

Ereata, T. 2003. Kiribati Country Report 2003, The Cadastral Template Project, used with permission.

Figure 10.4 Kiribati's approach to land parcellation, illustrated by its cadastral map, shows the land simply divided into strips with buildings and roads that are not contained within parcels, but splayed across them.

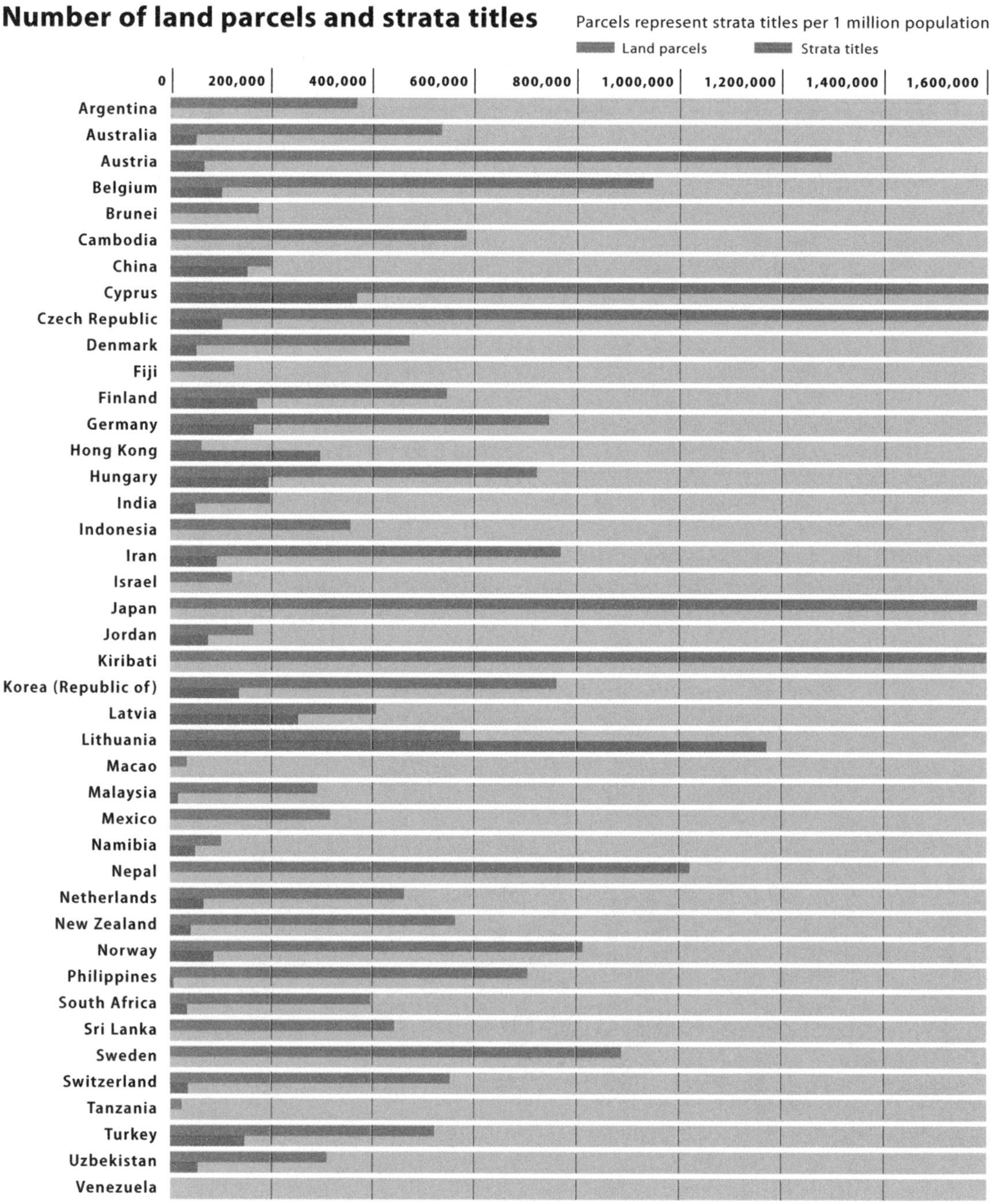

Number of land parcels and strata titles

Parcels represent strata titles per 1 million population

Land parcels   Strata titles

Figure 10.5 The number of land parcels in a country is usually significantly greater than the number of strata titles.

poor country are contributors. The low ratio of strata titles per million in Nepal is because its 5,000 strata titles are houses, not the true definition of strata titles. These houses are part of the country's cadastral register, which includes assets (such as buildings, etc.) on the property.

Hong Kong and Lithuania have the lowest ratio of regular parcels to registered strata units. In Hong Kong, this is because of high urbanization and issuing strata titles to buildings, engineering utilities, and premises respectively. In Lithuania, this may stem from the Communist era, when the rural area was divided into huge cooperatives. Similarly, Indonesia and Nepal have the highest parcel per strata values. The weighted average of this ratio was 4 regular parcels per 1 registered strata unit. Most of the countries with strata parcel data have a ratio of between 2.5 and 20 regular parcels per registered strata unit (Rajabifard et al. 2007).

PERCENTAGE OF PARCELS REGISTERED (INDICATOR 4)

Averages (with respect to total parcel numbers across all countries in the sample) of parcels that are legally registered and surveyed; legally occupied but not registered or surveyed; and informally occupied without legal title in urban and rural areas (indicator 4) are shown in table 10.6.

TABLE 10.6 – AVERAGES OF PARCEL REGISTRATION DATA			
	Legally registered and surveyed (%)	Legally occupied but not registered or surveyed (%)	Informally occupied without legal title (%)
Urban	84.5	11.5	4
Rural	77.7	16.4	5.9

According to the table, the informal occupation of land is 1.5 times more common in rural areas as in urban areas. The level of legally registered and surveyed parcels is also higher in urban areas (84.5 percent) than in rural areas (77.7 percent). Although the average of informal occupied land is only 4 percent–6 percent, it is as high as 38 percent for urban and 70 percent for rural regions in Namibia (figures 10.6 and 10.7). Other countries with exceptionally high levels of illegal settlement are Indonesia (60 percent in rural areas, 10 percent in urban areas), the Philippines (25 percent urban, 5 percent rural), and South Africa (20 percent urban, 5 percent rural). The percentage of land legally occupied but not registered or surveyed is also quite high across both urban and rural areas, at an average of 14 percent.

In terms of registration and surveying of land, the nations faring the worst are Cambodia, Indonesia, Japan, and Namibia. Only 10 percent of rural and 18 percent of urban parcels are legally registered and surveyed in Cambodia, which is largely a result of the post-Khmer Rouge rebuilding phase. Japan also has low levels of registration (18 percent urban, 46 percent rural), though the reason is unclear. Japan has the highest total number of land surveyors out of all the nations in the study. The large number of surveyors may indicate an attempt to rectify the lack of legally registered and surveyed parcels; this theory is supported by the systematic approach Japan is taking to the establishment of cadastral records. Indonesia has a total of 20 percent of

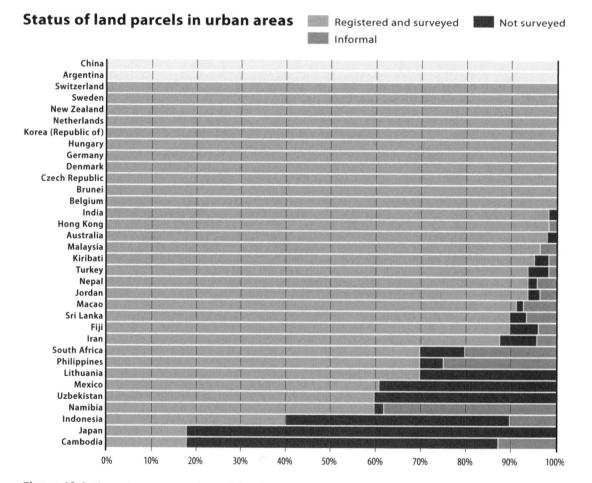

Status of land parcels in urban areas   Registered and surveyed   Not surveyed   Informal

Figure 10.6 The vast percentage of parcels in urban areas in most countries are legally registered and surveyed. The next group are countries with a large number that are legally occupied but not registered or surveyed. A much lesser percentage of parcels in urban areas are informally occupied without legal title.

rural and 40 percent of urban parcels registered and surveyed. The official reason for the low percentage of rural parcels surveyed and registered is predominantly because of the occupation of land by illegal settlers.

The nations with total coverage of their cadastral records (100 percent legally registered and surveyed) include Belgium, Brunei, the Czech Republic, Denmark, Germany, Hungary, South Korea, the Netherlands, Sweden, and Switzerland.

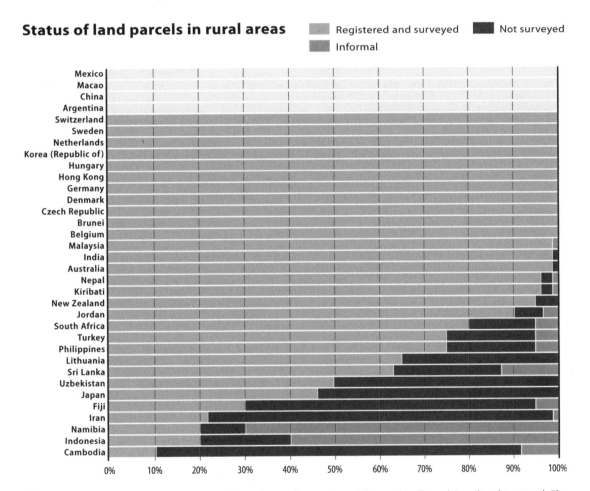

Figure 10.7 The vast majority of parcels in rural areas in most countries are legally registered and surveyed. The next greatest number are parcels that are legally occupied but not registered or surveyed. Indonesia and Namibia, on the other hand, have a great number of parcels in rural areas that are informally occupied without legal title.

PROFESSIONALS—SURVEYORS AND LAWYERS (INDICATORS 5, 6, AND 7)

The total number of professional full-time equivalent surveyors (within the cadastral system) per million persons and number of lawyers is shown in table 10.7. This is related to indicator 5, which relies on knowledge of the number of professional surveyors operating within each country. For example, Australia has approximately 3,500 professional land surveyors; however, the percentage of time that these surveyors commit to cadastral matters is approximately 30 percent. This gives a total of 1,050 full-time equivalent active cadastral surveyors out of the 3,500 professionals. This figure was then used to calculate the number of full-time equivalent surveyors per million. Some anomalies in the data are apparent from the extremely high value for Japan. The country report (Fukuzaki 2003) suggests that there are 201,351 professional surveyors in Japan but does not state the percentage of time they commit to cadastral matters, and there is no reason given for this large number of surveyors.

If a country has a low number of professional surveyors, it should also have a low figure per million people. However, although Kiribati has only five professional surveyors, they commit 100 percent of their time toward cadastral matters. As a result, the population of only 80,000 has a very healthy 62.5 full-time equivalent surveyors per million. The average value from the thirty countries with sufficient data was 58, but when the average was weighted by population, this value dropped to 25.5 surveyors per million.

Germany has 223 professional full-time equivalent surveyors per million people—the highest of the thirty jurisdictions. A high professional surveyor ratio may be an indication that the country relies on a very labor-intensive system or else a very sophisticated system. This is because the number of surveyors needed within a cadastral system is highly dependent on the complexity of its land transfer process, especially whether a country uses resurveying on land transfer, and land subdivision.

Hong Kong has the lowest value once again, but most likely, it is because of the high level of strata units. South Africa, the Netherlands, and China also have very low surveyor ratios, which could be caused by a number of factors. In many cultures, land transfer is not as common. Other influential factors include the amount of land transfer taxes that may be a disincentive to formal land transfer. To a significant extent, the number of surveyors needed within a cadastral system is related to the level of land transfer, as the bulk of surveyors' duties traditionally come about as a result of transactions in some form, usually subdivisions.

TABLE 10.7 – NUMBER OF FULL-TIME EQUIVALENT PROFESSIONAL LAND SURVEYORS PER MILLION POPULATION

Country	Total no. of professional land surveyors	committed to cadastral matters	–> full-time equivalent per 1 million population	Total no. of professional lawyers/ solicitors	committed to cadastral matters	–> full-time equivalent per 1 million population
Australia	3,500	30% (1,050)	54.7	12,000	30% (3,600)	187.5
Belgium	950	90% (855)	85.5	1,400	95% (1,330)	133.0
Brunei	50	80% (40)	114.3	10	60% (6)	17.1
Cambodia	220	100% (220)	18.3	?		
China	30,000	15% (4,500)	3.8	n/a		
Denmark	300	40% (120)	22.6	1,000	30% (300)	56.6
Fiji	49	90% (44)	56.9	250	15% (38)	48.4
Germany	23,000	80% (18,400)	223.0	10,000	60% (6,000)	72.7
Hong Kong	20	50% (10)	1.5	2,000	33% (660)	98.5
Hungary	1,700	50% (850)	84.2	2,000	60% (1,200)	118.8
India	30,000	30% (9,000)	8.8	100,000	30% (30,000)	29.2
Indonesia	5,600	100% (5,600)	23.9	2,000	100% (2,000)	8.5
Iran	10,000	90% (9,000)	137.4	40,000	30% (12,000)	183.2
Japan	201,351	n/a		2,000	90% (1,800)	14.2
Kiribati	5	100% (5)	62.5	10	95% (10)	111.8
Korea (Rep. of)	6,324	100% (6,324)	133.8	4,106	50% (2,053)	43.4

Continued on next page

Continued from previous page

TABLE 10.7 – NUMBER OF FULL-TIME EQUIVALENT PROFESSIONAL LAND SURVEYORS PER MILLION POPULATION

Country	Total no. of professional land surveyors	committed to cadastral matters	--> full-time equivalent per 1 million population	Total no. of professional lawyers/ solicitors	committed to cadastral matters	--> full-time equivalent per 1 million population
Lithuania	550	80% (440)	127.2	no data		
Macao	n/a			n/a		
Malaysia	300	75% (225)	9.0	7,000	70% (4,900)	196.6
Mexico	10,000	90% (9,000)	92.3	5,000	80% (4,000)	41.0
Namibia	20	80% (16)	8.9	40	90% (36)	20.0
Nepal	2,000	60% (1,200)	51.3	2,000	60% (1,200)	51.3
Netherlands	40	100% (40)	2.5	1,806	60% (1,086)	67.6
New Zealand	600	30% (180)	45.0	3,000	?	
Philippines	12,800	5% (640)	9.1	2,000	35% (700)	10.0
South Africa	860	10% (86)	1.9	2,000	70% (1,400)	31.3
Sri Lanka	1,500	55% (825)	42.9	4,000	65% (2,600)	135.1
Sweden	600	100% (600)	66.7	30	100% (30)	3.3
Switzerland	500	80% (400)	55.1	700	70% (490)	67.5
Turkey	150	50% (75)	2.2	10	50% (5)	0.1
Uzbekistan	2,000	80% (1,600)	61.5	100	small	

Note: All these tables are prepared from data collected from answers to questionnaires filled out by many contributors. Some errors and omissions may have occurred.

The number of full-time equivalent lawyers or solicitors per surveyors, and vice versa, is variable, reflecting the different functions and roles played by these professionals in different countries.

The overarching issue of capacity is clear, especially for developing nations. A lack of education for prospective cadastral surveyors exists in many Asian and Pacific Island jurisdictions because of the small size of island nations and their respective cadastral systems (indicator 7). Many countries (such as Kiribati) have no formal training programs and must send any prospective surveyors to educational facilities in neighboring countries. There is also no guarantee that individuals who undertake training overseas will return to work in their home country. Greater emphasis on educational capacity building is needed in the Asia and Pacific region, especially by encouraging cadastral education courses in regional universities such as the University of the South Pacific in Fiji.

The objective of this analysis of key data from the project is not to try to investigate specific countries in detail, or to get precise comparisons, but to show the variability and complexity of cadastral (and land administration) systems around the world. While no two systems are the same, these comparisons provide a better understanding of the issues involved in reforming or improving systems.

ANALYSIS OF CADASTRAL ISSUES AND CHALLENGES (INDICATOR 8)

In order to meet the needs of an information society and the evolving relationship of people to land, LAS throughout the world need to continuously change and adapt. In this regard, indicator 8 highlights some of the cadastral reform issues that countries are facing. Analysis of these issues shows they are grouped according to a country's level of development.

Digital cadastral mapping issues are seen across countries with either a low development level, such as Namibia, or the emerging capitalist nations in Eastern Europe, such as the Czech Republic and Hungary. Similar countries, such as Uzbekistan, also lack coordination in cadastral issues principally because separate management of land registration and cadastral surveying agencies causes conflict and anomalies in land information systems. The major need for these countries, however, is building capacity, including better educational facilities, access to funding and financial support, training and requalification of surveyors and other staff, and better management of cadastral projects and initiatives.

At the next level of development, newly industrialized countries, such as Indonesia and India, suffer from a lack of cadastral infrastructure. This infrastructure needs to be continuously built up by each country, particularly through developing educational facilities and professional bodies.

Well-developed countries, such as Australia, Japan, Switzerland, and Sweden, need maintenance systems for their cadastral infrastructure, a comprehensive system to support a transparent and equitable land market, and international compatibility within European nations.

Historically, developing countries face issues that have been resolved by more developed countries. While the resolutions of a decade ago may no longer be preferred solutions, the ability of nations generally to learn about these strategies is one of the major outcomes of the cadastral template project.

Constant reengineering of cadastral systems can be seen through the wide-ranging list of current initiatives being undertaken by various countries at all development levels. These reform strategies involve major initiatives in capacity building. This includes both capacity building in LIS through the use of technology (creating online registration information and making cadastral data and maps available over the Internet) and institutional initiatives, especially increasing coordination, cooperation, and communication among cadastral organizations. Many countries are also reforming land law and cadastral legislation. Namibia, for example, is concentrating its reforms on low-income communities in an attempt to deal with a large percentage of illegally occupied land.

Developed countries are also engaging in reengineering the process but are broadening the role of the cadastre into a spatial environment using SDI to tackle issues within the "triple bottom line" objectives (economic, environmental, and social) of sustainable development.

10.4 Improving capacity to make global comparisons

The analysis of the PCGIAP–FIG cadastral template project is just one example of ongoing initiatives aimed at a better understanding the complexity of cadastres and LAS. This example provides insight into the diversity of systems used by countries at all stages of development and illustrates the major historical sources of LAS. The template project is not a substitute for detailed national case studies. However, it provides a useful introduction to national systems, and then into comparative understanding. This project and similar exercises are important initiatives in developing best practices and improving systems.

These activities are only one part of the growing analytical literature and evaluation of systems that are discussed later in this book. Recently, the World Bank Doing Business and Development Reports focused on land and processes related to transfer and development, delivering comprehensive statistical and international comparisons on an annual basis. The Internet sites of independent aid agencies and NGOs also offer comparative information. The increasing use of varying sources of information for innovative cross-country comparisons is a new and welcome feature of the literature. It is now possible, for instance, to combine a country's standing on the "perception of corruption" ladder devised by Transparency International with the cost of registering an urban property taken from the Doing Business reports and the quality of public institutions cited in the World Bank Economic Forum proceedings to provide revealing comparisons of how LAS work in various countries (Proenza 2006). Comparisons of LAPs in small nations are also appearing, encouraging forthright analyses of project successes and failures (Bruce et al. 2006).

Part 4

Implementation

While Part 3 explores the components of building modern land administration systems, Part 4 reviews implementation strategies, including the major chapter on the land administration toolbox. Central to implementation is capacity building and institutional development. The book argues that this is the most important aspect of building sustainable systems. Chapter 11 introduces the modern capacity-building concept and distinguishes it from capacity development. Capacity-building issues in land administration are reviewed, including the need for broader institutional capacity in land management. Chapter 11 concludes by reviewing the need for appropriate education and training in land administration.

The land administration toolbox is described in the central chapter of the book. All the tools needed to build LAS are described in chapter 12 with a description of their use. The tools are divided into three groups: general tools, professional tools, and emerging tools. Most of the general tools such as land policy tools, governance and legal framework tools, land market tools, and capacity and institution-building tools, are covered in the preceding chapters of the book. The professional tools such as tenure tools, registration system tools, titling and adjudication tools, land unit tools, boundary tools, and cadastral surveying and mapping tools are discussed in depth in this chapter. The review of land administration tools concludes by introducing the emerging and increasingly important tools such as pro-poor land management tools, noncadastral approaches and tools, gender-equity tools, and human-rights tools.

Part 4 concludes with an introduction to project management and evaluation. Chapter 13 looks at the management processes required to design, build, and manage LAS in the context of the "project cycle." It describes the tools used to assist project management and monitoring such as SWOT analyses, Fishbone charts, Logical Framework Analysis, and Gantt charts. A reengineering framework is introduced to provide the context for LAS projects and to explore cadastral reform strategies. Chapter 13 concludes with a review of strategies to evaluate and monitor LAS, including benchmarking.

Chapter 11
Capacity building and institutional development

11.1 The modern capacity-building concept

11.2 Capacity development

11.3 Capacity-building issues in land administration

11.4 Institutional capacity in land management

11.5 Education and training in land administration

11

CAPACITY BUILDING IS ONE OF THE KEY ISSUES for designers of land administration systems. Traditionally, capacity building focused on the short term by means of staff development through formal education and training programs to meet the lack of qualified personnel. But capacity-building measures must be seen in the wider context of developing and maintaining institutional infrastructure in a sustainable way. Only then can capacity needs be met and adequate responses at the societal, organizational, and individual level be made.

The wider concept also diagnoses a serious lack of institutional capacity in many countries to undertake land administration activities in an adequate and sustainable way. Especially in developing countries and countries in transition, national capacity to manage land rights, restrictions, and responsibilities is not well developed in terms of mature institutions and the necessary human resources and skills. Moreover, in developing countries, there are often two systems of knowledge and production that exist in parallel: traditional and modern. When new knowledge is not integrated into traditional knowledge and production systems, it fails to be useful, despite its potential.

11.1 The modern capacity-building concept

The term capacity building is relatively new, emerging in the 1980s. It has many different meanings and interpretations, depending on who uses it and in what context. It is generally accepted that capacity building as a concept is closely related to education, training, and human resource development (HRD). When this conventional understanding broadened to a more holistic approach, societal, organizational, and individual aspects were clearly identified (UNDP 1998; Enemark and Williamson 2004):

♦ **The broader system/societal level:** The system or enabling environment level is the highest level within which capacity initiatives may be considered. For development initiatives that are national in context, the level covers the entire country or society and all the components that are involved. The dimensions of capacity at a systems level may include policies, the legal/regulatory framework, management and accountability perspectives, and available resources. For initiatives at the sectoral level, only the relevant components are included.

♦ **The entity or organizational level:** An entity may be a formal organization such as government or one of its departments or agencies, a private-sector operation, or an informal organization such as a community-based or volunteer group. At this level, successful approaches to capacity building include the role of the entity within the overall system and its interaction with other entities, stakeholders, and clients. The dimensions of capacity include mission and strategy, culture and competencies, processes, and infrastructure.

♦ **The social group or individual level:** Capacity assessment and development at this third level is considered the most critical. This level addresses the need for individuals to function efficiently and effectively on a professional basis within the organization and within the broader system. HRD is about assessing the capacity needs of people and addressing the gaps through adequate measures of education and training and continuing professional development (CPD) activities. The dimensions of capacity should include the design of educational and training programs and courses to meet the identified gaps within the skills base and to provide the appropriate number of qualified staff to operate the systems involved.

Capacity building is not a linear process. Whatever the entry point and the issue in focus, it is frequently necessary to incorporate the conditions and consequences at the upper or lower level as well. Capacity building should be seen as a comprehensive methodology aimed at providing sustainable outcomes through assessing and addressing a wide range of relevant issues and interrelationships.

Strategies for capacity assessment and development can be focused at any level, provided they are based on a sound analysis of all relevant dimensions. Capacity issues are often first addressed at the organizational level. Organizational capacity—such as the capacity of the national cadastral agency or the cadastral infrastructure and processes—is influenced not only by the internal structures and procedures of the agency, but also by the collective capabilities of the staff on the one hand and a number of external factors on the other. These external factors may be political, economic, or cultural issues that constrain or support performance, efficiency, and legitimacy. The whole level of awareness of the values of LAS also comes into play. By taking this approach, capacity measures can be addressed in a more comprehensive societal context.

Capacity development takes place not only within individuals, but also between them and in the institutions and networks they create—through what has been termed the social capital that holds societies together and sets the terms of these relationships. Most technology-building cooperation projects, however, stop at addressing the individual skills and institution building—they do not consider the larger societal level (UNDP 2002).

In terms of land administration projects, capacity must be seen as a development outcome in itself and distinct from other program outcomes—specifically, building technical and professional competence in specialist fields through HRD activities. Education and training measures become a means to an end. The end then is the capacity to achieve the identified development objectives over time—particularly to establish and maintain national land administration infrastructure for sustainable development (Enemark and Williamson 2004).

11.2 Capacity development

WHAT IS CAPACITY?

Arguably, many donor projects in land administration, especially before the mid-1990s, had a rather narrow focus on access to land and security of land tenure. The focus was on doing the project, including mapping, adjudication, and registration, and developing the necessary capacity for managing the processes within the system. It did not consider the wider land administration infrastructure or land policy issues. Institutional issues were therefore

Figure 11.1 Capacity building was the major issue in creating new land policy in Malawi that includes formalization of customary tenure.

addressed mainly in response to this narrower perspective. Capacity is now seen as a wider concept. The United Nations Development Programme (UNDP) offers this basic definition:

> *"Capacity can be defined as the ability of individuals and organizations or organizational units to perform functions effectively, efficiently, and sustainably." (1998)*

This definition has three important aspects:

1. it indicates that capacity is not a passive state but a continuing process;

2. it ensures that human resources and the way in which they are utilized are central to capacity development;

3. it requires that the overall context within which organizations undertake their functions will also be a key consideration in strategies for capacity development. Capacity is seen as two-dimensional: capacity assessment and capacity development.

Capacity assessment is the essential basis for formulating coherent strategies for capacity development. It is a structured and analytical process that assesses the various dimensions of capacity within a broader systems context, as well as evaluating specific entities and individuals within the system. Capacity assessment may be carried out in relation to donor projects — e.g., in land administration — or it may be done via an in-country self-assessment. Capacity assessment was the key issue in building new LAS in Malawi a decade ago (figure 11.1).

Capacity development goes beyond HRD, because it emphasizes the overall system, environment, and context within which individuals, organizations, and societies operate and interact. Even if the focus within an organization is on the capacity to perform a particular function, the overall policy environment must, nonetheless, always be considered to ensure coherence of specific actions with macrolevel conditions. Capacity development does not, of course, imply that there is no existing capacity; it also includes retaining and strengthening the existing capacities of people and organizations to perform their tasks. These ideas led to the UNDP offering an even more complete definition of capacity development:

> *"... the process by which individuals, groups, organizations, institutions, and societies increase their abilities to perform core functions, solve problems, and define and achieve objectives; and to understand and deal with their development needs in a broader context and in a sustainable manner." (2002)*

The new approach to capacity development is also influenced by today's globalization of the acquisition of knowledge. Capacity development is arguably one of today's central development challenges, since continuing social and economic progress will depend on it. This understanding of the new capacity-building approach is shown in table 11.1.

TABLE 11.1 – THE NEW CAPACITY-BUILDING APPROACH (ADAPTED FROM UNDP 2002)		
CHARACTERISTICS	**CURRENT APPROACH**	**NEW APPROACH**
Nature of development	Improvements in economic and social conditions	Societal transformation, including building of "right capacities"
Conditions for effective development cooperation	Good policies that can be externally prescribed	Good policies that have to be homegrown
The asymmetric donor-recipient relationship	Should be countered generally through a spirit of partnership and mutual respect	Should be specifically addressed as a problem by taking countervailing measures
Capacity development	Human resource development combined with stronger institutions	Three cross-linked layers of capacity: societal, institutional, and individual
Acquisition of knowledge	Knowledge can be transferred	Knowledge can be acquired
Most important forms of knowledge	Knowledge developed in the North for export to the South	Local knowledge combined with knowledge acquired from other countries—in the South or the North

BARRIERS TO CAPACITY BUILDING

LAPs often have failed to meet the overall objective of building a sustainable national land administration infrastructure. To a large extent, this is because of the complexities involved in addressing national land administration issues. This is not a criticism of these projects—the economic driver has a high priority in developing countries, and only recently has the capacity-building aspect developed into a comprehensive and sustainable methodology. To address these problems, an equal partnership must be built between doing the project and building the capacity to sustain the project. The past decade of experience delivers a clear lesson: Capacity building must be a mainstream component that is addressed up front, not as an add-on in donor projects related to building and improving land administration infrastructure in developing or transition countries. The same lesson applies to national efforts at building and upgrading LAS.

The project in Malawi on building capacity for implementation of land management illustrates best practice (Enemark and Ahene 2003), although it was never fully realized. Land policy reform requires a long-term vision and commitment. In the case of Malawi, project completion was estimated to take fifteen to twenty years. The process was initiated in 1995 by the World Bank, which provided support for land policy reform and a strategic action plan aimed at creating a modern environment for protection of property rights, facilitating equitable access to land for all, and encouraging land-based investment. Institutional reform and capacity building were keys to implementation of the policy. The project included a number of components such as drafting a new land law that formalized customary land law; initiating pilot district land registration, including mapping and demarcation; instituting rural/urban land-use planning and development controls; and engaging in a land resettlement project. The deficit of qualified personnel was addressed through developing an integrated curriculum at the certificate, diploma, and bachelor levels. The implementation was initiated in 2001 by placing the issue of capacity building up front in project design. Unfortunately, the project was not fully realized because of changed priorities within one of the donor countries.

Donors, in general, will often have a long-term vision of what they want to achieve. At the same time, however, they will have to account for the progress of the project to their constituencies and superiors at home. This tends to shape the project in a "manageable" way by using deliverable goals for accountable short-term achievements (such as the number of parcels registered, number of training courses provided, and so on) while the long-term goals (such as building the institutional capacity, and designing and implementing tertiary educational programs) are more difficult to turn into visible tangible activities. This kind of accounting management will work as a self-justifying system that pumps huge amounts of money into developing countries. At the same time, consultants have a strong interest in maintaining the status quo and have little incentive to criticize the basic system since, if they do, they will risk being replaced by more compliant staff. Donors have addressed these problems to some extent. However, many of the fundamental issues remain, though they can be addressed by this new approach (UNDP 2002).

11.3 Capacity-building issues in land administration

THE CONCEPTUAL FRAMEWORK

The land management paradigm drives land administration into a cross-sectoral and multidisciplinary approach that embraces technical, legal, managerial, political, economic, and institutional dimensions. To be effective, capacity-building measures must reflect all these dimensions and include assessment and development at all three levels: societal, organizational, and individual. An appropriate conceptual framework that is capable of supporting analysis of all the dimensions of building sustainable land administration infrastructure to support a broader land policy agenda (table 11.2) is therefore essential (Enemark and Williamson 2004).

TABLE 11.2 – CAPACITY BUILDING IN LAND ADMINISTRATION		
LEVEL	**CAPACITY ASSESSMENT ISSUES**	**CAPACITY DEVELOPMENT OPTIONS**
Societal level	Policy dimension Social and institutional dimension System dimension Legal and regulatory dimension	Land policy issues Land administration vision LAS Land tenure principles Legal principles
Organizational level	Cultural issues Managerial and resource issues Institutional issues and processes	Institutional infrastructure SDI Professional institutions
Individual level	Professional competence Human resources needs Educational resources	Education and training programs CPD programs Virtual programs Education and research center

GUIDELINES FOR SELF-ASSESSMENT OF CAPACITY NEEDS

The multilevel capacity-building framework applies to donor projects dealing with land reform and the design and implementation of LAS to secure rights in land, facilitate an efficient land market, and ensure effective control of the use of land. However, there is also a demand for a framework or guidelines that enable countries themselves to assess the capacity of their systems and identify specific needs for capacity development. These needs may then be met by specific capacity development measures, even with limited financial resources.

The Land Tenure Service in the Food and Agriculture Organization of the United Nations (FAO) developed guidelines for self-assessment of capacity needs (FIG 2008a). The guidelines serve as a logical framework for addressing each step, including land policy, policy instruments, and a legal framework; business objectives and work processes; and needed human resources and training programs. They propose a number of questions to be considered reflecting a best-practices approach. For each step, the capacity of the system can be assessed, and possible or needed improvements can be identified.

These guidelines aim to provide a basis for in-country assessment of capacity needs in land administration, especially for developing countries. The government may form a group of experts to carry out the analysis as a basis for political decisions regarding any organizational or educational measures to meet capacity needs. Of course, individual countries face specific problems that may not be addressed in these guidelines. Hence, they are meant as a tool for structured and logical analysis of capacity needs by posing the right questions rather than providing the right answers.

MAINTAINING SUSTAINABILITY AND CONTINUITY

A major problem in most LAPs is that they focus on the project itself rather than the long term. The sustainability of the system is often only sporadically addressed. Ensuring sustainability and continuity and developing a corporate memory within the country of land administration experience are essential for maintaining viability.

It is generally accepted that appropriately educated personnel and HRD are the keys to sustainability of land administration reform projects. To achieve this objective, it is essential to build up resources to support an ongoing HRD strategy and corporate knowledge in land administration. At the same time, tertiary and technical education programs must be balanced. Usually, technical education is best undertaken by the implementing agency or government technical institutes, while objective policy and technical research and education are better undertaken at the university level.

Most LAPs would benefit widely from establishing a national education and research center in land administration. The center should act as an ongoing body of knowledge and experience in land administration and use the actual project as a long-term case study and operational laboratory. The center should provide educational programs and supervise establishment of educational programs at other institutions. It should interact with international academics and professional bodies to assist the development of local academics. The land administration center most likely would be established in an appropriate national university, possibly within a lead academic department such as surveying or geomatics, but in conjunction with law, planning, valuation, sociology, anthropology, and public-policy departments, where appropriate. The establishment of a center at a university would also capitalize on the independence and transparency that universities can provide.

11.4 Institutional capacity in land management

Land management, the process by which the resources of land are put into good effect (UNECE 1996), encompasses all activities associated with the management of land and natural resources that are required to achieve sustainable development. The organizational structures for land management differ widely among countries and regions throughout the world and reflect local cultural and judicial settings. The institutional arrangements may change over time to better support the implementation of land policies and good governance. Development of institutional capacity in land management implies adoption of long-term strategic actions and capacity-building activities. These include

- ◆ Establishing a strategic approach to donor projects and ensuring that capacity-building measures are addressed up front — and not as an add-on
- ◆ Developing country self-assessment procedures to identify capacity needs and lobbying for the establishment of the necessary measures of capacity development in terms of policies, legal framework, institutional infrastructure, and human resources and skills
- ◆ Promoting the creation and adoption of a comprehensive policy on land development and establishing a holistic approach to land management that combines the land administration, cadastre, and registration functions with the topographic mapping function

♦ Establishing a clear split of duties and responsibilities between national and local government (decentralization) and ensuring that the principles of good governance apply when dealing with RRRs in regard to land resources and land development

♦ Promoting an understanding of land management as a highly interdisciplinary endeavor, including policy measures that range over social, economic, environmental, judicial, and organizational areas

♦ Promoting an interdisciplinary approach to surveying education that combines both technical and social sciences and links the areas of measurement science and land management through a strong emphasis on spatial information management

♦ Establishing strong professional bodies, particularly a national institution of surveyors, that are responsible for the development and control of professional standards and ethics, enhancement of professional competence, and interaction with governmental agencies to develop the optimal conditions and services

♦ Promoting the need for CPD to maintain and develop professional skills and promoting the interaction among education, research, and professional practice

Adoption of a comprehensive policy on land management is crucial since this drives legislative reform, which in turn results in institutional reform and, finally, implementation with all its technical and human resource requirements. A good overall approach to institutional development

TABLE 11.3 – COMPREHENSIVE APPROACH TO INSTITUTIONAL DEVELOPMENT		
CAPACITY ASSESSMENT	**CAPACITY DEVELOPMENT**	**SUSTAINABILITY**
Are the policies on land management clearly expressed?	Adoption of an overall land policy	Instigation of a self-monitoring culture in which all parties—national and local government, NGOs, professionals and citizens— review and discuss progress and suggest any appropriate changes
Is the legal framework sufficient and adequate?	Design of a legal framework addressing the RRRs in land	
Are the institutions adequate, and are the responsibilities clearly expressed?	Implementation of an organizational framework with clearly expressed duties and responsibilities	Lessons learned need to be fed back into the process for continuous improvement
Are the guiding principles for good management well expressed?	Adoption of clearly expressed guiding principles for good governance	Implementation of adequate requirements and options for activities related to CPD
Are the human resources and skills adequate, and are the relevant education and training opportunities available?	Establishment of adequate and sufficient educational options at all levels	

(table 11.3) looks at the four steps that constitute good strategic management: where we are now; where we want to be; how we get there; and how we stay there (UNDP 1998). This approach is in line with the broad capacity-building concept, which aims to assess, develop, and sustain.

GOVERNMENT ORGANIZATIONS

Institutional development enhances the capacity of national surveying and mapping agencies and private organizations to perform their key functions effectively, efficiently, and sustainably. This requires that government and other stakeholders provide clear objectives; that these objectives be enshrined in appropriate legislation or regulation; and that appropriate mechanisms be put in place for dealing with shortcomings due to individual or organizational failure. This requires agreement among a wide range of stakeholders in both the public and private sectors, and it is no trivial task.

Organizational development enhances organizational structures and responsibilities and furthers interactions with other entities, stakeholders, and clients to meet the agreed objectives. This requires adequate resources in staff and financial outlays; a clear and appropriate organizational focus to meet the agreed objectives of the organization; and suitable mechanisms to turn the concept into delivery. These mechanisms include organizational structures, definitions of individual roles, and instructions for completing the activities.

The UK Public Services Productivity Panel (HMT 2000) developed a useful and succinct model that recognizes five key elements to ensure organizational success:

1. **Aspirations:** to stretch and motivate the organization
2. **Coherent performance measures and targets:** to translate the aspirations into a set of specific metrics against which performance and progress can be measured
3. **Ownership and accountability:** to ensure that individuals who are best placed to ensure delivery of targets have real ownership to do so
4. **Rigorous performance review:** to ensure that performance is in line with expectations
5. **Reinforcement:** to motivate individuals to deliver the targeted performance

Of course, defining and implementing the details in any one of these respects is a significant task, and all must be in place if the organization is to succeed. By establishing the appropriate mechanisms and measures, and continuously improving and expanding them, organizations can ensure that they effectively turn input into output and, more importantly, achieve the required outcomes (e.g., certainty of land tenure).

All organizations need to continuously develop and improve if they are to meet, and continue to meet, the needs of their customers and stakeholders. In the land administration field, there are many examples of underresourced organizations unable to respond effectively to stakeholder requirements, leading to a lack of access to official surveys and land titling (and thus to unofficial mechanisms being used or even a breakdown in efficient land titling). There is a need to provide appropriate assistance to these organizations, given the key role they play in national development, so that the necessary capacity can be built and sustained. Of course, the need for this capacity building must first be accepted by the funding bodies. A range of funding methods exists, including releasing internal resources, if suitable resources exist, or finding external support.

The overall approach to LAS project management is presented in chapter 13, "Project management and evaluation." Following is an example of the successful development of sustainable capacity in recent years in Swaziland (Mhlanga and Greenway 1999). Prior to 1995, the UK government had provided long-term support for Swaziland's Surveyor General's Department (SGD). The retirement of the expatriate then holding the position of deputy surveyor general created the opportunity for exploring other mechanisms for developing sustainable organizational capacity. The UK government agreed to fund a series of short-term consultant contracts to supplement the ongoing work of two expatriate technical cooperation officers. The series of visits (approximately twelve in all, involving more than ten different consultants but with continuity provided by an overall lead consultant) made good progress and allowed the department to feel confident, in 1999, that it could continue its work without the need for outside input. The consultations worked in a number of ways. The ability to provide a range of expertise was regarded as a strength of longer-term inputs. The work and outcomes included

- **A thorough review of the strengths, weaknesses, and external impacts on the SGD,** including interviews with a wide range of staff and other stakeholders (including senior officials, private-sector surveyors, and customers). From this review, a number of work packages was agreed on, and work progressed (with periodic review and revision of priorities) over the following four years.

- **Creation of a clear vision, mission, and aims for the SGD** to provide a clear focus for its work. This was shared with all staff in the SGD through a series of workshops and briefings. A key element was the assessment by senior managers of the department's performance in 1995 against each of the aims, providing a powerful means of focusing required efforts on improvement, alongside consolidating areas of good performance.

- **Creation of a business plan for the SGD** to ensure progress toward the vision and aims.

- **Fundamental restructuring of the SGD** with a change of managerial hierarchy, the deletion of old positions, and the creation of a range of new positions. The new structure supported career progression as well as effective delivery of the outputs required. Policies for staff development and retention were also developed and implemented. These, and related changes, came about through interactive workshops, so that senior staff felt strong ownership of the results and could effectively argue for them in discussions with the central civil service and with SGD staff.

- **Creation of revised policies to guide the SGD work plan**, including policies on survey control, map revision, map specification, and marketing (including pricing).

- **Implementation of clear performance measures**

- **Support for the completion of the cadastral database** and implementation of digital map revision systems.

The work in Swaziland reflects the breadth of organizational development set out in the capacity-building approach. Key lessons learned from the project were that long-term consultancy input can easily become counterproductive, where the individuals involved are drawn into line management roles, leading to a limited transfer of skills and therefore not providing sustainable capacity development. In contrast, short-term visits require local managers to focus on completion of agreed actions between visits. Another key lesson was that management confidence, as well as competence, is crucial to success—and that building this confidence is therefore a necessary element in successful projects. In addition, a clear progression from vision to aims to meeting objectives is essential for success.

This case study provides confidence that appropriate efforts can build the required capacity in a sustainable way—in this case, with limited local and external resources being available.

11.5 Education and training in land administration

National education endeavors in building land management skills generally fall short of needs. Even when a country's resources are reasonable, the same problem of a shortage in skills occurs. The following discussion on education and training uses the surveying profession as a case study. However, most of the conclusions and strategies are equally applicable to the other disciplines or professions involved in land administration.

Project design is partly to blame for the shortage of skilled personnel and for a narrow technical focus on its skills base. Project funds almost never go to education institutions but remain with land administration agencies or related agencies. They do not go to capacity building in professional bodies, although these institutions, through their members' and official activities, frequently block change and modernization. These attitudes are built by fear of the unknown and can be successfully alleviated by broadening the outlook of the organizations and their members. An education program, such as that recommended by FIG; membership in and encouragement to participate in international forums; and inclusion of the private sector in official study tours all help to create a more positive and engaged approach to change the view of management.

EDUCATIONAL TRENDS

The training of surveyors and engineers is changing in a variety of ways.

Management skills vs. specialist skills: Changes in the surveying profession and practice, especially the development of new push-button technologies, require the core discipline of management to be a basic element in today's technical education in surveying, engineering, and related fields. Traditional specialist skills are no longer sufficient or adequate to serve the client base. Surveyors, for example, need to have the skills to plan and manage diverse projects, including not only technical skills, but those of other professions as well. In short, the modern surveyor or engineer has to be capable not only of managing within changing conditions, but of managing the change itself.

Developments in technology take the skill out of measurement and the processing of data. Almost any individual can press buttons to create survey information and process this information in an automated system. In the same way, technological developments make GIS a tool available to almost any individual. The skill of the future lies in the interpretation of the data and in its management to meet the needs of customers, institutions, and communities. Therefore, management skills will be a key demand in the future world of surveyors and engineers.

Project-organized education vs. subject-based education: An alternative to traditional subject-based education is the project model where traditionally taught courses augmented by actual practice are replaced by project work assisted by related lecture courses. The aim of project work is to learn by doing, or "action learning." Project work is problem based, meaning that traditional textbook knowledge is replaced by knowledge to solve theoretical and practical real-life problems. Delivery of a broad understanding of interrelationships and the ability to deal with new and unknown problems are the desired outcomes.

In general, the focus of university education should be "learning to learn." The traditional focus on acquisition of professional and technical skills (knowing how) often implies an "add-on" approach, where for each new innovation, one or more courses must be added to the curriculum to address a new technique. Arguably, this traditional subject-based approach should be modified by giving increased attention to entrepreneurial and managerial skills and to the process of problem solving on a scientific basis (knowing why). This should provide skills that enable dealing with the unknown problems of the future.

Virtual academy vs. classroom lecture courses: There is no doubt that traditional classroom lecturing will be largely supported by, or even replaced by, virtual media. The use of distance learning and the Web supports integrated tools for course delivery, which may lead to the establishment of the "virtual classroom," even at the global level. This trend will challenge the traditional role of universities and even shift the focus from on-campus activities to a more open role with regard to serving the profession and society.

The computer cannot replace the teacher and interpersonal interactions, and learning processes cannot be automated. However, the concept of a virtual academy represents new opportunities, especially for facilitating learning and understanding and even widening the role of universities. Online courses and distance learning represent a key path toward programs for lifelong learning.

The role of universities will have to be reengineered based on the new IT paradigm. The key will be knowledge sharing. On-campus courses and distance-learning courses should be integrated, even if the delivery may be shaped in different ways. Existing lecture courses should always be available on the Web. Existing knowledge and research results should also be available and tailored for use in different areas of professional practice. All graduates would then have access to the newest knowledge throughout their professional life.

Lifelong learning vs. vocational training: There was a time when a land administration professionals were qualified for life, once and for all. Today, they must qualify constantly just to keep up. It is estimated that the knowledge gained in a vocational-degree course has an average useful lifespan of about four years. The concept of lifelong learning or continuing professional development (CPD), with its emphasis on reviewing personal capabilities and developing a structured action plan to develop existing and new skills, is taking on increasing importance. University graduation should be seen, realistically, as only the first step in a lifelong educational process.

THE EDUCATIONAL CHALLENGE FOR SURVEYORS

There is a growing need to redirect the focus away from the traditional surveying discipline toward a more managerial and interdisciplinary model. The strength of the surveying profession lies ultimately in its multidisciplinary approach.

Surveying and mapping are clearly technical disciplines (within the natural and technical sciences), while the cadastre, land management, and spatial planning are judicial or managerial disciplines (within the social sciences). The identity of the surveying profession and its educational base, therefore, should be in the management of spatial data, with links to the technical as well as social sciences.

In this regard, universities should act as the main facilitator in forming and promoting the future identity of the surveying profession to include both the technical and managerial aspects. The subject of GIS and, especially, the department managing geographical and spatial information should form the core component of the surveyor's identity. This responsibility or duty of the universities, then, should be carried out in close cooperation with the industry and professional institutions.

The challenge of the future will be to implement the new IT paradigm and this new multidisciplinary approach into the traditional educational programs in surveying and engineering. A future educational profile for this field should include the areas of measurement science and land management (including land economics) supported by and embedded in a broad multidisciplinary foundation of spatial information management. This profile is illustrated in figure 11.2.

Figure 11.2 The educational profile of the future for surveyors and engineers must be based on both technical and managerial expertise.

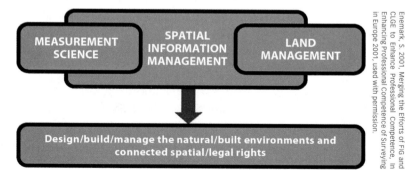

Enemark, S. 2001, Merging the Efforts of FIG and CLGE to Enhance Professional Competence, in Enhancing Professional Competence of Surveying in Europe 2001, used with permission.

This educational profile of the future was first introduced at the joint FIG/CLGE (Council of European Geodetic Surveyors) seminar on Enhancing the Professional Competence of Surveyors in Europe, held in Delft, the Netherlands, in November 2000. This seminar also included an overall assessment of the surveying programs in Europe (FIG/CLGE 2001).

PROFESSIONAL COMPETENCE

The term "professional competence" refers to an individual's status as an expert. This status cannot be achieved through university education alone, nor can it be achieved solely through professional practice. University graduation is no longer the passport to a lifelong professional career. Today, one must continually qualify just to keep up. The idea of "learning for life" is replaced by the concept of lifelong learning (figure 11.3). No longer can "keeping up-to-date" be optional. It is increasingly central to organizational and professional success.

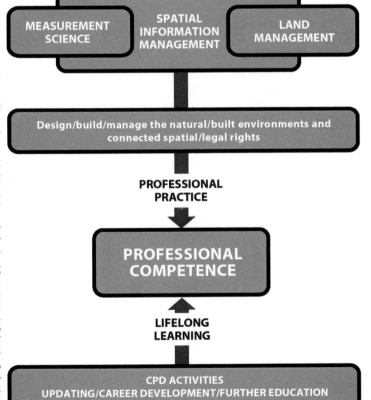

Figure 11.3 Professional competence is a mixture of university learning, professional practice, and continued professional development.

The response of the surveying profession, and many other professions, to this challenge has been to promote the concept of CPD as a code of practice to be followed by individual professionals on a mandatory or voluntary basis. Maintaining and developing professional competence is, of course, the responsibility of the individual practitioner. A personal strategy should be adopted and followed systematically. This, of course, relies on a variety of training options being offered by different course providers, including universities (FIG 1996).

An individual practitioner should be able to rely on a comprehensive CPD concept that is acknowledged by the profession and economically supported by the industry (public as well as private). Furthermore, the practitioner should have a variety of training and development options available to suit his or her personal plan of action. The options should be developed by the universities, offering, for example, one-year master's courses as part-time studies based on distance learning, and also by private course providers offering short courses for updating and just-in-time training. These options should be developed in cooperation among universities, the industry, and professional associations.

Furthermore, the individual practitioner should be able to rely on getting his or her professional competence recognized in a regional and global context. There is an interest in developing and extending such a global principle of Mutual Recognition of Professional Qualifications, though implementation is a regional policy issue. Mutual recognition allows each country to retain its own brand of professional education and training, because it is based not on the process of achieving professional qualifications, but on the nature and quality of the outcome of the process. In turn, this should lead to enhancement of the global professional competence of the surveying profession. The national professional associations, as well as the universities, should play a key role in facilitating this process (FIG 2002).

In short, enhancement of professional competence relies on an efficient interaction of education, research, and professional practice. This interaction is also the key driver to facilitate capacity building in LAS at all three levels: societal, institutional, and individual.

Chapter 12
The land administration toolbox

12.1 Using land administration tools

12.2 General tools

12.3 Professional tools

12.4 Emerging tools

12

12.1 Using land administration tools

THE TOOLBOX APPROACH

Given the variety of land administration systems, the toolbox approach offers a universally useful method of building or improving LAS. The toolbox approach allows a specific country or jurisdiction to select the most appropriate tools to meet its immediate and future needs. A systematic choice of tools is recommended. However, the range of tools changes constantly, according to emerging needs and new insights and technologies. International experience and best practices and land administration theory itself also keep evolving. A national land administration toolbox is always "a work in progress." It is always an unfinished, and even a never-to-be-finished, exercise.

While a definitive description of the full range of land administration tools is not possible, an overview can be provided to introduce the toolbox methodology. Each country has an existing land management system, whether it officially recognizes it or not. The point of the land management paradigm is to show how the processes used to deliver tenure, value, use, and development functions are part of a country's governance, whether or not these are undertaken systematically or as ad hoc responses to circumstances and irrespective of whether they are done through government, private businesses, or an amalgam of both. As countries seek to improve their land management capacity, the land administration system comes into focus as a more formalized entity. How to build and improve a country's LAS is therefore a recurring question. There is no single answer or recipe. There is, however, a reliable strategy that identifies what can be done and how to do it. This strategy requires that a nation investigate the use of existing tools within its system as well as other available tools that best suit its circumstances. In simple terms, this is the toolbox approach. Choosing the right strategies and tools is arguably the most important aspect of building LAS.

Tool selection is similar to selecting the right tools to repair a motor vehicle. Picking the right tools is also analogous to "cherry picking," or selecting the appropriate tools from a wide variety of options. The toolbox approach is universally applicable because each country needs to start from an analysis of its existing capacity when designing and improving LAS. Countries in similar stages of development and experiencing similar kinds of problems can achieve synergies by borrowing from each other and sharing their mutual experiences in selecting and experimenting with various tools. Within these open-ended opportunities, a set of best practices in LAS design becomes evident. The land administration tools and the options to implement them identified here are only a sampling of what's available. The selection indicates the range of ways to build, improve, or reform LAS. The range of land administration tools evolves and changes over time, reflecting the evolution of people-to-land relationships.

Historically, some of the tools have developed over many centuries and changed dramatically when printing became available, and again when computers were introduced. Most of the expertise in building and using these tools lies with the forty or so developed countries, which have gone through all stages of development and have flourishing land markets that took centuries to develop and refine. Contrarily, the greatest need for expertise lies in developing countries. Transfer of knowledge and tools, however, is difficult. In undeveloped countries, existing tools might be adequate for small-scale operations, but unable to deliver national-scale land management or tenure security. Paradoxically, the standard Western tools that work on a national scale tend to be too advanced and require substantial adaptation if they are to fit different circumstances.

When governments searched for national answers to LAS over the past twenty-five years, they used designs modeled according to the ideas of outside experts that were often difficult to integrate into the existing national skills base. Moreover, many governments made crucial land policy and administration decisions in the context of emergencies or crises, which discouraged exploration of various options or reflective consideration of the range of suitable tools. The processes, rules, and tenures adopted by developed countries service market-based economies. By contrast, developing countries found themselves making decisions and implementing LAS at a much earlier developmental stage. They had to grapple simultaneously with extreme poverty, homelessness, political disenfranchisement, and expanding slums. In these situations, the methods of providing basic housing for the poor were often informal and illegal. So, too, were the small-scale agrarian production methods that provided basic foods. These methods were so distant from formal methods delivered by highly refined Western tools that they need considerable adaptation to fit that model.

Analysis of international experience indicates that land management infrastructure works best when built from the ground up, with emphasis on opportunities for the enfranchisement of local peoples, validation of local pathways to local solutions, and extending sound local practices into formal national systems, especially for tenure, policy, and institutional arrangements.

Localization allows national history and experience to shape the choice of tools. Because of the need to start with what exists at ground level, there is no single approach to improving land administration by combining local pathways with modern land management tools. However, many strategies are available to land managers that incorporate and stabilize local practices and simultaneously permit economic growth, improvement of service delivery, and, where desired, modernism and working markets. While keeping it local is the starting point, the end goal is workable LAS that operate on a national scale and incorporate strategies suitable to the country context. Caution counsels against LAS solutions that are parochial, exclusionary, and excessively nationalistic. The central idea is to frame a system that can take advantage of the modern tools. In terms of the Millennium Development Goals (MDGs), systems relating to land should also reflect basic human rights, particularly in relation to access to land by women, tenure security for the poor, and provision of vital infrastructure in water and sanitation.

The tools described in table 12.1 represent the most recent developments in land administration theory. For example, they reflect the binary nature of land rights, restrictions, and responsibilities (RRRs). Rights simultaneously give opportunities to owners and restrict nonowners, who must respect the owner's realm of power and decision making in relation to land. In a similar vein, the general pattern of restrictions over land also involves dualities. A typical planning restriction limits opportunities of a landowner to use land for residential buildings only. The benefit of this restriction goes to the public at large in the orderly regulation of land uses and effective provision

TABLE 12.1 – THE LAND ADMINISTRATION TOOLBOX	
General tools	**1.** Land policy tools (chapters 1, 2, 3, 4, 5)
	2. Governance and legal framework tools (chapters 1, 2, 3, 13)
	3. Land market tools (chapter 6)
	4. Marine administration tools (chapter 8)
	5. Land-use, land development, and valuation tools (chapters 6, 7)
	6. ICT, SDI, and land information tools (chapter 9)
	7. Capacity and institution-building tools (chapters 11, 13)
	8. Project management monitoring and evaluation tools (chapters 10, 13)
	9. Business models, risk management, and funding tools
Professional tools	**1.** Tenure tools
	2. Registration system tools
	3. Titling and adjudication tools
	4. Land unit tools
	5. Boundary tools
	6. Cadastral surveying and mapping tools
	7. Building title tools
Emerging tools	**1.** Pro-poor land management tools
	2. Noncadastral approaches and tools
	3. Gender equity tools
	4. Human-rights tools

of services and roads. Well-organized restrictions are often implemented by a specific agency. Such agencies carry the administrative functions of organizing how the restrictions work and managing breaches. This world of restrictions has grown more complex since the 1980s as countries deal with chemical hazards, business compliance, safety standards, electricity wiring, large-scale plumbing and sewerage, building controls, and many other aspects of modern city and rural life. Future developments include energy-rating tax differentiation for houses and offices and other information involving even more intensive regulation of people and their activities on land.

The new technologies available in LAS allow management of these binary concepts. Historically, most LAS concentrated only on recording rights and interests. Owners' opportunities are identified by an owner's name and description of the interest. Increasingly, this narrow focus is expanding as modern and evolving systems place much more emphasis on the dual character of both rights and restrictions and move into recording the limitations affecting owners and property of all kinds.

Choosing the right land administration tools is not easy. Each tool must be compatible with all the others and with the needs and development capacity of the country. The tools must be chosen to help deliver the objectives of LAS, such as poverty alleviation, economic development, supporting social stability in postconflict situations, and sustainable development.

In developed economies, the activities surrounding the tools are highly professionalized. Production of paper trails and land development arrangements are commoditized by lawyers, notaries, and conveyancers. Formal identification of land is commoditized by surveyors. Mapping systems are produced by GIS experts and others. Bureaucrats and other professionals govern recording, registration, taxation, and the use of land to generate government funds. Valuers and planners are specialized professions. For developing countries, the trick is to deliver the certainty of tenure and services through appropriate professional standards without creating monopolies or high-cost services and without installing inaccessible or remote land management tools. The solutions must be available to the occupiers and owners to assist their land use, management, transfers by sale and on death, mortgaging, and other essential activities. The solutions must be simultaneously attractive to developers, who will provide the essential infrastructure for new parcels in the form of drains, roads, water, and utilities that add value and amenity. This is not "deprofessionalization." Rather, developing countries require coherent strategies to build up their skill base by starting and maintaining LAS processes at an achievable level, then design a betterment path that encourages opportunities to incorporate the more developed options to implement each of the tools.

In reality, the choices of appropriate tools are limited, or even restricted. The momentum of existing or legacy systems, rent seeking and corruption, the difficulty of implementing legislative reform, jealousies, silo mentalities, a fascination with the latest technologies (many being inappropriate), and a desire to replicate an inappropriate system found in a more developed country all work to constrain the choices. The solutions adopted tend to involve compromise.

Adoption of inappropriate tools has serious consequences. At best, poor and fragile LAS will result. At worst, the end result will compromise the system that existed at the start. Unfortunately, since 1975, cadastral, land titling, and land administration projects have produced more failures than the rare successes. The past decade has seen more successful implementations as the international community has come to better understand the complex issues involved in building LAS and their essential relationship with good governance in general.

This chapter pragmatically identifies land administration tools that most countries actually need for professionalized land administration purposes, but even this list will vary from time to time and place to place. Within the toolbox, three categories of tools are listed.

First are the general LAS tools that all countries require as basic infrastructure for LAS. General tools are vital for all government systems, not merely those that relate to land. Among these, risk management principles, business models and funding, and capacity building are universal governance tools. These must be included in LAS design. Professional land administration tools can only be fully understood in the context of the more general tools that form essential government infrastructure for delivery of services, including those related to land. These general tools are often forgotten by LAS project designers. They are summarized in the following sections but are discussed in detail in preceding chapters. The design of LAS also needs to include related tools to deliver land-use planning, development, and taxation systems, which are beyond the focus of this book. These related tools need to be integrated especially with both general and professional tools such as land policy, land market, strata and condominium, and tenure tools. The related tools are categorized as general tools in this structure and are described in dedicated texts relating to land-use planning, development, and taxation.

Next are the professional LAS tools that tend to be the domain of professionals in the land administration area. The order of the professional tools listed in the table is roughly according to when they need to be considered in the LAS design process, but logically, any system requires the entire contents of its toolbox (including the general tools) to be incorporated into a comprehensive design, irrespective of when any particular tool is implemented.

New and emerging LAS tools are then listed. These are being developed to meet the objectives of sustainable development, urbanization, and other critical needs. Many are still under construction rather than at the off-the-shelf stage of development. They illustrate the need to design LAS with sufficient flexibility to take up new directions and respond to new challenges, crises, and demands. The general, professional, and emerging tools are described in more detail in the following sections.

12.2 General tools

1. LAND POLICY TOOLS

There is great variation in what constitutes a land policy for a country. Some countries like Malawi (figure 12.1), Kenya, and Indonesia have a written land policy that directs policy and legislative formulation. UN–HABITAT, for example, has produced a booklet, "How to develop a pro-poor land policy: Process, guide, and lessons" (2007). Other countries have the components of land policies compiled in a statutory land code. Sometimes, references to land and property are included

Figure 12.1 Professional land administration tools are used in countries like Malawi.

in a constitution. In the Western world especially, few countries have a stated land policy. They may have an environmental policy, but not a land policy. However, in all countries, there are many statutes and regulations that refer to land administration, land management, and the regulatory controls on people-to-land relationships. Collectively, these could be considered a "land policy."

For many countries that are federations of states such as Germany, India, Canada, the United States, and Australia, the states or provinces control land matters and so would be the custodians of a land policy. However, there are often overarching land-related policies concerned with the environment, taxes, security, or health, for example, that override state land policies. In many cases, the national governments of these federated countries, for more than a century, ignored or kept at arm's length the large-scale land administration data held in their states and provinces, often because of perceived or real constitutional reasons. Today, however, in the modern world of ICT and virtual jurisdictions where spatial enablement is a key to many, if not most, government functions, they are rapidly coming to embrace this large-scale "people relevant" data as part of their national SDI (NSDI).

The overarching policy of sustainable development will need to be implemented in ways that suit national characteristics. Understanding these characteristics, especially those listed as follows, will assist the formulation of policy:

- The role that land plays in supporting sustainable development as part of the land management paradigm

- The major policy drivers affecting land policy: poverty eradication, income distribution, social justice, equitable access to land, and environmental management

- The extent of a land policy: Is it for the whole extent of the country (figure 12.2) or just for the parts of a country that are controlled by a specific government department? Are forest areas excluded from the policy? How will resources and land policies be integrated?

- The government institutions responsible for implementing land policy

- Methods of land distribution whether by market forces, by a centralist distribution system, or by informal social systems

- Announcements of fundamental rights to landownership and protection of property

- Who can own or buy land and controls on land speculation, land banking, and mass accumulations, if any

- The balance between state and individual or group rights to suit the national history and culture

- The need to build an efficient land administration infrastructure to support security of tenure and effective land markets

- The role of NSDI in supporting land policy

- Whether land administration infrastructure is centralized or decentralized

- Policies on land reform (which is a political process concerned with redistributing rights in land), noting that this is very different from land administration reform which is a process of improving LAS

- Land tax policies

- Fair compensation as a result of government resuming ownership of or acquiring land from private citizens

- The role that land administration data plays in spatially enabling a country

Figure 12.2 Highlighting its importance, the March 2007 special edition of the *Daily Nation* (Nairobi) focuses on Kenya national land policy.

Land policy creation and reform

Many of the ideas discussed in this book could be considered part of, or could contribute to, a land policy and then be considered as land policy tools. In particular, chapters 2, 3, 4, and 5 focus on land policy matters and tools. Following are key ideas, statements, or suggestions with regard to land policy formulation or the related land policy tools:

- ◆ A written and well-publicized land policy announces the nation's decision. However, most countries do not announce land policy in long complex documents and prefer to make short constitutional statements.

- ◆ International land policies are directed at informing international development agencies and creating evaluation processes for measuring their effectiveness. These international policies also provide leadership in formation of land policy at regional and national levels. Nations set their land policies in different ways, principally through the political processes used to formulate their constitution.

♦ A national constitution typically includes provisions about the roles of the state and individuals in relation to land. In centrally organized states (China, Vietnam, Laos), land belongs to the nation. For Western democracies, the role of the state is substantially reduced by constitutional announcements of fundamental rights to landownership and protection of property. The most common constitutional situation is a middle path that balances state and individual or group rights to suit the national history and culture.

♦ National governments directly or indirectly allocate land to specific users and uses. In land administration theory, each nation creates a unique range of land RRRs. Three major systems of land allocation or distribution systems are used worldwide: social allocation through relatively informal social orders, land market systems used in capitalist economies, and bureaucratic systems used by centralized economies. These systems run parallel with tenure or property types described earlier in the book.

♦ Some nations set land policy through specific exercises or processes. A national land policy forum with high-level government and professional participation is recommended for countries seeking a lasting and influential policy base (Dale and Baldwin 2000). The government of Saint Lucia in the Lesser Antilles, for instance, undertook a well-designed and extensive process of community consultation in establishing its land policy. Formal consultation processes, green and white papers calling for public comments on proposals, and public evaluation of contributions are typical ways for modern nations to set national or regional land policy. More commonly, countries rely on parliamentary processes to create legislative announcements of land policy, using the normal democratic processes to engage the public audience. South Africa, Timor-Leste, and other nations used public engagement processes to build their systems as part of national healing and identity building in post-traumatic political change. Public engagement processes help ensure that national land policies and administrative institutions reflect the ways people actually think about land.

♦ Whatever processes are used to establish formative and continuing land policy, coordination in announcement and application of the policy is essential for coherent land administration (UNECE 2004, 61). If ministries provide inconsistent advice to parliaments, and if ministerial functions clash or, worse, chase inconsistent national, regional, and local land policies, LAS will fail.

♦ Within a land policy, land administration is not land reform. If possible, land administration reform should be apolitical and concerned with putting in place an efficient land administration infrastructure to manage the people-to-land relationship. Land reform and land tenure reform are, by nature, political in character and typically involve redistributing land among various groups. The processes involved should be kept separate from the development of a land administration infrastructure. In general, the introduction or improvement of LAS should not change the land tenure relationships between people and land. In this sense, the land administration infrastructure provides an inventory of RRRs within a country. On the other hand, LAS are an essential part of projects to reform land tenure, and reform paths must potentially link into a national land administration system.

◆ In general, land policy should precede and determine legal reform, which in turn should result in institutional reform and, finally, implementation. The reality is that legal and institutional reforms are very difficult and require major political commitment. As a result, these functions and reforms should at least continue in parallel.

◆ Land policy decisions and land reform decisions should be kept separate from the management of LAS. An example is forestry and state lands, which should all be included or recorded in LAS, even though management and policy decisions with regard to these lands are usually the responsibility of separate agencies. On the other hand, the land administration infrastructure in a country will be critical to the implementation of any sustainable development or environmental management policies. The land administration infrastructure forms the foundation for implementing these policies. Thus, all national environmental and sustainable development policies should clearly articulate the role of land administration in implementing particular policies.

◆ Implementation of land policies requires a legal framework, which enforces the rule of law. The framework requires good laws, as well as legal institutions, professionals, and government officials who are versed in the law and a justice system that enforces the law. The legal framework is essential to ensure that landholders are secure in their occupation; they are not dispossessed without due process and compensation; and the land market can function with confidence and security. The legal framework is essential for security of tenure of landholders of all kinds, even those who rely on social tenures.

◆ A land policy should have a national focus. Land administration, cadastral systems, and land titling are national activities, not just rural ones. They are just as relevant to urban areas. Addressing urban poverty is as major an issue as rural poverty. Land administration reform is just as urgent in informal or squatter settlements in urban areas (and is often more urgent) than in rural areas. Cities are now recognized as the engine of economic development in developing countries. This is especially an issue from the perspective of social stability, environmental management, and sustainable development. At the same time, issues of addressing indigenous rights within a land administration infrastructure are just as critical as rural and social issues, although they require different strategies. More importantly, it is virtually impossible to undertake substantial land administration reform without considering all land—urban, rural, state, forest, marine areas, and indigenous land. A national approach is essential for land administration reform.

2. GOVERNANCE AND LEGAL FRAMEWORK TOOLS

The importance of good land management for overall civil peace is forgotten by many people privileged with living in countries that provide functional processes. For the majority of the world's peoples, land management capacity is threatened by lack of governance and poor legal infrastructure. The relationship between democracy and market systems is also assumed. There is a range of governance and legal framework tools discussed in chapters 2, 3, and 13. Suffice it to say, without good governance and appropriate legal infrastructure, efficient and effective land administration is not possible.

3. LAND MARKET TOOLS

The capacity of a land market to accelerate wealth creation in a country ensures that market-driven land policy and institutional design predominate in land administration theory and practice. The pertinent land market tools are discussed in chapter 6. Complex property markets can only exist when the administration systems are exceptionally robust and reliable. Formal and secondary markets require administrative infrastructure. The attraction the land market holds for land administration lies in the market's open-ended capacity to generate the funds needed to build and manage the country's infrastructure and institutions.

4. MARINE ADMINISTRATION TOOLS

The marine environment, particularly the coastal zone, requires special tools to administer the complex RRRs in these areas. Chapter 8 introduces the issues and tools to assist administration of the marine environment.

5. LAND USE, LAND DEVELOPMENT, AND LAND VALUATION TOOLS

A central theme of this book is the key role that land administration plays in the land management paradigm. Within the paradigm, land administration has four major components or dimensions of land tenure, land value, land use, and land development. However, the issues of tenure and related cadastral activities, such as land registration, are at the heart of building LAS and are the focus of this book. This is not to diminish the importance of land value, land use, and land development within the paradigm, but to acknowledge that these are three separate disciplines with unique and specialized tools. This book introduces the three areas and their interaction with the tenure function within land administration in chapters 6 and 7. A detailed discussion of their associated tools is available in the many professional texts on these disciplines. However, land administration theory acknowledges that effective LAS require all four areas to be effective and that processes in the latter three areas must utilize tools that are appropriate for different circumstances.

6. ICT, SDI, AND LAND INFORMATION TOOLS

Paper-based systems (figure 12.3) raise basic information issues about accuracy and reliability, as well as privacy. They can be used to compile general statistics about the number of parcels, titles, and transactions but remain very limited in their capacity to deliver information for coherent land management. The onset of computers placed an essential new demand on governments—to take a holistic approach to land information. Chapter 9 presents an overview of the impact, benefits, and opportunities for building LAS through ICT, SDI, and land information tools.

7. CAPACITY AND INSTITUTION-BUILDING TOOLS

A country's institutional, individual, and organizational capacity to manage land processes (figure 12.4) is the most expensive part of any land administration system. The normal way to nurture capacity is to build basic LAS infrastructure that recovers costs and generates research and training funds. At the initial stages of LAS development, the costs of training are beyond the capacity of the infrastructure and require heavy subsidy from general taxes and project funding. The development of well-trained technical and professional staff to conduct land-related processes is also a high priority.

Socially engaging the beneficiaries of any system in its design and management is the most effective and often the cheapest way to deliver capacity. Since engagement is also directly related to instilling confidence and participation in a formal system, strengthening capacity at the social,

Figure 12.3 Globally, most LAS are still paper based, even though they may be well organized.

institutional, and individual level is an additional positive outcome. Chapter 11 provides an overview of the required capacity and institution-building tools required to develop effective LAS. These tools include

- Concept of capacity-building
- Capacity assessment and development
- Guidelines for self-assessment of capacity
- Institutional capacity
- Education and research

8. PROJECT MANAGEMENT MONITORING AND EVALUATION TOOLS

A wealth of published material on the social, economic, and environmental consequences of land management and land administration is available worldwide such as described in chapter 10, "Worldwide land administration activities." However, there is little written on the principles and tools required to design, build, and manage LAS projects. Chapter 13, "Project management and evaluation," describes and discusses a range of project management, monitoring, and evaluation tools to support building LAS. These include

- SWOT analyses
- Fishbone and GANTT charts
- Logical framework, or LogFrame analysis
- Reengineering framework
- LAS design criteria
- Documenting key processes and practices
- Pilot projects
- Social and economic analysis (baseline and longitudinal studies)
- Community and stakeholder engagement (participatory development)
- Critical success factors
- "Project Cycle"
- Financial management
- Quality assurance
- Benchmarking
- Evaluation framework

Figure 12.4 LAS in less developed countries such as Vietnam often do not have the capacity to organize records or capitalize on digital opportunities.

9. BUSINESS MODELS, RISK MANAGEMENT, AND FUNDING TOOLS

Building LAS is big business for any government, irrespective of its geographic coverage, number of parcels, or whether it is a national or merely a local government initiative. Whether construction is undertaken incrementally or on a "once and for all" project, LAS will be a substantial part of government infrastructure and will demand the same levels of professionalism and risk mitigation that are applied to delivering large-scale, physical infrastructure projects. General principles of risk mitigation should therefore form an essential backdrop for project design (Matsukawa and Habeck 2007). Within these general principles, LAS involve special characteristics and opportunities.

A discussion of these issues and related business models, risk management, and funding tools cuts across many sections of this book and is not explored in depth. This section attempts to briefly discuss these issues and identify some of the associated tools.

Most of the processes in LAS, once established and institutionalized, are capable of generating continuing income streams. However, some processes, particularly initial registration of existing parcels and dispute management, are cost heavy and unable to be funded by those who immediately benefit, except where high-value land is involved. These processes need to be subsidized by other land-related processes and taxes. Each nation will have its particular capacity to attract donor funding, private-sector commercialization, and payments from users. The business models therefore need to be developed in the context of government budgets in the immediate and long term, remembering that imposing unrealistic initial price hurdles for essential processes will doom any project to failure.

Most LAS are so complicated that they depend on support from both the government and private sector. There are therefore opportunities to spread risks among taxpayers and professionals. The typical method of spreading risk among nongovernment businesses involves insurance. Insurance spreads risk by using the collective premiums (and other funds) to generate a pool out of which a stream of compensation (after profits and administration costs) is provided to those who incur the insured risks. Ideally, insurance is designed to pick up system failures, which should be controlled by a series of standard requirements:

- ◆ Licensing (with strong prudential standards and effective complaint and investigative capacity)

- ◆ Probity checks and financial guarantees

◆ Imposition of liability on people or businesses that create risks as a means of encouraging those undertaking the activity to reduce or eliminate risks

Options or tools for insurance suitable for LAPs include

◆ Models of state-operated, professional, service-based insurance (Denmark), which depend on high-integrity professional capacity

◆ Registration insurance programs (sometimes called a government guarantee), which provide a government guarantee of title. Particularly associated with Torrens registration systems, they typically delimit the situations covered, place upper limits on the liability of the guarantee fund by a cutoff point, require extensive administrative proofs of loss, and position the government fund as an insurer of last resort, available only after all other options are exhausted

◆ Private professional indemnity insurance for transaction and registration service providers as provided by notaries and lawyers in many countries

◆ Private-sector title insurance (which is used extensively in the United States) depends on financial costs and duplicate title verification and investigation systems

◆ Self-insurance or no insurance (historically used in nascent and developing systems)

Title insurance needs special precautions. Guaranteed land registration or Torrens systems use publicly funded pools (drawn out of taxpayers' funds or contributions of users) to cover errors, fraud, and other limited risks. These registration systems need to provide the guarantee of title, because they reverse the normal rules relating to forgeries and documents made beyond the formal power of statutory and other agencies. Usually, forged and *ultra vires* documents are void; that is, they have no effect. The forgery of an owner's signature on a purported transfer of land does not disturb ownership, and the innocent owner's position remains inviolate. However, guaranteed land registration systems change this result by allowing buyers—indeed, all persons intending to take an interest in a registered parcel—to rely on the register, provided they are not involved in forgery, and take the land in good faith.

Torrens systems allow the public to essentially take an instrument conferring an interest in land, without extensive investigation of the capacity or the identity of its originator; to accept even cursory identification and authority to deal (the State of Victoria in Australia, for example, requires only a signature before an adult witness); and to register that instrument. The public essentially relies on the registration program to confer the interest in land to them, even if the document they register is forged. This leaves the owner who suffered the forgery at risk of losing his interest in his home or farm—a risk that must be covered by the guarantee fund to ensure that public confidence in the system is maintained. These guaranteed systems are only sustainable when the criminal, corporate, and other laws are effective, reducing the number and significance of forgeries to manageable proportions. Alternatively, if a population is largely illiterate, use of title guarantees to manage forgeries needs to be supported by identification systems such as photographs of owners, thumbprints (not signatures), and combinations of proof to control incidences of forgery.

Decisions about what insurance systems to use in LAS are crucial. The key capacity is the ability to conduct a thorough, ongoing, on-site risk assessment and an assessment of local needs and capacities before designing suitable options (costs and benefits).

The funding tools for a LAP must be constructed so that the outlay in human and financial resources can eventually cover the geographic area and the realm of processes involved. If resources are too slight in the initial and middle stages, LAS will never succeed. For example, failure will occur if the per annum intake of parcels into the system is not equivalent to the number of new parcels created (Payne, Durand-Lasserve, and Rakodi 2007). Imposition of transaction fees high enough to reduce incentives to register, use of land taxes to fund non-land-related government expenditures, and shifting income benefits to one part of the constituency at the expense of another are all pitfalls that need to be anticipated and avoided. The business model should scrupulously reduce opportunities for graft and corruption, whether among government employees, contractors, or associated professions.

The lessons now apparent from the recent studies of the relationship between good governance and land administration indicate that the business model and funding arrangements for any discrete part of a land administration system must be applied to the related parts, so that LAS are treated holistically. Treatment of each part (a land registration program, a land tax system, or provision of sewers and drains) must be related to consideration of how its operations will affect related processes and parts. The most crucial issue is the creation of land information as a generic asset for government as a whole by effective management of all administration processes and their respective costs.

If a right is registered, the systems generally shift the rights in land to compensation opportunities, within highly prescribed and formal procedures. The existence of this insurance or assurance opportunity provides some degree of public confidence that their land is forever protected.

12.3 Professional tools

1. TENURE TOOLS

Of all the land administration tools, tenures are the most complicated, because they institutionalize the variety of ways people approach land, in informal and formal systems. Because the variety of tenures is so large, land administrators use general tenure categories as described in table 12.2.

Given the dimensions and varieties of tenures, classification for its own sake is a worthless endeavor. Instead, the approach to tenures here is functional and based on observing the

TABLE 12.2 – TENURE TOOLS	
TYPE	**DESCRIPTION**
Formal/informal tenures	**Formal tenures** are legally recognized and supported by organized LAS. **Informal tenures** are recognized by other normative systems in all countries. They can be formed by social norms (in cities, kampongs, favelas, forests, and even jails) or traditional and customary norms that exist in most countries, including in the South Pacific, Africa, South America, and developed countries. Informal tenures can mirror formal tenures but lack organization through legal records.
Customary, traditional, indigenous, and native tenures	"It is not easy to find a satisfactory formula that will adequately define 'customary' land tenure" (Simpson 1976, 223). It generally covers rights to use or to dispose use rights over land, which rest neither on exercise of brute force nor on evidence of rights guaranteed by government statute. These rights are recognized as legitimate by their community. The rules governing acquisition or transmission of these rights are usually explicit or generally known, though not normally recorded in writing. The social and spiritual relationships with land are just as important as the material ones. These tenures constantly evolve.
State ownership	All governments own land. This can be called state, crown, public, or national land. These parcels can include large areas in national parks, reserves, and a wide range of public facilities. Usually, roads are also owned by the state.
Private ownership	Private ownership systems underpin land markets and presuppose a theory of property. These rights have no time limit and last forever, in contrast to leaseholds. Private ownership can rely on • **allodial rights**, where individuals have the right to own land (in parts of the United States and European systems relying on the Roman/Dutch idea of absolute ownership) or • **freehold rights**, which are held by the crown or state and derive from feudal tenure systems. In modern economies, for all practical purposes, both can be called private ownership.

Continued on next page

Continued from previous page

TABLE 12.2 – TENURE TOOLS

TYPE	DESCRIPTION
Trust ownership	Land is owned by a person or entity on behalf of another, particularly in English-derived systems.
Common property or group tenure	Common lands or facilities are held by a group, sometimes under traditional use rights or a legal framework. Usually, the shares are not tradable or are traded only on condition, such as with the common facilities held in a condominium.
Leasehold (including rental arrangements)	An owner (including the state) can allow a person or entity to have possession of land, an apartment, or even a room, for a specific time (fixed term) or a time that can be fixed (for life). The periods are variable, depending on the needs of the parties. The owner retains a reversion that entitles him or her to possession at the end of the term and, meanwhile, to rents or other services from the lessee.
License	A license is similar to a lease. It can be proprietary in nature or merely contractual. It typically covers a specific activity, such as putting up a sign or grazing stock.
Occupation right	Squatters and others who possess land can be given some formal recognition of occupation, such as by antieviction laws.
Illegal squatting	Possession or occupation of land, without any legal entitlement. Within illegal settlements, groups will sometimes be highly organized by informal arrangements or tenures. In many legal systems, a squatter's occupation over a long period will turn into ownership through the doctrine of adverse possession.
Possessory tenures	Legal systems often recognize opportunities to acquire land through adverse possession, provided it is open and without violence. It is, by its nature, without permission of the owner.

processes used by a society to stabilize its access to land and resources. The classifications are therefore designed to assist discussion and frame useful generalizations. They are not intended to be political or to raise one tenure over another as a world's best practice. Indeed, most countries use all kinds of tenures simultaneously. Among the tenure types, the global trend is to move toward privatized individual tenures because of the popularity of land markets as the principal means of land distribution and management.

All tenure types can be more or less organized (figure 12.5). Therefore, levels of formalization of the tenures in national LAS crosscut the typologies. A country's capacity to integrate resource tenures, especially in mining and forestry, with land tenures is another, often ignored, crosscutting issue.

Figure 12.5 A tenure supporting a nomadic herding society in Mongolia shows the effects of overgrazing and desertification due to population pressures.

Tenures in land administration

Analyses of the components of tenure systems from the viewpoint of land administrators tend to concentrate on the commodification of land. The first foray at a land administration approach saw tenures as organizing rights (the right) in a physical parcel (the object) held by an owner (the subject) and enforced by the state (Kaufman and Steudler 1998). The "bundle of sticks" analysis, in which each opportunity of an owner was a component of an overall idea of ownership, was a common illustration. It led to ownership being broken down into component parts such as the right to sell, devise, exclude others, build, and so forth. However, this analysis tells only half the tenure story. Theories about property in general, especially those based on jurisprudential analysis of legal orders, extended the model of private rights, so that each right was seen as a relationship between the owner and other parties (including the state) in relation to a parcel of land. The crucial addition in this analysis is the part played by other parties in supporting the rights of owners and undertaking duties that demonstrate respect for this ownership. For successful land administration then, the infrastructure in any system should reveal the entire picture of the binary nature of rights (figure 12.6) and include the correlative duties that give effect to the rights—likewise with restrictions and responsibilities affecting land.

New tenure theory therefore accounts for the lost analytical half—the part that deals with articulation of the relationship between other parties (including the state) and the owner in relation to the land or parcel, not merely the relationship between the person and land. The

Figure 12.6 Informal and formal tenures are intermixed in the Philippines.

simplest and most influential definition of tenure, therefore, is that proposed by UN–HABITAT in its pro-poor land management texts: i.e., "the fact that other people believe the land you occupy and use is the land that you are allowed to live on and use" (UN–HABITAT 2004, 13).

This is essential for understanding how LAS might respond to the way people use land and the tensions that undermine their security of tenure. It focuses on the primary function of LAS—the building of respect among other parties for landownership irrespective of the kind of tenure. This contrasts sharply with the old way of seeing land administration functions as recording information about owners and their parcels. It also reflects the binary nature of RRRs. New tenure theory must also account for the increasing number and significance of responsibilities and restrictions generated through private-sector activities, the most important being the arrangements made by owners' corporations in relation to multioccupancy parcels and buildings and through private provision of essential infrastructure services.

Mature tenure systems need tools to deliver the functions identified in figure 12.7.

Mature tenure systems deliver the following six functions:

◆ **Articulation of rights:** Within any legal system lies a systematic and exhaustive account of the interests in land that a country makes available. The fundamental

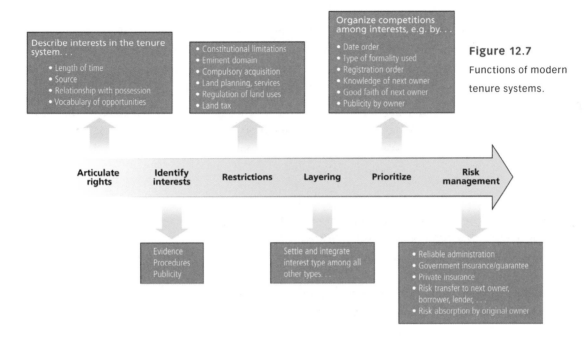

Figure 12.7
Functions of modern tenure systems.

questions of who can do what, when, how, and where are all addressed. The larger and more important interests (ownership) are expounded in detail, but mature systems also create lesser interests to cover local access, drainage, support, and building maintenance.

♦ **Identification of interests:** The formalization of interests allows strangers to understand the configuration of arrangements relating to land and provides objective evidence of these arrangements.

♦ **Restrictions on and responsibilities for land:** These restrictions provide a framework for overarching government and public intrusion into land uses and entitlements.

♦ **Layering of tenures:** Layering allows simultaneous arrangements to subsist in relation to land, resources, and water. These arrangements provide multiple opportunities to commercialize and utilize assets. Layering demands well-organized LAS. English-derived systems of common law recognize multiple opportunities for layering interests in land by using three simultaneously effective titles in one parcel: legal titles, equitable titles, and possessory titles. European systems based on Roman/Dutch law recognize only one owner at a time.

◆ **Prioritization of interests:** A system of priorities among interests within one tenure (for example, land tenures) and among interests held in diverse but related tenures (for example, land tenures and resource tenures) must exist. Where registration programs apply, most of the issues with priorities are settled automatically according to the order established in the system.

◆ **Management of risks:** Tenure systems allow for risk shifting and risk management. While most economic analysis is on the exploitation opportunities of the owner, the land administration perspective of tenure management is far more concerned with risk elimination, allocation, and management.

If all these systems work well, disputes about access to land and resources are minimized. Most existing LAS are undergoing continuous improvement in their capacity to produce these six tenure functions. Historically, most countries manage rights and restrictions through dichotomous and separate systems. In modern LAS design, this dichotomy will diminish, and the differences between public and private sources of land information will shrink in significance. Segregation of management of tenures in public and private land is no longer relevant, since sustainable development shows that all land, no matter what the tenure type, suffers the same problems of encroachment, salinity, desertification, water loss, contamination, weed infestation, and many other problems. Just as water management systems have replaced administrative boundaries with water catchment boundaries, all land tenures need to be reevaluated to comprehensively account for the economic, social, and environmental obligations of owners and create coherent restrictions and responsibilities (Raff 2003).

Given the land management paradigm, land management is more than a parcel-by-parcel exercise, because the various capacities of owners of interests in land and resources (especially those that raise tensions between landowners and miners, and landowners and foresters) and the overarching restrictions and responsibilities in the public interest will undergo constant renegotiation. The land registration model in this new environment will service the land information needs of a modern economy, not internal registry information needs. To achieve this, the concept of land registration will change. Increasingly, it will incorporate tenures of all kinds, including resource and marine tenures.

Delivering security of tenure
The sources of security of tenure, in a coherent order, are antieviction processes, remedies that protect and return land, sound and regular administration, and smooth transaction and inheritance processes. The ultimate measurement of when security exists lies not in the processes themselves, but in the confidence they provide for their beneficiaries.

In many parts of the world, these sources of security remain unavailable. In this regard, UN–HABITAT has proposed building a global monitoring system on tenure security. In monitoring secure tenure, the perception of protection against forced evictions remains a significant touchstone or measure. Moreover, perceived security is probably just as important as legal security in some situations, though not in formal land markets. The perception of security led to land improvement and credit use of land in slums in Buenos Aires, Argentina (`http://www.onderzoekinformatie.nl/nl/oi/nod/onderzoek/OND1310262/`).

2. REGISTRATION SYSTEM TOOLS

Registers of private rights in land and resources
Globally, national approaches to registration generally involve generating many independent registers to manage specific land and resources. These include land registries, mining interest registries, road registries, and building registries, to name a few. These registries are run by governments, with rare exceptions in countries with Latin influences. Where condominiums are popular, the operational data of their corporate owners tends to be managed by building administrators. This latter category carries increasingly significant information related to the management of buildings, including developments of vertical villages of up to 700 or more separate units. The larger building title systems require more management than a small township, because the density is far greater. In the modern world of spatial information, all these registries need to be built and operated in the context of achieving seamless treatment of interests, restrictions, and responsibilities and delivering comprehensive spatial information.

Within this general approach, a broad distinction can also be drawn between land and resource registries. The latter generally manage both commoditization of interests in the resource and opportunities to work or extract the raw material. The policing of work activities is integrated into the management of the right to undertake the work. This approach is not possible with land registries where private ownership is entrenched and constitutionally protected. Management of land-based activities additionally creates positive opportunities through licensing and other forms of business regulation. These land-based opportunities stand outside traditional analysis of LAS, though new spatial technologies are uninhibited by these historical classification barriers and are capable of managing all kinds of information and processes. In the technical environment, information about a permit to build, to operate a mine, or to run a hospital, hotel, or retail outlet is no different from restrictions on land in general or parcel information in particular. In the future, LAS will increasingly extend the range of information created by these processes for the benefit of both the public and government.

Special features of land registries

The distinction between deeds registries (in which deeds are made between an owner and prospective owner to transfer title, then registered in a government or private register) and land registries (which transfer title when an application is made by one owner for registration of the new owner, typically through a "land transfer" document) is now archaic. In practical terms, the best-run systems using either approach deliver the same kind of results of security of tenure; comprehensive capture of all changes in ownership; public transparency and accountability; and sensible solutions to the merging of text and cadastral data. Also, it is equally possible to have poor, inefficient, and ineffective title systems or deeds registration systems. Despite deed and title systems generally achieving these accepted best practices in land registration, it is still essential to examine the history of any local system in order to reengineer its components. Land registries are embedded in people's attitudes toward land, and changes in their processes must be made with care.

Every system differs. Table 12.3 contains generalizations about how the systems generally work. No one system actually does all the things described, but they all borrow from the characteristics. More importantly, while this table emphasizes the differences, the best systems achieve comprehensive, reliable, guaranteed, and inexpensive land administration for a variety of purposes, especially for contributing to sustainable development.

Many countries started with a deeds-based system and moved into title registration to achieve greater reliability and simplicity. However, title registration is not for novices. It is expensive to set up and hard to run. Some countries start with qualifications on the record, so that ownership is not guaranteed on some titles, and the survey plan is not guaranteed on others. Malaysia qualifies many titles as to survey as part of its successful program. Title registration demands that every transaction and social change in land ownership be recorded — otherwise, the system breaks down, with disastrous results. On the other hand, the deeds system is more reliable for the early stages of development and allows the public to have deeds produced to evidence transactions with some degree of commercial confidence, provided they have the funds to pay for professional services. Irrespective of the efficacy of the deeds registration program, external evidence of ownership exists.

One of the major issues with deeds is the role of notaries and lawyers. The title system offers some respite from title search endeavors. In deeds systems, these are onerous and involve searching the deeds themselves and the records in the public registers or depositories of the deeds. This sometimes gets so complicated that another set of extracts is created, which forms an additional search industry.

TABLE 12.3 – DIFFERENCES AMONG REGISTRATION SYSTEMS

ASPECT	DEEDS REGISTRATION	TITLE REGISTRATION
Legal origin	The deeds system is associated with Roman/Dutch law in Europe and early common-law conveyancing in England and its colonies.	The old Hanseatic city-state system developed in Germany and spread into the New World, where systems broke away from the cumbersome deeds conveyancing to simple land registration systems.
Cultural origin	Now used in Latin-culture countries in Europe (France, Spain, Italy, Benelux), South America, parts of Asia and Africa, and extensively in the United States.	Common-law countries and new emerging nations. German style found in Germany, Austria, Switzerland, and Nordic countries.
Legal consequence	Title is a concept that is transferred when the deed is executed. The title therefore runs through the "chain of deeds," and each deed must be perfect enough to transfer the title up the chain.	Title is transferred only if and when the document is registered. The title record determines the title. Any unregistered documents can create rights among the parties, but they do not affect the land.
Concept of title	Title exists in law and is transferred through deeds. The deeds remain essential evidence of landownership.	Title exists in the register. The official record is compelling evidence of ownership.
Searching	All the deeds making up the title need to be searched, as do their registered copies.	Only the last entry of the owner on the official record needs to be searched, not the documents.
Positive or negative	A pure deeds system is negative, though highly developed systems offer a fail-safe system and registration creates the positive impact of transferring title. The deeds themselves, not their registration, are primary evidence of title. Registration gives a higher degree of evidentiary protection, against unregistered deeds.	The title system is positive, meaning that the titles are the proof of ownership. No other evidence of title is needed.
Parcel identification	Identification is achieved in many systems by a text description in the deed, often referred to as metes and bounds, or sometimes with a map sketch. The boundaries and area are not guaranteed.	Initial registration involves establishing surveyed or (fixed) boundaries, or general boundaries, for the parcel. The parcel is geometrically identified with its related parcels, usually by the reference to, and incorporation in, the cadastral map. The boundaries and area are not guaranteed.
Role of the cadastre	The cadastre identifies the land for taxation purposes and is not necessarily based on cadastral surveys.	The cadastre identifies the land for title purposes. Boundaries are reliable and can be reestablished.

Continued on next page

Continued from previous page

TABLE 12.3 – DIFFERENCES AMONG REGISTRATION SYSTEMS

ASPECT	DEEDS REGISTRATION	TITLE REGISTRATION
Administration system	Generally, the deeds are copied, and copies are held in "land books."	The record is held on a single page or in a digital file referring to the parcel.
Actors	Lawyers or notaries are usually essential. Deeds registrars check and manage filing and recording of the deeds in the books.	Often, lawyers and surveyors are required. In the best systems, individuals can do their own conveyancing. The land registrars check and record the information in the documents as well as social transitions affecting the land.
Agencies	Registry offices are typically set up or overseen by local courts.	Registration or land title offices are typically set up under an administrative arm of government.
Registration	Involves lodging a copy of the deed in an official book or collection. Administration requires a complicated system of cross-referencing of parties' names, parcel identifiers, and deed numbers to retrace the history of the land.	Involves recording land transactions in the order in which they are lodged at the land title office on a single page, or single computer file. This page or file is called the "title," and registration is simply recording the transaction on the title.
Forgery	Forgery breaks the "chain of title," so that all later deeds are ineffectual.	Forgery by a person seeking registration is ineffective. The forger cannot get title. But all other people not party to the forgery can rely on registration of the forged instrument to gain a title for themselves.
State insurance and guarantee	There is no guarantee of title by the registration system.	The title is normally guaranteed by the state. Hence, the administration system must be very reliable.
Private and professional insurance	Professionals always carry insurance to protect their clients against failures in their work. Notaries carry insurance and can provide professional guarantees. In other places, notably the United States, private insurers sell insurance cover against failure of the system.	There is no need for private insurance of the title, but private cover can sometimes be offered to protect people against restrictions and responsibilities outside the title system affecting the land. Lawyers carry insurance against losses they or their staff cause.

No matter whether the registry is deeds or title based (figures 12.8, 12.9, and 12.10), its relationship with parcel maps and the cadastre generally is a product of its culture, capacity, and approach to mapping. Many systems start with rudimentary sketch plans or approximate charting maps and generally improve on their mapping capacity. For modern systems, the cadastre, or accompanying map of the parcels, is essential as the means of ensuring continuity among the

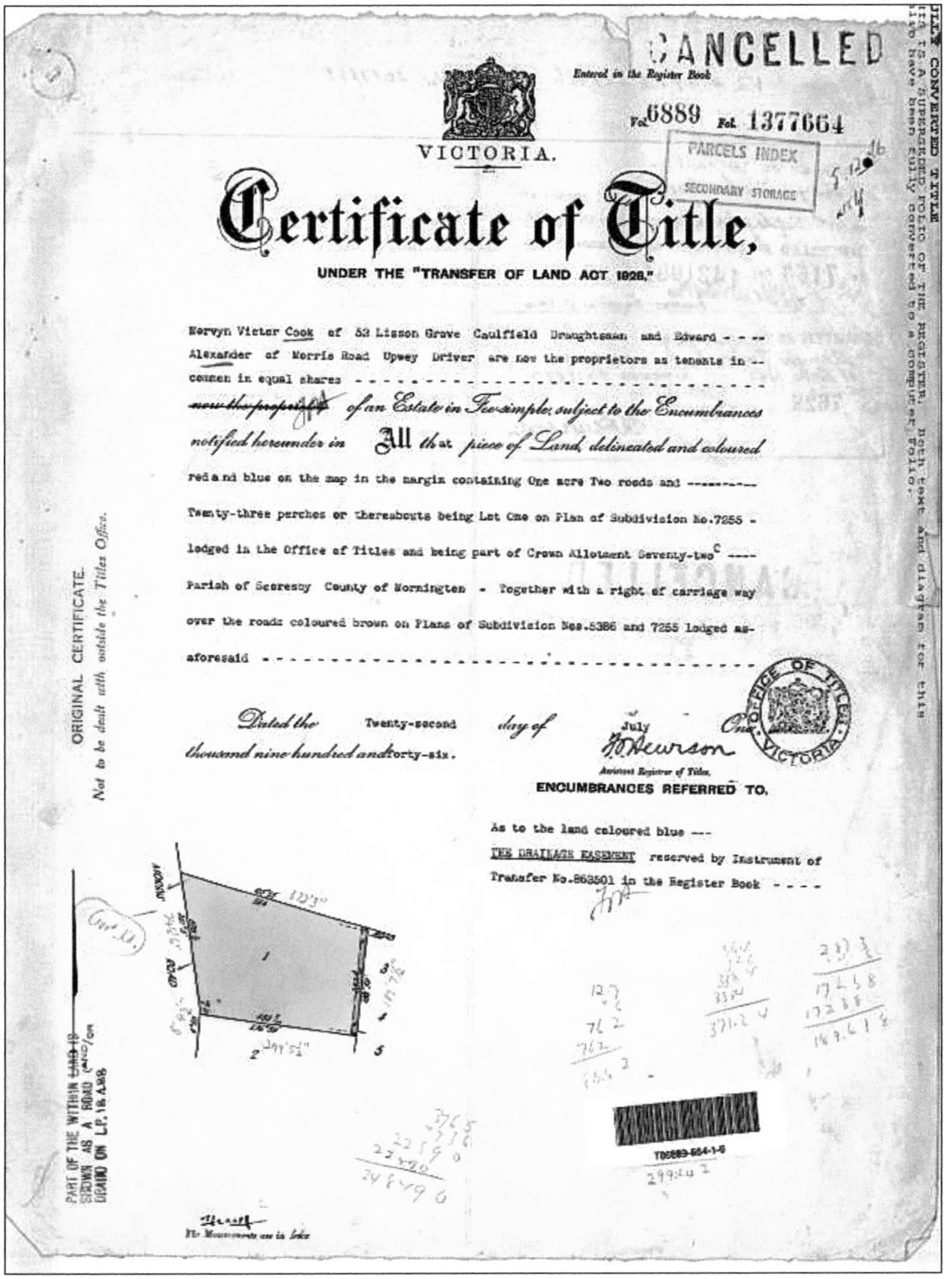

CANCELLED

Entered in the Register Book

Vol. 6889 Fol. 1377664

PARCELS INDEX

SECONDARY STORAGE

VICTORIA.

Certificate of Title,

UNDER THE "TRANSFER OF LAND ACT 1928."

Mervyn Victor Cook of 53 Lisson Grove Caulfield Draughtsman and Edward ---- Alexander of Morris Road Upwey Driver are now the proprietors as tenants in -- common in equal shares ---------------------------------

~~now the proprietor~~ of an Estate in Fee-simple, subject to the Encumbrances notified hereunder in All that piece of Land, delineated and coloured red and blue on the map in the margin containing One acre Two roods and ---------- Twenty-three perches or thereabouts being Lot One on Plan of Subdivision No.7255 - lodged in the Office of Titles and being part of Crown Allotment Seventy-two^C ---- Parish of Scoresby County of Mornington - Together with a right of carriage way over the roads coloured brown on Plans of Subdivision Nos.5386 and 7255 lodged as-aforesaid ------------------------------------

Dated the Twenty-second day of July One thousand nine hundred and forty-six.

Assistant Registrar of Titles.

OFFICE OF TITLES VICTORIA

ORIGINAL CERTIFICATE.

Not to be dealt with outside the Titles Office.

ENCUMBRANCES REFERRED TO.

As to the land coloured blue ---

THE DRAINAGE EASEMENT reserved by Instrument of Transfer No.863501 in the Register Book ----

PART OF THE WITHIN LAND IS SHOWN AS A ROAD ON LP.16,486

Figure 12.8 A traditional Torrens title certificate from Australia shows the owner, encumbrances, and parcel location and dimensions.

Department of
Sustainability
Victoria and Environment

Copyright State of Victoria. This publication is copyright. No part may be reproduced by any process except in accordance with the provisions of the Copyright Act or pursuant to a written agreement. The information is only valid at the time and in the form obtained from the LANDATA REGD TM System. The State of Victoria accepts no responsibility for any subsequent release, publication or reproduction of the information.

REGISTER SEARCH STATEMENT   Land Victoria   Page 1 of 1

Security no : 124026210747L   Volume 10272 Folio 612
                                               Produced 04/06/2008 10:09 am

LAND DESCRIPTION

Lot 2 on Plan of Subdivision 341712K.
PARENT TITLE Volume 09140 Folio 443
Created by instrument PS341712K 25/03/1996

REGISTERED PROPRIETOR

Estate Fee Simple
Joint Proprietors
    DAVID JOHN WILLE
    CHRISTINE ELLEN WILLE both of 46 ILLUKA CR. MT WAVERLEY 3149
    U236468Q 28/05/1996

ENCUMBRANCES, CAVEATS AND NOTICES

MORTGAGE V628816W 08/09/1998
    CITIBANK LTD

COVENANT (as to whole or part of the land) in instrument A428828

    Any encumbrances created by Section 98 Transfer of Land Act 1958 or Section
    24 Subdivision Act 1988 and any other encumbrances shown or entered on the
    plan set out under DIAGRAM LOCATION below.

DIAGRAM LOCATION

SEE PS341712K FOR FURTHER DETAILS AND BOUNDARIES

ACTIVITY IN THE LAST 125 DAYS

NIL

The following information is provided for customer information only.

Street Address: 2-3 YERONGA AVENUE UPWEY VIC 3158

STATEMENT END

Figure 12.9 A typical printout from a digital title register in Australia shows the registered owner, encumbrances, and activity within the previous 125 days.

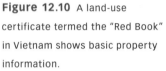

Figure 12.10 A land-use certificate termed the "Red Book" in Vietnam shows basic property information.

titles as described in the register and shown in the plan and map, usually by developing a system of unit identifiers. The surveying and inspection processes are simultaneously developed to ensure that these descriptions reflect what appears on site. Surveys are the most reliable method, albeit expensive, of ensuring that the parcel map represents on-the-ground reality. Registration systems generally do not guarantee these relationships. Only the rarest and most expensive systems can provide firm, almost legal guarantee of actual compliance with the parcel map, for example, in Hamburg, Germany. The common solution to compliance is reliance on well-qualified cadastral surveyors and the quality control systems they employ, in addition to in-house checking systems in cadastral offices. There is a wealth of literature on land registration, often with a book(s) written for each jurisdiction for the most developed systems (see Simpson 1976).

Major distinctions, however, exist in the detail of working systems. The kind of interests that can be registered; how subordinate interests, such as usufructs, easements, covenants, charges, and leases are managed; the system that has priority; the formalities attending registration; insurance coverage (if any); and risk management systems vary remarkably. Treatment of adverse possession of whole and part parcels, inheritance, communal titles, building titles, and resource titles also varies from system to system. For example, in Australia, the State of Victoria's Torrens system allows adverse possessors to part parcel (in other words, the boundaries can move over time), while in the neighboring State of New South Wales, this is not permitted. This difference has a significant effect on the operations of the system. Within these local differences, land registration best practice is now internationally agreed. Registries, irrespective of their historical antecedents, should achieve similar results of tracking the history of ownership and interests in land parcels, protecting titles, and delivering transaction efficiency (World Bank Doing Business reports). The way these are achieved generally involves moving to agreed land registry best practice while adapting these standards to suit local circumstances.

Positive registries and indefeasibility

A positive registration system like the German, English, and Torrens styles confers title. A negative registration system like the French, Latin, and U.S. styles of deeds systems does not determine title but only provides evidence of transactions. In the most positive systems, no interest in the land is permitted "off the register" (as a general principle). Only when the interest is recorded is it protected and, in Torrens systems, the interest cannot be taken away without the owner's permission. In Torrens systems, these standards are rigidly enforced. Thus, Torrens systems are successful in countries where fraud, forgery, and corruption are tightly controlled and where governance capacity is high. Countries unable to achieve these results will need to use registration systems that are less positive, and even negative. In effect, registration adds little to the effectiveness of the land transaction. Over time, as administrative capacity develops, systems can move to more positive outcomes.

Overriding interests

All titles to land must be able to be overridden in special circumstances, even though they are indefeasible. The range of circumstances varies and can include adverse possession of whole and part parcels, land taxes, tenants' rights, easements, fraud and forgery, land recovery or acquisition by the state. The legislative framework establishing these exceptions must be clear and easy to administer.

Accuracy and completeness

Land registries should record all the major events relating to the title to a parcel. They need to track transactions, gifts, changes through social events including marriage and death in so far as titles are altered, and actions by security holders. They should allow historical searches, not merely verify current ownership and interests.

Priorities among claims to land

The registry system needs to attract immediate registration of any dealing in land. Most do this by offering priority to people who claim the land in order of registration. These systems both attract and reward prompt registration. The need to make registration compulsory by some penalty system is thus reduced or entirely eliminated. Many systems are unable to attract registration of derivative (later) transactions, even when registration itself is valued. Systems are much more robust if registration of these derivative transactions is subsidized and encouraged.

The relationship between parcel maps and text information

A register logically needs physical information about the land — the where. Typically, this is stable over the life of the parcel. Who owns what, when, and how they obtained it, however, are ever-changing details. In paper-based systems, the map and the text information are kept separately, though some systems depend on a "parcel file," which keeps the sketch plan or survey plan on the front page and details of ownership over time inside the file. Most offices maintained these separations when they converted to digital systems. This separation of maps and text is then compounded by further institutional separation in many nations. The historical silo approach in land administration means that maps are kept in one agency (a cadastral agency) with the land books or text information kept in the land register, often associated with a court system. This was the approach used in many European countries and variously in the United Kingdom and United States. This silo approach is no longer best practice. The development of the multipurpose cadastre helped integrate the two registration functions.

Authentic registers

Registers of people, cars, securities, shares, land, resource licenses, mining rights, and so on are part of the machinery of all democratic governments. These registers are increasingly important for national and regional management, and their conceptual nature is changing dramatically. The idea of making a register "authentic" (to use the European term) or "official" is now well established. It involves a national decision to use one agency as the determinative source of information. This agency creates the information once, and it is used many times throughout government. The information is nationally reliable, and increasingly internationally reliable, especially in EU countries. An authentic register is designated by law as the sole officially recognized register of the relevant data to be used by all government agencies and, if possible, private organizations throughout the country (Van der Molen 2005). Basic spatial information including parcel information falls naturally into the concept of authentic registers — it is impossible to manage modern government efficiently without it and far too expensive for a nation to maintain separate datasets in multiple and independent agencies.

As an example, the Netherlands is developing six "key registers": the register of popular census, the register of business entities, the register of land, the register of the cadastre, the register of geographic information at a scale of 1:10,000, and the register of buildings and addresses. Now, it is working on a register of car license plates (rather than cars), a register of social securities, a register of incomes, and a register of real estate values. It plans a register of noncitizens, as well as a register of geological and soil data. At the periphery is a possible register of large-scale topographic data, though the needs for agricultural subsidies in the European Union make the 1:10,000 small-scale register essential. The Netherlands situation in land management illustrates how planned national responses to this type of information can deliver capacity to manage land in a nation where up to 60 percent of the land surface is below river and sea levels. For other countries with less dramatic land management demands, these registers are expensive to run, but their capacity to return value to the nation is clear.

In building national authentic registers, the principle of subsidiarity in land information is vital. This ensures that collection and maintenance of data is at the source level, without compromising access.

Ideally, international agencies and national governments will recognize the need to create authoritative and reliable metadata registries. Thus far, land registries tend to stand aside from these developments and to concentrate on improving internal systems. The international interest in e-conveyancing reflects countries' concentration on the immediate horizon of reaching transaction efficiency. While e-conveyancing initiatives are commendable, they need to be constructed in a way that delivers national transaction information in a spatially enabled government.

3. TITLING AND ADJUDICATION TOOLS

The objective of titling and adjudication tools is to incorporate property as described by land parcels into formal LAS. The process of either converting informal tenures to formal tenures or the initial alienation of land from government by formal titling, registration, and adjudication can be sporadic or systematic, or a combination of both. Each approach has its strengths and weaknesses. The approach adopted is dependent on many factors, which include availability of technology, the process objective and availability of government funding, or whether the government wants the process to be self-funded. It is worth remembering that usually the major cost associated with land titling is the associated surveying and mapping.

Today, most land titling or land administration projects adopt a systematic approach as epitomized by the well-known Thailand Land Titling Project (Angus-Leppan and Williamson 1985). These projects systematically cover a country or jurisdiction village by village, town by town, map sheet by map sheet, municipality by municipality, or state by state. Typically, they use aerial photography to assist in compiling the inventory of land parcels (or in other words, creating the cadastre) through an on-site adjudication process. The land titling proceeds area by area, with the title documents often prepared in a local field office and distributed to landholders. Usually, this process is accompanied by programs to build land titles or land registration offices, as well as capacity building to maintain the new system. Once complete coverage is obtained, future processes (transfer, lease, mortgage, inheritance, and so on) are done sporadically to maintain the currency of the register.

There are two sets of tools, with some tools more common. When land is already settled, the tools document existing ownership. When land is government owned or controlled, the tools govern release into private ownership, usually termed alienation of crown or national land. Most of the large land titling projects focus on documenting existing occupation, with this being the focus of most discussion. However, a surprising number of historical and current projects or components of projects involve survey and distribution of government or national lands for distribution to private citizens. Within these two types, the tools are again grouped into systematic and sporadic tools.

Alienation of state or national lands

Alienation of state lands or grants to citizens for private use was the preferred form of settlement in the New World of countries like Australia, the United States, and Canada. The U.S. Public Lands Survey is the classic example where the government surveyed vast tracts of land systematically prior to alienation in the nineteenth century. On the other hand, countries like Australia took a more sporadic approach to alienation in the nineteenth century. In recent times, land redistribution in a similar manner assisted the eastern and central European countries to move from a centrally organized economy to a market economy in the 1980s and 1990s. Many of the large cooperative farms were divided up for private ownership. Similar processes occurred after independence from colonial reign, when large colonial farms were redistributed under land reform programs, particularly in South America and Africa. But even today in countries like Mongolia (figure 12.11), large tracts of state lands—or in Asian and Pacific countries, areas of forest land—are alienated for private use. At the same time, almost every country has a process to alienate state lands where they are no longer of use for government purposes. Whatever the political or policy reason for getting land into private ownership, the technical tools for adjudication, surveying, and mapping are similar.

Systematic and sporadic land titling

Systematic land titling occurs through government intervention to systematically identify, adjudicate, and issue deeds or titles for all land parcels in a region or area. It can be in an informal urban settlement or a formal well-established rural area. The process is the same and usually is undertaken through a government-funded project where all eligible landholders are given security of tenure and appropriate documentation to support their claim. Virtually all the large land administration or land titling projects worldwide since the 1980s have used

Figure 12.11 Once
Mongolia alienated state lands
in a traditional nomadic society
and bestowed private rights,
fences were erected.

systematic processes. These projects require a large injection of funds that can range from less than $100 million to more than $1 billion.

Sporadic land titling is rarely part of a modern titling project. In most developing countries, it has its history in a colonial past. It is designed for individual landholders who wish to individually pay to have their property surveyed and titled. The attraction for governments is that there is not a requirement for a large and expensive project; however, the hidden costs to sporadic titling are significant. After an area, jurisdiction, or country has been titled systematically, then all future dealings and subdivisions are done sporadically since that is the nature of individual landownership. Land market processes consist of individual activities relating to land parcels, whether to transfer, mortgage, lease, inherit, subdivide, and so on. By their very nature, these are individual activities that occur irregularly. Within the context of LAS, they are sporadic activities.

Table 12.4 compares systematic and sporadic land titling.

4. LAND UNIT TOOLS

Effective and efficient LAS require unique identification of each individual land unit within the system, whether those units are parcels, properties, easements, or any other interests in land. This issue was also discussed in section 5.1, "Designing LAS to manage land and resources," where the cadastral units of parcels, properties, and entities were described. Simply, a land administration system, or a cadastre, cannot be built without an effective parcel numbering system.

TABLE 12.4 – COMPARISON OF SYSTEMATIC AND SPORADIC TITLING

ACTIVITY	SYSTEMATIC TITLING	SPORADIC TITLING
Policy objective	Usually part of a large government initiative to deliver secure titles or tenures to a wide cross section of the population at little or no cost to landholders.	Usually part of a user-pays government initiative that allows individual landholders to gain title or more secure tenure at their own initiative and cost.
Legislative basis	Requires a legislative and regulatory environment—however, often not as detailed as for sporadic titling. Usually requires less precision (graphical survey and mapping approaches are more easily used) and is more often undertaken fully by government officials.	Usually requires a more detailed and rigorous legislative and regulatory environment to control the activities of private surveyors. In addition, more detailed and specific regulations and oversight are required because of the greater precision required in the surveys.
Cost	Systematic titling costs the government significantly more initially, but landholders usually bear a minimal cost. Systematic titling is an investment by a government to improve security of tenure, growth of land markets, knowledge of who owns what and where, and documentation of state lands, etc.	Little or no cost to government (at first glance), other than maintaining the records, checking the quality of the information, and maintaining the quality assurance of the private surveyors (but often this is cost recovery as well).This usually requires government to support a university education for land surveyors. Overall, this is usually an expensive process for landholders and generally restricted to the wealthy. It is definitely not a pro-poor initiative.
Capacity requirement	Since government officials are usually involved and graphical surveying and mapping is used, less training and education are required to do the simple adjudication, identification, and mapping.	Since private surveyors are often used, government boards are required to ensure standards. Educating and training professional land surveyors also usually requires a university degree with all associated government costs.
Control surveys	All titling requires some mechanism for relating each land parcel spatially. However, when graphical approaches are used based on maps or aerial photographs, the survey control can be much sparser and is really only required to identify the maps or aerial photographs. This can be as approximate as showing a 1:1,000 scale land parcel map on a 1:100,000 scale topographic map sheet.	Sporadic cadastral surveys are ideally coordinated surveys that must be connected to national control surveys. Today, this is relatively easily done by using satellite positioning (GPS). However, it is possible to do a precise and accurate cadastral survey of a land parcel in isolation from other surveys. This requires considerable expertise and training to ensure there are no overlaps and that the parcel can be charted approximately to other parcels and surveys. Many countries started their systems this way, including Australia and parts of the United States and Canada, as well as many countries in Africa and Asia.

Continued on next page

Continued from previous page

TABLE 12.4 – COMPARISON OF SYSTEMATIC AND SPORADIC TITLING

ACTIVITY	SYSTEMATIC TITLING	SPORADIC TITLING
Mapping	The key advantage of systematic titling is that it allows the use of low-cost graphical approaches using large-scale orthophoto maps, rectified photomaps, unrectified aerial photomaps, and satellite photomaps. However, some small-scale mapping system is required to show the relationship of all the individual large-scale maps. In addition, it allows the use of low-cost satellite positioning technologies (GPS) to assist in boundary location. This mapping has the added advantage of providing government with an excellent basis for land management. One disadvantage in many developing countries is the restriction of community access to large-scale maps by the military, thereby disallowing this low-cost approach.	Sporadic titling actually does not require a basemap, although it is highly desirable. It is the actual function of the field survey that creates the map. However, once field measurements have been obtained, usually as coordinates, it is very easy to input into digital systems and maps based on GIS technology and to accurately record parcel boundaries.
Charting	All titles and parcels need to be charted on a map in order to show the location of each parcel in relation to other parcels. It is possible to easily identify and chart each land parcel by using a systematic approach based on some form of aerial photomap.	It is much more difficult to create a picture of all land parcels in a neighborhood without a large-scale basemap to chart each land parcel. This can lead to overlaps between land parcels and inappropriate practices.
Adjudication	Systematic adjudication is much more efficient and usually more equitable than sporadic adjudication. When a whole area is adjudicated, the process is widely communicated, the adjudication is usually led by a senior person in the community, and the process is very public and transparent. In addition, if some form of aerial photomap is used, the different properties and boundaries are usually easily identified.	The adjudication function for each land parcel is usually the same, but it is much more difficult or even impractical since all neighbors need to come together to agree on the boundaries—often an almost impossible task.

Continued on facing page

Continued from previous page

TABLE 12.4 – COMPARISON OF SYSTEMATIC AND SPORADIC TITLING

ACTIVITY	SYSTEMATIC TITLING	SPORADIC TITLING
Cadastral survey requirement	Both graphical and field survey approaches can be used, but by far the most common approach is a graphical approach using general boundaries. Ground surveys are often used to fill in gaps where boundaries are not visible from the air.	Fixed mathematical boundaries (bearings and distances) are the norm for sporadic titling. However, some graphical boundaries can be used—for example, for natural or water features.
Boundary marking and delimitation	If a graphical approach is used and the boundaries are visible from the air, such as rice paddy dikes, boundary marking is often not required. However, landowners sometimes prefer an easily placed boundary marker. If markers are placed, the distances between markers can be measured, but there is no necessity to accurately determine the bearings of the boundaries—the physical features are the legal boundaries.	Sporadic surveys of fixed boundaries require all parcel corners and boundaries to be marked and reference marks placed on key corners, especially where parcels are surveyed in isolation. These are vital to reestablish the boundaries if some of the corner marks are missing or lost. If all corners are coordinated on some local, or ideally national, coordinate system, the coordinate system is, in effect, the reference network. Landowners still like to see all boundaries marked, so that they can fence their property or know what is theirs to use.
Boundary reestablishment	In a graphical system based on an aerial map, it is very easy to reestablish boundaries. In most cases, the boundaries are clearly visible on the ground.	In a fixed-boundary system where the boundaries are described mathematically, it is usually necessary for a professional land surveyor to reestablish boundaries and mark the corners. Some systems, for instance in New South Wales in Australia, require most boundaries to be resurveyed by a professional land surveyor every time the land parcel is transferred.
Private- or government-sector involvement	Systematic titling is almost always instigated and controlled by government on a project basis. Usually, government officials are involved with specific activities often contracted to the private sector.	Sporadic titling can be done by either the government or private sector. However, it is increasingly common for the private sector (both survey and legal) to be involved. Sporadic systems in many countries promoted the growth of a private surveying sector, where surveyors are licensed by the government.
Title delivery	Government needs to proactively and systematically distribute new landholder documents, titles, or deeds to landholders. However, often landowners do not pursue their documentation once the land titling project has been completed and their land has been "titled."	Since titling is initiated and paid for by landholders, they are usually eager to obtain their deeds or title documents as soon as possible, and government does not need to intervene to encourage derivative registrations.

Continued on next page

Continued from previous page

TABLE 12.4 – COMPARISON OF SYSTEMATIC AND SPORADIC TITLING		
ACTIVITY	**SYSTEMATIC TITLING**	**SPORADIC TITLING**
Appeal processes	Most land titling has an appeal process, usually to the courts. However, there is far less chance of a dispute in a systematic titling process since the adjudication of boundaries is a well-organized, public, and negotiated process. Landholders negotiate and compromise, knowing that the titling process will pass them by if they delay, and a disputed resolution in the future will be very expensive and slow.	Well-developed cadastral systems with trained and experienced land surveyors generate few disputes. In the rare disputes, either the land surveyor, an appropriate government official in a land registry, or a chief government surveyor will resolve the matter. Any rare residual cases involve the courts. Some countries—for instance, Denmark—give the primary legal responsibility to resolve disputes to professional land surveyors and negate the use of courts unless as a last resort.
Additional activities	Systematic titling requires many processes and results in many documents and maps being produced in a geographic location relatively quickly. New local offices are required to manage and store all the records, and new government officials need to be trained to manage the records so that changes in ownership, subdivisions, mortgages, and inheritance can be recorded and kept up-to-date.	Typically, no new government resources are required since the number of new titles, survey plans, and maps increases slowly.
Advantages	Low initial cost for landholders. Government can effectively manage land since it knows who owns what and where it is. A vital end product is a complete map of all land in a jurisdiction or country to support land management and sustainable development. A systematic approach is much more equitable than a sporadic approach, especially when the basemap is used for land taxation, to identify encroachments, or other planning or environmental controls. Titling is completed relatively quickly (15–20 years is possible for most countries).	Low cost to government (initially). Individual landholders, if they have the financial resources, can obtain title to their land relatively quickly.
Disadvantages	High initial cost is borne by government. The system requires a major project and usually considerable donor or external expertise.	This involves high cost to landholders. It demands a high level of surveying expertise. Titling may not be completed within 100 years or more, if ever. The national parcel map remains impracticable, with severe consequences for national land policy and management.

The unique identification of land units has been important since the earliest cadastres and LAS. Today, computerization of LAS and the increasing importance of land information underscore the issue. In North America, a particular focus on land unit identification (for example, Moyer and Fisher 1973; Ziemann 1976) compensated in part for the lack of organized cadastres, as compared to Europe. The importance of land parcel identifiers was evident from North American conferences, symposia, and workshops in the late 1960s as summarized by David Moyer and Kenneth Fisher (1973). These conclusions are equally relevant today:

> "The chief obstacle to records improvement was the lack of a common system of land parcel identifiers and that any universal system must be compatible for application to other land-related records (land use, ecology, etc) as well as to land title records (those defining an interest in land and identifying the owner or holder of a security interest)."

Land units and identifiers are strongly influenced by legacy systems in existing LAS; thus, lack of consistency in both units and identifiers remains an issue. A clear description of the hierarchy of ownership units is needed within a framework that recognizes the peculiarities and requirements of each country or jurisdiction. There are two forms of parcel identifiers.

The first is a descriptive identifier based on a sequential list of cadastral maps (such as parcel XXX, map sheet reference YYY together with a reference to administrative units such as villages, municipalities, towns, provinces, or states) or plans (parcel XXX, cadastral or survey plan YYY, again with the addition of administrative units), or an identifier that relates to a physical location, such as a street address (XXX Smith Street) together with suburb, locality, municipality, county, and state identifiers, either in numerical or descriptive form. For example, each administrative unit may be given a numerical identifier in addition to its name.

The second form of identifier is to give the parcel a geocode or spatial identifier. That is a geographic coordinate from a mapping system (or a latitude or longitude), usually for the centroid of the parcel. However, as long as a parcel is shown on a map that is part of a state or national mapping system, it can be located spatially by graphically scaling its geocode off the map, albeit at varying degrees of accuracy depending on the accuracy or scale of the map. Quite often, a geocode is given in addition to the descriptive identifier. UNECE (2004) provides many examples of these identifiers.

Each parcel identifier should be simple and easy to use, especially by nonprofessionals, and is typically a parcel number on a cadastral or survey plan or map or a street address. However, street addresses are most prone to error, even though they are the most common identifier used by the public and wider government. Consequently, the primary identifier adopted in LAS is usually one that is created as part of the formal subdivision process, which in turn is part of the land registration system.

Geocodes are not essential for an efficient and effective system of parcel identifiers. However, they are increasingly significant, because they permit a greater multipurpose use of land parcel data. For example, a high-integrity geocoded national address file (G-NAF) based on the legal integrity of the land registry and cadastral mapping system, such as developed by the Public Sector Mapping Agency (PSMA Australia Ltd.), can be considered one of the ultimate developments of parcel identifiers for multipurpose uses (PSMA 2008). Only development of spatially enabled identifiers such as G-NAF can achieve the vision of a spatially enabled society. In particular, this enablement allows individual jurisdictions to fully integrate land administration and cadastral activities, or at the very least the datasets produced by these activities, with data produced by geographic or topographic activities. This integration remains problematic for many countries in both the developed (Korea and Japan, for example) and developing world, where institutional government silos continue to operate. Integration remains one of the major challenges facing many countries that seek the advantages of spatial enablement.

A parcel numbering system must relate to all land parcels in a jurisdiction, whether the tenure of those parcels is private, government, common, or communal. The system must also identify individual public thoroughfares as parcels, since many lanes, roads, and highways are owned or controlled by different organizations such as municipalities, regional, or national governments or increasingly by private interests such as toll roads.

Importantly, the land parcel framework is the basis of a national ability to manage the relationships of people to land. The framework needs to be sufficiently complete and easy to use, so that easements, restrictions, buildings, land-use controls, and a wide range of other RRRs can be related to the land and managed effectively. A complete cadastre with an effective parcel identification system is essential to capacity and delivery of sustainable land policies.

Buildings pose special problems for LAS. When buildings have individual identifiers, or parcels are created in 3D, or parcels are created by strata or condominium subdivisions, it is highly desirable that those building identifiers also relate to individual land parcels that can be identified on a cadastral map. Often, these issues are discussed under the topic of "3D cadastres" (see, for example, Stoter and Van Oosterom 2002).

The most comprehensive review of real property units and identifiers is the 2004 UNECE study "Guidelines on real property units and identifiers." Any reengineering of LAS should make reference to these guidelines.

The UNECE hierarchy is a more detailed description of real property units than was summarized in chapter 5, "Modern land administration theory." However, chapter 5 identifies and describes the biggest issue in developing LAS: that is, managing the different characteristics of parcels and properties (described as a BPU). The number of principles, suggestions, and recommendations in the UNECE guidelines are helpful in reengineering LAS.

A global approach to best-practice land unit principles follows:

- **Land as a whole:** Land should be treated as a whole, allowing building rights to be a subset of the rights that are associated with the land. All land parcels, be they private, government, roads, common property, or communal, should be included in the cadastre—in other words, the cadastre should be a complete representation of interests in land within a jurisdiction. There should be no gaps.

- **Definitions in law:** Definitions of the parcel and the BPU should be contained in the land law. The land law should define the extent of ownership vertically and horizontally, both on dry land and for land under or over water. Special legislation is needed to cover the management and responsibilities of apartments and the common areas in a condominium.

- **Parcel identification:** Each parcel should have unique ownership or homogeneous real property rights or clearly defined managers such as the nation-state in communist systems or government agencies for state and public lands. The physical extent of parcels may be defined by survey or by physical features on the ground; the legal extent is defined by real property rights or land-use rights. Parcels should change only through the process of law.

- **Referencing systems:** The land parcel referencing system should be based on the needs of users with the data in the register compiled on the basis of the land rather than the owner. The same parcel referencing system should be used in land books, in the cadastre, and in municipalities, so that real-property-related data can be easily integrated.

- **Referencing identity:** The reference that identifies a parcel should be unique. Two parcels should not have the same reference, even when they are located in different districts or municipalities.

◆ **Reference permanency:** References to basic property units and parcels should be permanent over time. Ideally, political or administrative jurisdictions should not be part of the parcel identifier, because these may change — for instance, when municipalities are amalgamated. Ideally, the parcel identifier should not include the name of political or administrative jurisdictions, because these may also change. For the same reason of permanence, the parcel number should not be used as part of the building identifier.

◆ **Address system:** A national standard for street (postal) addressing should be established. Street addresses and apartment numbers should be designed primarily to support the process of finding the relevant feature in the field — for example, by supporting the delivery of goods and services at that address. They should be treated as attributes of parcels in cadastral registers. The address system should include a postal code or postcode that can be used for mail sorting and delivery and by commercial companies for marketing products and services and for data analysis.

◆ **Geocoding:** Geographic coordinates of real property boundaries and any point that represents the middle of a parcel (together forming a geocode) should be recorded in the register as attributes of the parcel. GIS technology may then be used to search the data files on a location basis.

5. BOUNDARY TOOLS

Boundaries are fundamental to land administration and civil peace. Without a system to equitably and transparently create, describe, and mark boundaries that are accepted by society, whether in formal or informal systems, disputes and eventually civil unrest, and even war, can result. Therefore, LAS must have a system of creating, describing, and marking parcel boundaries.

Equally important, LAS need a system of resolving boundary disputes, usually between neighbors and most commonly as a result of encroachment. It is preferable for disputes about formal boundaries to be resolved by administrative methods following good governance principles, such as tribunals, appointed officials such as assessors or surveyors general, or, in some countries, land surveyors, even though judicial processes are usually available as a last resort. Unfortunately in many developing countries, boundary disputes can only be resolved by the courts, with the result that judicial systems become clogged by relatively minor disputes that can take years, and sometimes decades, to resolve, if ever.

The term "boundary" refers to either the physical objects that mark the limits of a parcel, property, or interest in land or an imaginary line or surface marking or defining the division between two

Figure 12.12 The world's greatest boundary monument is the Great Wall of China. This general boundary is man-made.

legal interests in land. Boundaries are defined by laws and regulations, with many variations across countries and even states or provinces within a country. For example, a landowner may refer to a fence, a hedge, or a wall, and say, "That is my boundary." This statement can influence a third party if it reflects the legal definition in the jurisdiction of what constitutes "a legal boundary." In many jurisdictions, fences, hedges, or walls may have legal standing in marking or identifying the boundary, while in others, they may have no legal status at all. In still others, they may play some, though not a legally defining, role in boundary determination.

A land administration system requires a boundary system underpinned by law that defines, describes, and relates every boundary to the ground. There are many options to create, describe, and mark boundaries on the ground. Typically, boundaries are identified on the ground by monuments, where a monument is any tangible landmark that indicates a boundary. A monument may indicate the boundary itself or the end or turning point of an artificial line describing the boundary; it may not be on the boundary but reference a boundary corner mathematically. Monuments may take many forms (figures 12.12, 12.13, 12.14, and 12.15). They can be natural features or artificial marks that meet prescribed regulations for marking boundaries.

Figure 12.13 An on-the-ground boundary mark with a metal pin is used in Switzerland.

There are two broad categories of boundaries—fixed boundaries and general boundaries—described as follows:

- **Fixed boundaries** are where the precise line of the boundary is determined by legal surveys and expressed mathematically by bearings and distances, or by coordinates. These boundaries are also referred to as artificial boundaries. A fixed boundary is usually marked on the ground by monuments, such as concrete posts, iron pipes, wooden pegs, steel rods, or marks in rock or concrete. Boundaries are usually determined and marked by a land surveyor who is registered or licensed by the state to undertake cadastral surveys. Fixed boundaries are the most common form of boundary in the developed world and are found in most jurisdictions worldwide.

- **General boundaries** are where the precise line on the ground has not been determined, although usually, it is represented by a physical feature, either natural or man-made, such as a fence, hedge, ridge, wall (in a strata or condominium parcel), ditch, road, or railway line, and shown graphically on a map—normally, a large-scale topographic map, as is the case in the United Kingdom.

Many general boundaries are referred to as natural boundaries, where the defining physical feature is a natural feature rather than man-made, such as a ridgeline, the centerline or bank of a river or stream, or various forms of seashore boundaries such as the mean high water mark. The law surrounding these forms of general boundaries can be complex. In this case, the boundary may be able to move with time (called an ambulatory boundary, or in the case of a boundary

Figure 12.14 A concrete boundary mark that numerically identifies the mark is used in the Philippines.

adjoining water, a riparian boundary). Sometimes, the boundary may be fixed in the position of the natural feature at the time of creation, although this is less common. General boundaries are best known from the boundary system in the United Kingdom, but in reality, general boundaries can be found to a lesser or greater extent in most, if not all, LAS worldwide.

It is possible to precisely define a general boundary if required, such as the center of a fence, ditch, hedge, or face of a wall; however, this rarely occurs. There also can be many difficulties in defining the precise boundary, since there can often be disagreement as to where the precise boundary is. But this is also the strength of the general boundary concept, since it does not create disputes over small boundary movements and leaves the precise boundary undetermined.

There is another category called **approximate boundaries**, where the position of the boundary has not been determined, although the general location of the parcel or property or interest in land is determined, and usually, but not necessarily, shown approximately in a graphic manner on a map. Generally, approximate boundaries are not as precise or as accurate as "general" boundaries. An excellent example of approximate boundaries is the system of qualified titles used by Malaysia, where the title to the parcel is determined, but the actual boundaries are not. In this case, the parcel or property has a full legal title guaranteed by the government, but the title is "limited as to boundaries." The system allows for titling without delay and at lower cost, postponing the formal surveying of boundaries until the owner requires it, the land is developed, or the government wishes to upgrade to fixed boundaries.

Approximate boundaries have similarities to general boundaries. However, it is usually possible to define a general boundary precisely if required, but usually there is no indication of what constitutes an approximate boundary or its location. For example, in Malaysia, the qualified title may refer to a parcel that is part of a row housing development where each parcel is fully occupied by a structure—either the house itself or a brick or concrete wall. The title refers to the identifier of the parcel, sometimes simply on an architect's plan or its street address. Approximate boundaries, such as used in Malaysia, can play a useful role at the early stages of establishing a land market in a country going through rapid economic development. However, they have significant limitations as LAS mature.

Fixed boundaries, general boundaries, and approximate boundaries all have a role to play in the land administration toolbox, and all have unique strengths and weaknesses.

The methods or tools used to mark boundaries, usually called survey, corner, or boundary marks, are extensive. Marking principles and options include

- ◆ Marking corners using wooden pegs, steel pipes, steel rods, concrete blocks, stones, metal pins, and even bottles
- ◆ Indicating the boundary by using marks drilled or chipped in stone, concrete or walls of houses, marks made on trees, or trenches dug in soil
- ◆ Using marks as reference marks—here, the pipes, pegs, and marks are usually not on the boundary but are related to the boundary mathematically or by offsets
- ◆ Using parcel corner identifiers, usually a numerical identifier, shown on the cadastral survey or map and sometimes shown on the boundary mark
- ◆ Identifying the professional land or cadastral surveyor who placed the mark, often with the license or registration number of the land surveyor inscribed on the mark (common in the United States)

Marking land parcels has two roles: first, to define the parcel on the ground, and second, as evidence for reestablishment of boundaries at a later time. It is important to remember that the basic principle for boundary redefinition is that monuments placed as the boundary corner will be given precedence over measurements when there is disagreement. There is significant legal precedence in many countries to support this principle.

Figure 12.15 In Athens, Greece, most high-rise buildings and condominiums rely on general boundaries such as the condominium walls themselves.

There are two categories of survey marks: visible and hidden marks. Visible marks are there to visibly identify the boundary or identify a reference mark. Hidden marks are often buried marks that usually only a land surveyor can locate. These are used to help reestablish boundary marks since the visible marks, being on the boundary, often are removed, or lost or destroyed while building fences, buildings, or structures on the boundary. They are also often lost during wildfires and floods.

Boundary principles and practices continue to evolve, even in highly developed systems. Most LAS include both fixed and general boundaries. The most common general boundaries outside the United Kingdom are either those in condominiums or strata subdivisions in buildings, where the physical walls of the condominium unit denote its boundary, or natural or riparian boundaries.

A question often asked is "Can boundaries move?" Again, this depends on the law of the jurisdiction. In some jurisdictions, boundaries can effectively move, while in others, they cannot. There are two forms of boundaries that can "move." First, riparian or natural boundaries in some jurisdictions can move, provided the movement is through imperceptible accretion or erosion that meets specific legal criteria. The second case is where the jurisdiction allows adverse possession for part of a parcel in a fixed-boundary system or in a system that is based on general boundaries. This is also called prescription and can occur when a jurisdiction gives legal sanction to encroachment on a neighboring parcel after a set period of time, provided certain legal criteria are met. In the case of general boundaries, the boundary can move as a result of small or imperceptible movements of the physical feature, but again, rules apply. Needless to say, the area of law surrounding prescription, adverse possession, and encroachment, by either fences or buildings, is complex and requires expert understanding of the laws and practices within each jurisdiction (see Park 2003).

Another issue concerns whether boundaries are guaranteed—for instance, as with land titles in a title registration system. Again, there is not a simple answer. On one hand, no land administration system legally guarantees "metes"—that is, the actual mathematical coordinates or dimensions that describe a boundary—and none should guarantee the area of a parcel. However, though rare, there are some cadastral survey systems, such as in Hamburg, Germany, that are so accurate and precise that the dimensions that determine the boundaries effectively guarantee the boundaries. On the other hand, what is often guaranteed are the "bounds" of a parcel. That is, the laws of the jurisdiction will guarantee that a particular parcel bounds a certain street or neighboring parcel.

Creation and marking of boundaries and redefinition of land boundaries, usually by professional land surveyors, are usually complex processes surrounded by extensive laws, regulations, and practices. Most jurisdictions have rules, regulations, and government directions that describe all these processes. Those used in the Danish system are outlined in table 4.5. At the same time, many countries, such as the United States, Australia, and Canada, have extensive and detailed books and manuals that describe both the legal and practical interpretation of the regulations based on court cases and practical case studies.

In most countries, the creation, determination, or marking of boundaries can only be undertaken by a trained government surveyor or a private-sector land surveyor who has been licensed or registered to act as an agent of the state to undertake boundary surveys.

Every boundary system has its strengths and weaknesses. For example, the choice to use fixed or general boundaries involves weighing pros and cons similar to those involved in the choice between systematic and sporadic land titling discussed earlier. The major weaknesses involved in choosing boundary systems are that the choices are often influenced by local history, usually as the result of a colonial system. Using low-cost, low-technology techniques or fast and efficient approaches to creating, defining, and identifying boundaries are sometimes strongly opposed by professional vested interests because of a real or perceived loss of income derived from using an expensive and slow alternative. The choice of an appropriate "boundary tool" is rarely a simple technical decision.

6. CADASTRAL SURVEYING AND MAPPING TOOLS

Importance of making the right selection
Some rudimentary systems that record rights in land and support aspects of land markets (such as certain deeds registry systems or local village systems and the qualified titles used in Malaysia) do not require formal surveying and mapping systems. However, all systems eventually require the ability to spatially identify land parcels and interests in land in order to reduce boundary disputes, promote security of tenure, support effective land markets, and, in a broader sense, support the sustainable-development objectives of economic development, environmental management and social justice. Without using surveying and mapping tools within LAS, it is difficult, if not impossible, to use the concepts of place and location in an unambiguous manner.

The Thailand Land Titling Project provides an example. In the northern province of Chiang Mai (one of seventy-two provinces in Thailand) in the early 1980s, very rudimentary LAS existed, with little or no spatial integrity. The local judiciary estimated that there was about one murder per month (and many more serious criminal activities) resulting from boundary disputes. Today, by benefit of the project, with most of the private lands surveyed and included in LAS, boundary disputes that result in serious criminal activity no longer occur.

An appropriate surveying and mapping system within LAS will undoubtedly deliver many benefits, but it must be acknowledged that this is the most costly component of any system or LAP, at both the initial phase as well as the ongoing maintenance phase. As a result, the choice of the surveying and mapping tools used in LAS or a LAP can mean either success or failure. There have been some notable successes in land titling or land administration projects in developing countries, but there have been many more that have either failed or are only partly successful at best. A major contributing factor to lack of success for many projects is the poor choice of surveying and mapping tools.

There are examples around the world where the most appropriate boundary system involves using general boundaries (see "Boundary tools," also in this section) as part of a systematic titling process where the key surveying and mapping tools are based on the use of aerial photomaps or orthophoto maps. However, sometimes the military in these jurisdictions simply prevents access to maps by the nonmilitary arms of government. Some countries (such as Mongolia, even though there are many others) still struggle with decade-old security regulations that are now anachronisms in the age of satellite mapping and similar technologies. These outdated policies effectively force a jurisdiction to adopt an expensive fixed-boundary system where boundaries are created by ground survey methods and create insurmountable problems if the country has neither the resources nor the capacity for this kind of system. Thus, a land administration system is doomed before it starts.

The surveying and mapping tools used in one country are rarely fully transferable to another because of local circumstances. The tools needed for one area within a country (such as urban, peri-urban, rural, or customary areas) are often not appropriate for the country as a whole with the result that different tools are required within different parts of a country. Simply, cadastral surveying and mapping technologies must be adapted to suit different land tenures and different parcel boundaries, as well as systematic and sporadic approaches to titling and adjudication.

Cadastral surveying and mapping

"Cadastral surveying" is the process of creating, measuring, and marking boundaries on the ground (figure 12.16); preparing cadastral survey plans of those activities for the purpose of reestablishing boundaries; and recording the boundaries on an aggregated (cadastral) map. Often, at the early stages of developing a land administration system, the cadastral survey plans are either attached to a deed or title and filed as part of the deeds/title register, or shown or "charted" at various levels of accuracy on a "charting map." These charting maps often have low spatial accuracy and are often at a small scale. In many cases, the individual land parcels in the cadastral survey plan are not charted or plotted, but are identified by reference to the cadastral survey plan (often a plan of subdivision). In some countries (such as parts of Australia and the United States), basic valuation maps were used in the past for charting cadastral survey plans. A "cadastral map" is usually built when the parcels in a cadastral survey plan are plotted to scale on a map and the map is kept up-to-date. However, in many European countries, the cadastre was originally created from a complete cadastral map normally encompassing an individual village, parish, or jurisdiction. When the cadastral map is kept in digital form and updated digitally, it is often referred to as a "digital cadastral database" (DCDB). Once all land parcel data is in a DCDB, it is possible to use Web services for collecting and transferring land information and to move into an e-government environment for managing land processes. The accuracy of this cadastral map, compared with other spatial information, ensures that the spatial enablement of government systems can be achieved or pursued.

LAS that rely on up-to-date cadastral maps increasingly take on multipurpose roles. Very quickly, the benefit of the DCDB in support of a multipurpose role in government and society outweighs its initial benefit of supporting security of tenure and simple land markets.

Figure 12.16 Surveyors in the Philippines conduct cadastral surveying and marking.

Unfortunately, many government officials and professional land surveyors have difficulty understanding this change in emphasis, with the result that use of multipurpose cadastral data slows down. In developed LAS, the primary role of cadastral surveying is to describe and identify land parcels for inclusion in the cadastral map or DCDB with the secondary purpose being security of tenure and supporting simple land markets.

Different forms of cadastral surveying and mapping
While cadastral surveying can use a range of graphic tools, the most common methods rely on bearings and distances or coordinates, or both, for measuring and recording boundaries. Within this mathematical approach to cadastral surveying, there are a number of conceptual forms that are independent from the technology described in later detail. There are two fundamental approaches to determining the location of land parcels by cadastral surveying and mapping. The first is the European approach, where complete cadastral maps, even if just "island maps" covering a specific village area, parish, or jurisdiction, are the primary focus. In this case, cadastral surveys always have to refer to the cadastral map as the primary source of identification. In Torrens and related systems, the primary focus for parcel identification is the cadastral survey plan and associated cadastral survey. It is only in the mature stages of jurisdictions that adopt this latter approach that the cadastral map is compiled at a later stage of development. See the article by Ian Williamson and Stig Enemark titled "Understanding cadastral maps" (*Australian Surveyor* 41, No. 1 (1996): 38–52). Whatever the form chosen, it is desirable that the cadastral surveying and associated boundary marking is only done once. Following is a brief historical

evolution of many cadastral surveying and mapping systems. This detail is justified recognition that many countries still exhibit some of these systems and that usually the cadastral surveying and mapping component is the most expensive aspect of a land administration system.

- **Isolated survey system:** In the early stages of LAS, such as in parts of Australia, Canada, and elsewhere, cadastral surveying was performed within an isolated survey system. That is, parcel boundaries were measured and marked to a high level of accuracy and precision and related only to neighboring parcels and boundaries. Corners were referenced to reference marks in case the corner mark was lost or destroyed. The key aspect of isolated surveys is that they float in isolation, even though they are connected to neighboring parcels where possible. In most cases, the survey is oriented to north by a magnetic compass; however, the accuracy of this meridian determination is only approximate. An important aspect of isolated surveys is that they are sufficiently accurate and precise, so that a surveyor only has to find two undisturbed marks within the survey to effectively reestablish the survey. The resulting isolated surveys are usually charted approximately on a charting map. In rural areas, this may have been a small-scale topographic map or a parish or county map. In urban areas, the charting map was often whatever was available, such as a valuation map or even maps showing water or sewage services. Charting maps of this kind were used in the nineteenth century, but in the late twentieth century, more accurate and complete cadastral maps were needed. These were often produced by physically scaling and fitting cadastral survey plans to a topographic map.

- **True meridian:** Some jurisdictions improved on the isolated survey approach by requiring all cadastral surveys to be based on the true meridian (or true north) — that is, zero degrees being north, even though the survey remained essentially an isolated survey. This was done by observations of the sun or stars to accuracies of about a minute of arc. These surveys were then much easier to reestablish or connect to neighboring surveys. The procedures also greatly facilitated the charting and plotting of cadastral survey plans on a charting or cadastral map. While these systems and isolated surveys used the same method to compile their cadastral map, the processes benefited from having all parcels based on the true meridian.

- **Local plane coordinate system:** Many jurisdictions around the world, in countries as far apart as Malaysia, Switzerland, and Hawaii in the United States, improved on this approach by adopting a local plane coordinate system for a village, town, or region where the origin (0,0) was a trigonometric station near the center of the region. This allowed easy adoption of a true meridian at this point and the use of a localized plane coordinate system, even though the accuracy of the trigonometric coordinates of the central control station is variable and not designed for large-scale cadastral use. There are many advantages to this approach in that cadastral boundaries of parcels could be relatively easily plotted on a local cadastral map, but there are also significant deficiencies. For example, the farther a survey is from the origin, the effects of convergence and scale make it difficult to relate coordinates from one local origin to another. Fitting together cadastral parcels at the extremities of each local plane system is difficult. Interestingly, in some of these systems, such as in some states of the United States, zero degrees was south, not north. These systems served security of tenure and simple land market requirements well but created difficulties when the cadastral maps were to be integrated into a national mapping system or a national SDI. However, a benefit of this approach was that relatively accurate cadastral maps were developed as surveys were carried out. By the mid- to late twentieth century, these local coordinate systems and maps needed to be integrated with a state or national mapping system. Different techniques were employed, but the most common involved identifying key monuments on major block corners in the local system, using satellite positioning (GPS) or other techniques to give the monument a position in the new state or national system, and then transforming all the local coordinates into the new system. The many variations of this approach are, in principle, all fundamentally the same.

◆ **Coordinated cadastral survey:** In the next evolution of cadastral surveying, the surveys are based on coordinates and use coordinates as the fundamental focus instead of bearings and distances, although all coordinated cadastral surveys obviously use a combination of coordinates, and bearings and distances. This requires the jurisdiction to have a comprehensive, sufficiently dense control network based on a jurisdiction-wide map projection suitable for cadastral purposes. Historically, this involved a breakdown of the classic geodetic control system from first-order control down to third- or fourth-order cadastral control. In the latter part of the twentieth century, GPS was used to place this control where the concept of orders became meaningless in practice. In coordinated cadastral survey systems, cadastral surveying is a relatively simple exercise, especially when the accuracy of the cadastral control and coordinated boundaries is capable of delivering field survey accuracy (within two centimeters or better in local control and boundary corners). However, within many coordinated cadastral survey systems, particularly those that evolved from isolated survey systems, coordinates are simply the method used to create and mark boundaries and to prepare cadastral survey plans. In these cases, the cadastral survey plans are still the final record used to redefine or plot cadastral parcel boundaries. Cadastral maps that are based on an aggregation of coordinated cadastral plans are still significantly less accurate than the actual cadastral survey or cadastral survey plan.

◆ **DCDB:** The next evolution involved moving from a jurisdiction-wide, coordinated cadastral survey system to a survey-accurate and complete DCDB, even though it may still be supported by coordinated cadastral survey plans. This is usually termed a coordinated cadastre. Within a coordinated cadastre, all coordinates are survey accurate, and the DCDB effectively becomes a continuous cadastral survey plan. Since the mid-1990s, survey-accurate cadastral maps, while still not the norm, became increasingly common within Europe, some parts of North America, in the urban areas of New Zealand, and in some parts of Australia (particularly the Australian Capital Territory). Now, it is more common for a city or local government to decide on economic grounds to replace their graphically accurate cadastral map with a survey-accurate cadastral map or, in effect, a coordinated cadastre. This trend is justified by the ease of checking subsequent cadastral surveys, the ease of carrying out cadastral surveys, the ease of undertaking engineering design for roads and other works, and the ease of maintaining the DCDB. Many private land surveyors in developed countries upgrade their local graphic DCDB for their own use for the same justifications. There is considerable debate in some countries as to whether a dense network of ground marks is required to support a coordinated cadastre or whether a sparse network of ground marks can be expanded by GPS for use as cadastral control. The choice is complex and involves technical and other issues in order to determine the best form of control in each jurisdiction for cadastral surveying.

◆ **Legal coordinates:** Arguably, the ultimate evolution of a cadastral survey system is when the coordinates have legal status and are guaranteed in the same way that a title is guaranteed by government in certain LAS. However, this is a controversial ideal since it overrides the basic legal principle of monuments over measurements with respect to boundary determination as described earlier. Many jurisdictions would oppose legal coordinates since it goes against many aspects of land law and the manner in which society operates. To date worldwide, no system has adopted guaranteed coordinates; however, some jurisdictions have sufficiently accurate and precise systems that the status of legal coordinates is arguable.

This summary of the different forms of cadastral surveying and resulting cadastral maps is an oversimplification. In reality, LAS at the state or national level evolve their own peculiarities. However, the final objective of LAS is the same—that is, to use a cadastral surveying system to create and maintain an accurate representation of cadastral parcels on a cadastral map that is part of a state or national system and a component of the jurisdiction's SDI. This is an ongoing technical challenge in both developing and developed countries as they adopt Web-based systems and increasingly operate in an e-government environment.

Different cadastral surveying tools

There is an enormous range of cadastral surveying technologies in the toolbox, each having its own strengths and weaknesses. They are discussed under graphic and numerical categories. The choice of which tool to use will be influenced by many issues, including the development and capacity of the country or jurisdiction, the type of boundaries, the availability of technology, and the strategies adopted within LAS to establish and maintain the system. While there is an inevitable and increasing trend to use technology and associated numerical means, these are by no means the most appropriate tools for all situations. Following is a brief overview of most of the tools. Details of these tools and techniques, and their strengths and weaknesses, are available in a large number of surveying textbooks and manuals available worldwide in most languages.

Graphic tools: Most cadastral surveying and mapping systems started using graphic approaches. They are usually simple, low cost, reliable, and often surprisingly efficient when used by well-trained personnel.

- **Plane table:** LAS around the world typically were originally based on plane table surveys (such as in India and Korea and many of the European systems in the late eighteenth and early nineteenth centuries). It is still used in some countries.

- **Orthogonal method:** Use of optical squares and measuring tapes can be surprisingly effective. These were used historically in countries such as the Netherlands but are rarely used today except as a means to quickly relocate old corners and monuments in preparation for a cadastral survey to reestablish existing boundaries.

- **Stadia:** Use of the two stadia hairs in a theodolite or alidade to measure distances by observing a distance on a vertical staff is effective for producing a graphic cadastral map. This was used historically in some parts of Europe.

- **Photogrammetry:** This can be used in two ways for cadastral surveying and mapping purposes. By far, the most common use is to produce a photomap that is then used to identify features on the ground (roads, ditches, houses, trees, and so on) that either represent a boundary or are related to a boundary. The other way to use photogrammetry is to determine the actual numerical coordinates of parcel corners. This approach is rarely used. Use of photomaps as a graphic base to identify, adjudicate, and plot cadastral boundaries can be extremely effective, especially when there are sufficient boundary features identifiable on the map. It is quite compatible with the concept of general boundaries. This tool is common in systematic land titling. The tools or options are as follows:

- **Traditional topographic mapping** can be effective since the maps are true to scale, but it can also be expensive to produce the initial topographic maps. However, these maps can be part of the mapping series of a country and used for multiple purposes.

- **Orthophotography** is aerial photography that has been incrementally rectified. Usually, orthophoto maps are more useful than topographic mapping, especially in hilly or mountainous terrain since they help identify boundary features. Orthophoto maps also have many uses.

- **Rectified photomaps** can be just as effective as orthophoto maps, especially in relatively flat terrain, and have many uses.

- **Nonrectified photomaps** are a low-cost alternative but have all the inherent distortions of aerial photos. They can be very useful in identifying parcels and boundaries for land-use certificates and other occupational tenures.

- **Satellite mapping:** Until recently, satellite mapping was on too small a scale to be effectively used for cadastral purposes. However, the new generation of high-resolution satellite images has a resolution of better than one meter, and they are increasingly becoming a viable option. Again, there is the option of either rectified (true to scale) or nonrectified satellite maps that have the same advantages or disadvantages as photomaps from aerial photos. Sometimes, satellite mapping can be more expensive that aerial photo mapping, but the multipurpose roles of these tools can reduce their cost. At the same time, aerial photomaps are increasingly becoming available in products that are available over the Internet. For example, for much of Australia, both high-resolution imagery and the government-produced cadastral map are freely available, and adding the street address of any property in Australia will bring up both the image and cadastral map.

Numerical tools: Increasingly, numerical tools are the tool of choice, since in most cases, the data can be obtained or converted to digital form and used in an ICT environment. However, there are still many reasons for using numerical cadastral surveying tools to simply determine measurements to facilitate surveying and to determine and mark boundaries. Many of these

Figure 12.17 A surveyor uses a digital theodolite for cadastral surveying in Tokyo, Japan.

techniques play a role in updating and upgrading the cadastral mapping systems as discussed in the following section on DCDBs. The primary role of all the numerical tools is to determine a coordinate value for each boundary corner or the distance and bearing of each boundary.

- **Polar method (compass, theodolites, tapes, EDM, Total Station):** The polar method is the most common tool used in cadastral surveying. It effectively measures a bearing and distance from one point to another. The technique is just as useful for isolated cadastral surveys as for coordinated cadastral surveys. While a compass and tape is the simplest form of polar surveying, it has been replaced today by digital theodolites (figure 12.17) with integrated electronic distance measurement and even integrated satellite positioning using GPS (generally called Total Stations).

- **Offset methods (optical squares and tape):** While offset methods are still used for marking cadastral surveys and locating lost corner marks, the method is rarely used today in cadastral surveying.

- **Photogrammetry:** Again, while photogrammetry has the ability to measure coordinate values of boundary corners, it is rarely used.

◆ **GPS (GPS as a measuring tool as distinct from GPS as a network):** GPS is rapidly becoming a mainstream tool for cadastral surveying, especially in a real-time kinematic mode where one-centimeter accuracy is now available. GPS can be used in this mode for all forms of cadastral surveying, from measurements to setting out boundaries. The only difficulty is that some cadastral surveying regulations do not yet recognize GPS measurements because of issues of traceability of the accuracy of the measurements.

◆ **Digitizing and scanning:** Scanning or digitizing paper maps is a common tool to convert an analog cadastral map to digital form. Most LAS around the world have, or are using, some form of scanning or digitizing to upgrade LAS (figure 12.18). The one weakness in the resulting map (and data) is graphic accuracy. As a result, LAS often need to convert all the old analog survey measurements to digital form so that the system can move fully into an ICT environment.

DCDBs

A cadastre can be conveniently defined as a parcel-based and up-to-date land information system (LIS) containing a record of interests in land (e.g., RRRs). It usually includes a geometric description of land parcels linked to other records describing the nature of the interests and ownership or control of those interests and sometimes includes the value of the parcel and its improvements and planning controls. A DCDB is the representation of the geometric component of a cadastre in electronic format.

A DCDB usually consists of the following data:

◆ Parcel boundaries
◆ Parcel identifiers
◆ Easements

And sometimes additional components such as

◆ Property boundaries
◆ Building footprints
◆ Street addresses
◆ Administrative boundaries
◆ Valuation data
◆ Other land-use features (gardens, roads, railway lines, forests, etc.)
◆ Planning zones and land uses

Boundary coordinates in a DCDB can serve two roles: first, to support the establishment of the DCDB, and second, to facilitate boundary reestablishment or redefinition.

A graphically accurate DCDB is one where the coordinates of all corners are determined graphically, usually by digitization of original hard-copy plans and maps or by inputting data from isolated surveys.

A survey-accurate DCDB is based on coordinates determined by ground surveys, which are used to define, describe, and redefine parcel boundaries. For all practical purposes, the coordinates in this DCDB are the true coordinates—the result is a fully coordinated survey system as part of a coordinated cadastre.

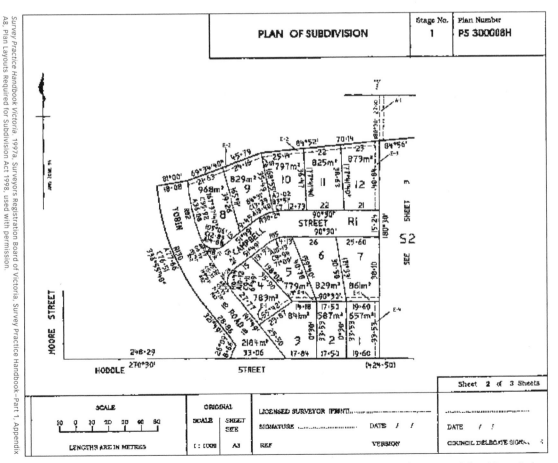

Survey Practice Handbook Victoria, 1997a, Surveyors Registration Board of Victoria, Survey Practice Handbook–Part 1, Appendix A8, Plan Layouts Required for Subdivision Act 1998, used with permission.

Figure 12.18 A typical plan of subdivision in Australia resulting from cadastral surveys of fixed boundaries.

The reality is that many systems are a mix of both graphic and survey-accurate data. Importantly, the design of the system has significant implications for upgrading the accuracy of the system as discussed as follows.

DCDB as an important component of an SDI

A DCDB is a form of infrastructure but not a complete SDI and is often referred to as the parcel or property layer within an SDI. However, the DCDB forms an integral component of large-scale SDIs, where the cadastre is a fundamental dataset (figure 12.19). By using the geodetic network and cadastre as fundamental datasets of an SDI, it is possible to build and integrate other spatial information datasets, particularly administrative, address, and utilities layers.

Updating DCDBs refers to those processes that ensure that all new and existing legal subdivisions are recorded—i.e., that the cadastral map or DCDB is up-to-date. Specifically, this should include

 ◆ Recording all new legal subdivisions
 ◆ Ensuring map completeness

Upgrading DCDB activities (an improvement, not maintenance) can include the following:

 ◆ Increase in accuracy
 ◆ Inclusion of survey measurements
 ◆ Alignment of cadastral features with topographic features
 ◆ Changes in the data model
 ◆ Generation of topological structures
 ◆ Inclusion of a historical layer
 ◆ Creation of unique identifiers for spatial entities

Considerations in selecting the most appropriate surveying and mapping tool

As is evident from this discussion on cadastral surveying and mapping tools, a wide range of choices need to be made in order to establish the best set of tools for a specific jurisdiction. Choices are also influenced by existing laws, institutions and processes, historic and colonial influences, and even cultures. The different technologies can be rated on matters of accuracy, simplicity, cost, efficiency, utility, and flexibility as done by P. F. Dale and J. D. McLaughlin in their book *Land Information Management* (1988). However, a broader evaluation is required that can take in the following considerations:

 ◆ The form of boundary is obviously central. For example, are general boundaries or fixed boundaries the primary choice?

- In major LAS projects, the land titling strategy is central to many of the choices of tools and is particularly influenced by whether a systematic or sporadic strategy is adopted.

- Land use and people-to-land relationships can have a significant influence on the choice of tools. Is the cadastral surveying for high-value, high-rise buildings in the center of a city or for low-value mountainous or desert regions? Or is the cadastral surveying for new high-value subdivisions for expensive homes or to determine initial tenure for occupants of informal slums?

- Availability of technology is a key issue. For example, is digital theodolite technology, together with all the software to process and use the data, available and well maintained in the country or region? If the technology fails, can it be easily repaired at a reasonable cost?

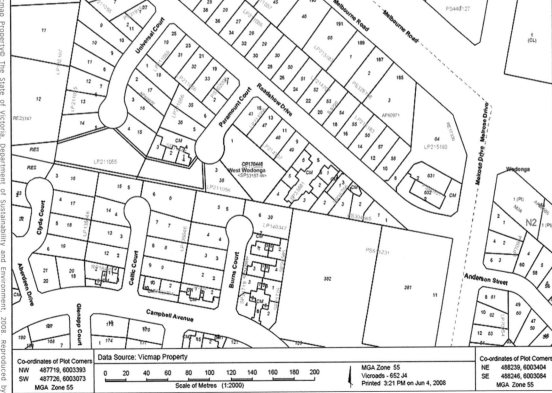

Figure 12.19 A page from the cadastral map of Australia's DCDB shows property in detail.

THE DIGITAL CADASTRAL DATABASE—THE CASE OF DENMARK

As described in chapter 2, the Danish cadastre was established in 1844 for collecting land taxes from the agriculture holdings based on the quality of the soil. The resulting property framework from the enclosure movement (see figure 2.9) formed the basis for the new cadastral maps established in the early 1800s. These maps were surveyed by plane table at a scale of 1:4,000. Each map normally includes a village area and the surrounding cultivated areas. As a result, the maps were "island maps" and not based on any local or national grid (figure 12.20). These old analog maps have been maintained over time with subdivisions and cadastral alterations.

The process of digitizing analog maps was undertaken in two stages. First, state control points and cadastral surveys connected to the national grid were entered into the map to form a "skeleton" cadastral map. In urban areas, about 40 percent of the boundary points were entered this way and in rural areas about 20 percent. Second, the remaining parcels were inserted by digitizing the analog map and fitting these into the skeleton map by transformation (figure 12.21). Identified elements in the digital topographic map were also used to support the transformation.

Using this process, the accuracy of the boundary points in the resulting digital cadastral map may vary considerably, ranging from a few centimeters in some urban areas to several meters in rural areas. Therefore, the digital cadastral map will not be totally consistent with a digital topographic map.

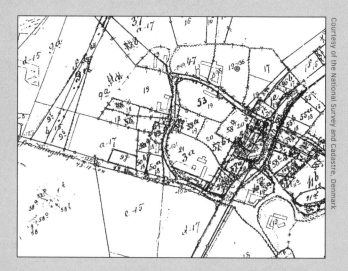

Figure 12.20 Part of an analog cadastral map from 1983 as used, rectified, and updated over a period of about 100 years. The map is an "island map" and is not linked to a national grid network. The display is difficult to interpret, and even though the map was redrawn in 1984, it is not efficient for daily land-use administration.

Courtesy of the National Survey and Cadastre, Denmark

The accuracy of the boundary points relates to the way they are established in the map. This information is therefore attached to the boundary point in the database. Other metadata includes information about the type of boundary and the file number from the cadastral archive of the National Survey and Cadastre of Denmark.

In the analog cadastral maps, new boundaries were adjusted graphically to the position of existing boundaries. This process is reversed in the digital map, where new cadastral measurements are used for adjusting the position of existing boundaries. This dynamic process will ensure an ongoing improvement of the accuracy of the DCDB.

The DCDB may also be improved by upgrading certain areas—e.g., in relation to major land development projects. This process includes a new transformation of the existing boundary points based on identification and positioning of a range of boundary points within the area.

The DCDB includes a number of problems deriving from the history of the old analog map and the process of computerization. Successful use of the DCDB depends on the degree of educated use of the map.

In summary, Denmark's establishment of a DCDB has provided the opportunity to combine the cadastral identification with the topographic information to support efficient management of land RRRs in a sustainable way.

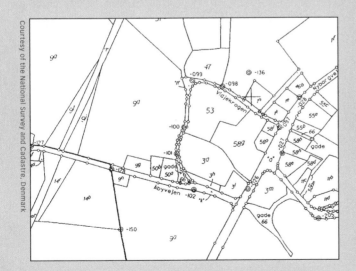

Courtesy of the National Survey and Cadastre, Denmark

Figure 12.21 The digital cadastral map from 1993 shows the same area as figure 12.20, but the map is now linked to the national grid as a "frame map" showing only the current cadastral situation. The boundary points, shown by circles, are established in the map using control points and cadastral measurements. The digital cadastral map is thus tailored for integrated land management.

◆ Obviously, cost is a central issue including the cost of labor. Is it better to employ teams of low-cost surveyors to do plane table surveys, or are fully digital technologies affordable and maintainable? While it may be possible to obtain the latest technology from an aid project or international development loan, are the organizations getting this high-technology equipment able to maintain it in a reliable and accurate manner?

◆ In the end, a decision on cadastral surveying and mapping tools comes down to the socioeconomic development and capacity of the country or jurisdiction. The choice of technologies and tools must be in keeping with what is affordable, what can be maintained and serviced, and importantly, what educational and training capacity is available to use the tools effectively.

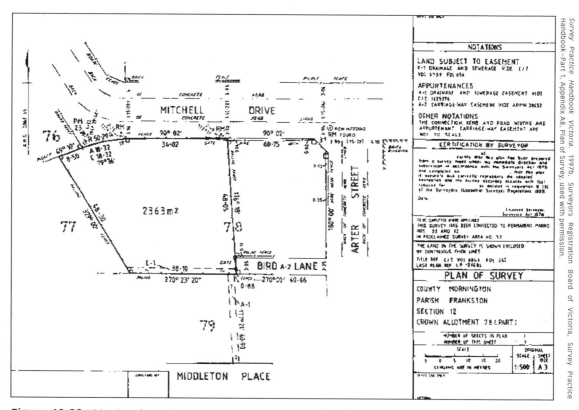

Figure 12.22 This plan of a cadastral survey in Australia uses fixed boundaries.

Survey Practice Handbook Victoria, 1997b, Surveyors Registration Board of Victoria, Survey Practice Handbook–Part 1, Appendix A8, Plan of Survey, used with permission.

The resulting cadastral survey system in each jurisdiction will be different. However, there will be similarities worldwide in the resulting cadastral survey plan and resulting field notes. Figures 12.22, 12.23, and 12.24 are just a few examples.

7. BUILDING TITLE TOOLS

Building tenures

As land becomes scarce and transportation costs rise, buildings, rather than surface land, are a major development opportunity for commercial, residential, and even industrial land uses. Historically, building access in capitalist systems was managed by leases and in central economies by bureaucratic allocation. Both economic systems sought to handle building access and management as closely as possible to the way on-the-ground land access is managed and new

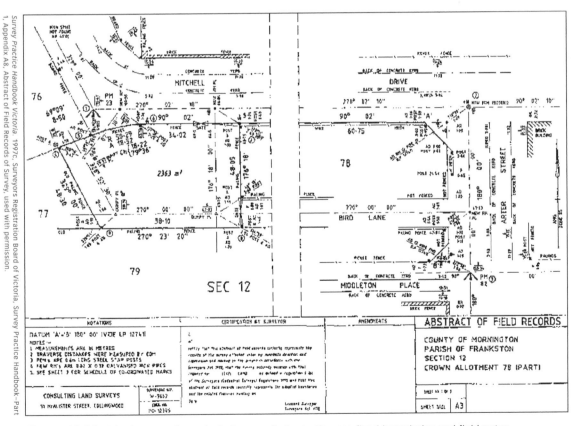

Figure 12.23 This abstract of a cadastral survey in Australia uses fixed boundaries and field notes.

forms of tenure developed. In capitalist systems, "flying freeholds," or condominium titles, were invented, which had the same characteristics as the title to a land parcel. In centralist systems, use rights, occupancy rights, or other entitlements gave stronger protection against bureaucratic interference. The condominium style of building ownership (figure 12.25) is best practice for mortgage financing, building management and replacement, transaction efficiency, and engaging condominium owners in management processes.

Building titles that support land parcels and buildings characterized as being purpose built for strata or condominium use are popular and flexible. Such parcels provide housing for millions and a large number of business facilities. Developments range from the very simple and small scale to the large vertical villages supporting multiple uses and hundreds of unit owners. These units, buildings, and management systems form a significant part of a nation's real property assets. They are increasingly used to create facilities attractive to modern commercial, industrial, and residential uses. Because these assets are multiply owned, they increase the flexibility and options for those seeking secure housing, workplaces, or property investments. They are

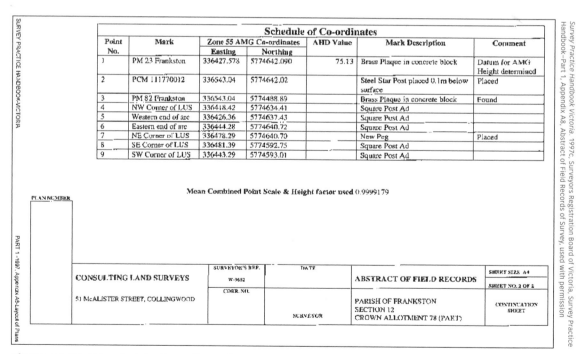

Figure 12.24 This schedule of coordinates from a cadastral survey in Australia uses an accompanying abstract of field notes.

contrasted with the large buildings and retail centers that are typically owned and managed by investors or property trusts, where occupation by retailers and manufacturers is through a lease or some form of derived or dependent title.

A basic model for building titles is derived from a New South Wales strata title developed in the early 1960s before computerization became a part of surveying and land registration. The achievement of the NSW model is the conversion of a raw land subdivision system to a subdivision of buildings comprising multiple stories and multiple purposes. The model is emulated, with various changes, throughout the world. Its generic qualities are most recently described in the Guidelines for Ownership of Condominium Housing produced by UNECE in 2003. The essential ingredients of the model are separate titles to units, separate title and clear common ownership of shared property and facilities, a management system combining the owners into

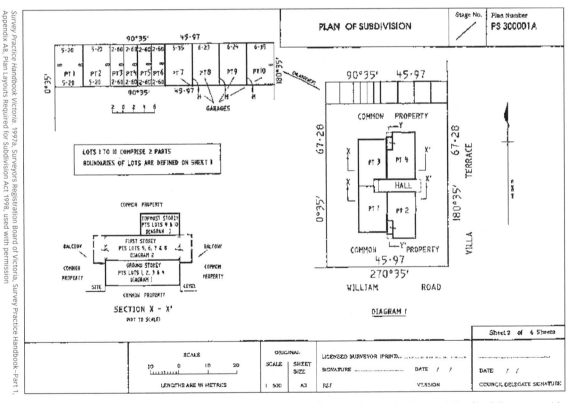

Figure 12.25 A plan of subdivision for strata titles in Australia uses the condominium style of building ownership.

TABLE 12.5 – BUILDING TENURE TOOLS	
TOOL	**DESCRIPTION**
Community living	The building owners are a group that live in close proximity, similar to the inhabitants of a village. They share responsibilities and facilities and need to restrain their behavior in consideration of the needs of their neighbors. To achieve cooperation, the building manager needs open communication systems, newsletters, complaint systems, and dispute-handling systems.
Ownership of units	Each unit is owned separately and traded as a real estate asset in the national land market.
Titles	Each unit has a separate title, conferring the same legal right as in an ordinary land parcel.
Common property ownership	The common property, including the building itself, land under the building, and the air above, needs a separate title that is clearly owned. Ownership must belong to all current unit owners, without imposing administrative processes of data entry to track changes in unit ownership. Two mechanisms are available: the creation of an owners corporation or creation of common ownership that automatically shifts shares according to the transfer of the units.
Distribution of equity, costs, risks, and profits	The strata or condominium organization of the building includes a schedule of entitlement of some kind to allocate the owners' shares in the common property and for payments to cover annual and unforeseen costs. The entitlements are often calculated on the basis of approximate value and size of the units.
Management of the building	Buildings are complex and need constant maintenance. The owners corporation or common owners usually appoint a manager and a small management committee of owners.
Bylaws and rules	Each building is unique and will have rules that apply to its owners and occupiers. These rules might have a statutory status or be registered in the land registry office. The legislation will have model bylaws that can be modified to suit the owners.
Quality controls	The building needs a plan for maintenance throughout its life, including replacement of large features such as the roof, staircases, and water pipes.
Boundaries	Best practice is to use a general boundaries system, so that the internal walls of the unit define its extent. No benefit is derived by precisely measuring a unit and showing the results on a survey plan. The outer boundary of the development forms the parent parcel. The boundary of each unit in the building itself must be defined, depending on the distribution of responsibility among the owners corporation or entire group and individual owners. There are several basic approaches. First, if the unit boundary is the inner walls, ceilings, and floors, the entire group of owners is liable for anything that goes wrong within the walls, pipes, and wires. Second, if the boundary of the apartment is the outside walls, the individual owner may bear these risks. Third, if the boundary is in the middle of the walls, determining responsibility may be difficult. Legislative provisions may clarify these issues.

Continued on facing page

Continued from previous page

TABLE 12.5 – BUILDING TENURE TOOLS	
TOOL	**DESCRIPTION**
Total destruction or demolition	Buildings do not last forever. City planning sometimes requires demolition, and sometimes buildings deteriorate to the point of condemnation. The title system allows for sale, replacement, and rebuilding options and sets up a voting system so that most of the owners are able to manage the process. A requirement for an absolute majority will paralyze redevelopment.
Insurance	The building and its use will impose risks on the owners. These risks must be insured against. The largest risk is injury to members of the public who enter the building and its common facilities. The owners as a group can collect insurance premiums and negotiate a single policy to cover public liability and other building risks—for example, damage by storm, fire, or water. Otherwise, each owner needs separate public liability and building insurance, and the owners as a group need insurance for risks associated with common property. All owners need separate insurance for their fittings and fixtures.
Accounting and financing	Funds for rates, insurance premiums, management expenses, common facilities, and utilities are essential. Therefore, the strata or condominium organization of the building must be supported by methodical bookkeeping and rules about handling of public money. Finances for large buildings will include substantial sinking funds to cover extensive repairs and things such as elevator maintenance. Sometimes, the building owners as a group will be a taxable entity.
Disputes	A system for handling disputes and complaints needs to operate within the strata or condominium organization itself. For intractable disputes, an independent dispute resolution system is necessary. This can be the standard courts or a specialized building dispute tribunal.

a legal organization capable of appointing a manager of the building, an accounting system that facilitates cost sharing, and a disputes system. See table 12.5 for more detail on building tenure tools. For countries with land registration, including Australia where eight Torrens systems manage building titles (though some multioccupancy buildings rely on company share scheme titles and other out-of-date systems), these titles need to sit satisfactorily within systems originally devised for managing vacant land.

The legal and practical arrangements for buildings must maximize cooperation and minimize disputes. The tools in table 12.5 are generally used because experience suggests they work well; however, each country will adapt these tools to suit local needs. The key is to establish systems that ensure good management of common facilities and the building itself, not only for its occupiers and owners, but for members of the public. One of the greatest confusions arising with buildings stems from people thinking of them in terms of their physical features, whereas the asset is the cube of airspace identified by the walls, ceilings, and floors of the unit. Although the model of private ownership applies, the style of living requires each owner to accept mutual responsibilities and obligations that are significantly different from those of an owner of a stand-alone residence. Management of modern buildings requires a range of competencies, and in many cases, this means that professionals need to be employed.

Countries with highly developed LAS and effective property markets use a model for titling buildings that provides three separate administrative frameworks (Stoter 2004, 7). While these three models are similar to and build on accepted LAS framework, they are not usually built in less developed countries:

- **Juridical framework:** the legal status of stratified properties and particularly the RRRs of their owners

- **Cadastral framework:** the capacity of the plans of the entity to be stored in and related to other parcels in the land administration system, particularly the land survey system

- **Technical framework:** the system architecture (computer hardware, software, and data structures) supporting cadastral registration

Destruction by accident or demolition of the building is the major issue with the longevity of these titles. In English-based systems, the estate or interest in the strata land exists even though the building is destroyed. In other systems, the interest is lost. In either system, an owner of a building that's destroyed is faced with the costs of redevelopment or sale of the site at land value. Either way, the interest in the building needs to be managed to preserve value and insured to cover accidental or other destruction. The overall laws governing building developments should provide for destruction either by choice or through necessity following deterioration of its fabric.

3D Cadastre

The two-dimensional, standard survey plan needs substantial modification to provide sufficient detail about a building's upper levels. Using multiple pages to represent the various levels works for smaller developments, but as developments become more complicated and multiuse, the readability of plans diminishes. The solution is development of a genuine 3D cadastre, capable of depicting height as well as length and width. While computer-assisted design systems support architecture and engineering (with sufficient accuracy for even the most technically complicated designs), surveying standards as yet do not permit a 3D model (Stoter 2004). It would take years for most countries to develop appropriate and usable systems.

12.4 Emerging tools

The need for new tools

Since the Millennium Development Goals were identified, much more attention is being paid to acceleration of delivery of security of tenure for countries whose land administration capacities are nonexistent or merely inadequate. The focus is on tools for poverty alleviation rather than market effectiveness. The urgency of the situation is pressing. The traditional tools of titling and formal land administration cannot be implemented quickly enough to answer immediate needs. Given the sheer numbers of people and parcels, the world's existing and even potential capacity to create formal LAS through land titling and administration cannot deliver extensive security of tenure, let alone sustainable development objectives. Though the traditional tools, including large-scale land titling, are important (Biau 2005), too much is required for many scenarios. The global number of slum dwellers may double to 2 billion unless mitigating action, including stabilizing rural land use, is undertaken in the next thirty years (Augustinus, Lemmen, and Van Oosterom 2006).

This reality does not undermine the idea that formal systems are needed. They are. But these traditional tools need to be augmented by newer tools. Intermediate tools capable of stabilizing land use are essential to bridge these ever-increasing gaps. Quicker, more deliverable, flexible, and scalable tools are needed, as initiatives in developing countries tend to start at the ground level and work upward (Van der Molen 2006). Simultaneously, countries need to build robust national LAS capable of absorbing these emerging tools.

There is no blueprint for doing this, but it is now accepted that land should be delivered with sanitation services, water access, and appropriate buildings for living and working, while simultaneously managing less densely settled land (forests and farms) so that resources are preserved. The Global Land Tools Network (GLTN) is the focal point for organizing these activities in the context of urban land. Its outputs include Islamic land tools, gendered land tools, and pro-poor land tools, now in various stages of development. International aid agencies and UN agencies are undertaking substantial work on these emerging tools and on increasing the capacity of formal systems to absorb these initiatives. The search for practical tools is shared by the Commission on Legal Empowerment of the Poor (UNDP 2008).

1. PRO-POOR LAND MANAGEMENT TOOLS

New land administration theory distinguishes pro-poor from land market tools, not because of antimarket sentiments but simply to focus on the poorest of the world's people and their security of tenure. In general, the differences between pro-poor and land market tools are easy to explain. One set depends on social systems, the other on legal systems, as shown in table 12.6. Thus, established LAS in the twenty-first century need to meet two different sets of deliverables: support for market-based land distribution tools and tenures and for social tools and tenures (figure 12.26).

The emerging tools aim generally at bringing social systems into the formal system and increasing the flexibility of formal systems to permit incorporation. For example, allowing land registry agencies to undertake simple recording of all kinds of paper evidence about land

TABLE 12.6 – CHARACTERISTICS OF PRO-POOR VS. MARKET TOOLS		
DELIVERABLE	**PRO-POOR**	**LAND MARKET**
Access to land	Socially derived system	Legally derived system
Sources of authority	Social system	Legal system
Sources of protection	Social practices	Legal rights
Disputes	Local system and authority	National and highly formalized system
Formalities	Low, secondary evidence such as oral, ceremonies, etc.	High, using formal documents
Starting point	Secure access	Secure rights
Evidence	Observable practice, oral	Formal documents and registration
Transition	Inheritance systems	Transaction systems
Boundary delineation	Observable and practical: levees, paths, marked trees	Formal systems: surveys and maps
System	Social husbandry devices	LAS
Cognitive capacity	Socially internalized	Market understanding

Figure 12.26 Customary tenure in Ghana requires pro-poor land management tools.

interests in a spatially enabled system or map for very low fees would allow poor owners with access to registration offices to build up their evidence of ownership.

The new tools also need to meet anticipated problems, which vary according to culture and geography. For instance, in Asia and the Pacific region (APR), the range of problems includes

- ◆ **Land-use planning:** Very poor capacity to manage land-use planning and land development is common. Most LAPs concentrate on delivery of land through titling, while the planning systems remain separate.

- ◆ **Land acquisition:** In Vietnam and Indonesia, major litigation and disputes are about land acquisition (taking) and payment of value. China is also showing a similar pathology. Land is acquired from farmers and reallocated to commercial or residential use with a concomitant increase in value. Farmers receiving low-end or even derisory compensation are unhappy with the allocation of development rights and commodification of these rights into profit streams by the land taker or developer.

- ◆ **Separation of land and buildings:** The separate treatment of land and buildings might make sense at an initial stage of normalization as, for instance, in Timor-Leste's initial property law, but it raises problems for LAS that are compounded when markets develop.

◆ **Informal land markets:** Informality in APR countries never disappears. Informal markets can be economically dramatic: Witness the Hanoi market, where prices are "as high as Tokyo." But they lead to a lack of government engagement to capture transaction and other taxes. APR registration programs fail in particular because informal practices survive without the capacity to manage dual systems (figure 12.27).

◆ **Oral transactions:** Land transactions in Asia are often undertaken on good faith and familiarity, not formalities and registration. In legal systems where oral transactions are the norm (as in Indonesia), no registration program can reflect reality.

◆ **Inheritance:** Programs of formalization in APR countries are generally unable to coherently capture land changes on death.

◆ **Land hoarding and speculation:** The fear of land hoarding leads to restrictions on landownership by foreigners and corporations and controls on individuals and families. In countries with relatively poor administrative capacity, these are ineffectual and create opportunities for informed land grabbers (Lohmann 2002; Leonard and Ayutthaya 2003). For economists, these restrictions and controls represent a point of tension between market ideals and government intervention. The balance of national opinion in APR countries is squarely in favor of more and effectively implemented restrictions, rather than fewer.

◆ **Confusion of land administration initiatives with land reform:** Even the most conservatively and technically designed LAP has political consequences and is difficult enough to manage. Using LAPs to run land reform initiatives is much more difficult (Bledsoe 2006).

African countries especially need pro-poor tools to address chronic land management and administration problems. Indeed, most of the pro-poor tool initiatives are sited in African countries.

The development of pro-poor tools involves many hundreds of people and agencies—indeed, so many are involved that singling out ones for mention is done only to introduce the Internet sites where much of the work is documented. Among many sites, those that receive attention include sites of the UN agencies, particularly UN–HABITAT, the World Bank, and FAO; ITC; the Lincoln Institute of Land Policy in Cambridge, Massachusetts; international development agencies such as Norwegian Aid, SIDA, GTZ, DFID, GRET, IIED; expert agencies such as Netherlands Kadastre and many others; and professional organizations such as FIG, CASLE, and others. Contributions by academic researchers to development of pro-poor tools are extensive.

Figure 12.27 Informal settlements in Kenya pose a challenge to land registration programs.

New Web sites are developing constantly. An overview is available on the UN–HABITAT site for the GLTN (`http://www.gltn.net/`)

The stage of development of any particular tool reflects the emerging interest in delivering systems for tenure security that are precursors to national efforts. The core values behind pro-poor tools are governance, equity, subsidiarity (the principle that management of land should fall to the lowest possible level of appropriate competence), affordability, systematic large-scale approaches, and gender equity. Most of the tools are "under construction" compared with the refined technical tools familiar in developed economies.

Social tenure models

The integration of customary or informal land rights into formal systems and the delivery of secure tenures for the millions of people whose tenures are predominantly social rather than legal is the focus of the social tenure domain model, now under development.

> *"Rights such as freehold and registered leasehold and the conventional cadastral and land registration systems, and the way they are presently structured, cannot supply security of tenure to the vast majority of the low-income groups and/or deal quickly enough with the scale of urban problems. Innovative approaches need to be developed."*
> *(UN–HABITAT 2003)*

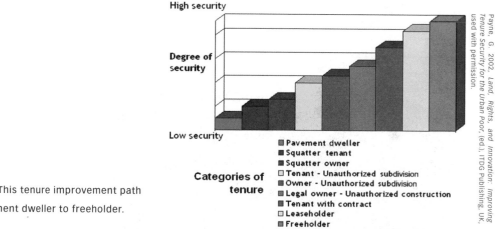

Payne, G. 2002, *Land, Rights, and Innovation: Improving Tenure Security for the Urban Poor*, (ed.), ITDG Publishing, UK, used with permission.

Figure 12.28 This tenure improvement path goes from pavement dweller to freeholder.

A solution to this problem may be found in the Social Tenure Domain Model (STDM) originally developed as the Core Cadastral Domain Model (CCDM). The key issue here is that, in the technical field, there is often an insufficient focus on pro-poor technical and legal tools. In the development of the CCDM, efforts are being made to avoid such criticism; a lot of useful functionality has been developed, but the name of the model, class names, and terminology used are still too much aligned with formal systems. For that reason, the social tenure domain is being proposed as the next step for research, which could be a specialization of the CCDM based on domain-related terminology (Augustinus, Lemmen, and Van Oosterom 2006).

Savings and other cooperative schemes
Communities the world over are striving to create their own solutions to land problems. Setting up local cooperatives to produce salable products, savings schemes to allow a group of innovative people to save and purchase land, small loans on microcredit to finance productivity initiatives, and many other mechanisms are appearing. Of these, one of the most successful involves

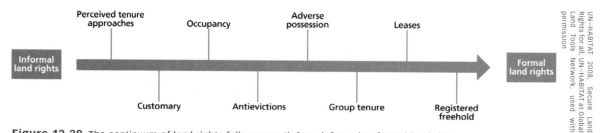

UN–HABITAT 2008, *Secure Land Rights for all*, UN–HABITAT at Global Land Tools Network, used with permission

Figure 12.29 The continuum of land rights follows a path from informal to formal land rights.

Figure 12.30 Nomadic or customary tenure is the norm in Mongolia.

savings schemes among groups of women to facilitate purchase of land or a house in common, then sharing the residential opportunities. In some scenarios, this might move into separate parcels. Namibia and Kenya are developing innovative savings plans of this kind.

Scaling up

While many of the tools are experimental, some have gained a foothold, particularly those focusing on activities in the lower-end tenures of leases, or even squatting. G. Payne (2002) and UN–HABITAT (2003, 2008) use a betterment path for urban areas (figures 12.28 and 12.29), ranging from pavement dweller through squatter, to ownership. Scaling up means formalizing land arrangements to give the beneficiaries greater security of tenure. This process does not mean that all societies will develop freehold tenure systems (figure 12.30). Each step in the process can be formalized (figure 12.31), with registered freeholds offering a stronger protection than at earlier stages.

The development of a similar, scaling-up process in the context of rural tenures is much harder, because the penetration of formal systems has been slower. One attempt to develop a scaling-up chain, using possession as a starting point, is shown in figure 12.32.

Figure 12.31

Street traders in Thailand exhibit both informal and formal occupation.

2. NONCADASTRAL APPROACHES AND TOOLS

Land management and administration in Africa has suffered from lack of capacity and poor governance. In most African countries, these systems are operating below sustainable levels with some even at crisis level. National cadastres remain a distant goal. There are positive signs, however. In Africa, conventional land administration tools are seen as remnants of a colonial heritage, and new responses to organizing land administration have begun to focus on inclusion of traditional tenures and systems that predate the colonial era. The misfit between the conventional tools of individual ownership and precisely surveyed boundaries and the customary tenures that embraced flexible responses to climate, water availability, and grazing opportunities is stark. The customary forms of tenure evolved, including flexible tenures, certificates of occupancy, village title systems, local land management systems, and many others. The design of many of these flexible options, however, remained formalistic, and their actual implementation often failed (Land Equity 2006, 103; Payne, Durand-Lasserve, and Rakodi 2007, 27).

Some of these emerging tools have moved from cadastral to noncadastral approaches, a trend made possible by the increased usability and utility of spatial information systems and satellite images. The inclusion of social tenures in the standard cadastral models is also improving options for inclusion of nonstandard land interests in formal administration systems (Augustinus,

Lemmen, and Van Oosterom 2006). These noncadastral tools include a move to land management—e.g. for slum upgrading—to manage conflicts or to allocate land to internally displaced persons and refugees. Support is therefore required for non-parcel-based object identification within LIS. None of this is expected to be easy. The extension of land administration to areas of slums and customary communities, with overlapping and non-polygon-shaped rights and claims, is an unexplored challenge given the complexity in local and customary land practices that defy ordered mapping approaches.

A major new focus is on land management tools, rather than titling tools. Land management is especially relevant for groups whose distance from land markets, preference for communal rather than commercial values, and hope for self-determination require inventive responses by governments and agencies that seek to help (Toulmin and Quan 2000). The land administration approach to group land management is now well established with more than thirty years of theory and practice (Lavigne Delville 2002b). Spatial technologies, combined with high-resolution satellite images, are the foundation of attempts to create coherent administration systems within these fluidities. The fundamental concepts revolve around creating autonomy for

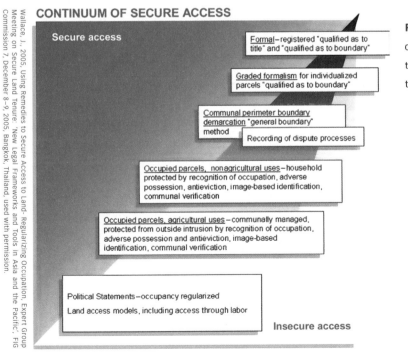

Figure 12.32 The continuum of secure access to rural land goes through the stages from informal to formal occupation.

Figure 12.33 Gender issues are central components of customary tenure in Malawi.

the local land group through a boundary identification system that keeps strangers out and allows internal management systems to evolve as needs arise.

3. GENDER EQUITY TOOLS

In the years since the development of the land administration toolbox, there have been calls within the literature (Haldrup 2002; UN–HABITAT 2006c; FIG 2001; World Bank 1996) for "gendered" tools. The establishment of a gendered toolbox would serve to compile the array of strategies that address gender equity at the grass-roots level. While efforts in this regard are being made by the GLTN, the United Nations, the World Bank, and many others (figure 12.33), a complete, well-defined, and widely accepted gendered toolbox still does not exist. Meanwhile, a concerted effort is being made to introduce gender responsiveness to standard land administration tools (GLTN/FIG 2008). Furthermore, existing gendered tools do not fit in with the usual tools used to deliver security of tenure. Specifically, gendered tools have struggled to deal with the evolution and dynamic nature of the social and cultural norms of different countries (Payne 2004; Schech and Haggis 2002).

Women of the West took years to achieve the gender equity they currently enjoy: They can own land and raise capital against that land in their own capacity, without the signature of a male authority figure such as a brother, father, or husband. These achievements took more than a century to achieve and required cooperation from male politicians and power holders outside the immediate family.

For women in developing countries, the problems are profound and will take longer to address. Security of tenure for women remains a pro-poor goal, given their responsibilities for child nurturing and raising and their contributions to food production. The tools used to drive gender equity will vary according to the national scenario. If the land is predominantly titled, reforms will focus on the titling system by gendering the large-scale national system. However, most countries with gender inequity lack rudimentary titling systems, and the tools come from outside LAS.

A preliminary attempt at developing gender land tools is shown in table 12.7.

TABLE 12.7 – GENDER LAND TOOLS	
TYPE OF TOOL	**TOOL**
Land administration tools	1. Careful and targeted land titling 2. Joint titling and shared tenure (UN–HABITAT 2005) 3. Cooperative purchasing of land 4. Inclusion of women's names on utility bills 5. Recognition of agricultural and domestic labor
Law-related tools	1. Inheritance rights for women 2. Guaranteeing land rights to children 3. Legislating intrahousehold rights 4. Constitutional changes 5. Elimination of legal restrictions on women owning land 6. Streamlining land-related agencies
Economy-related tools	1. Microloans through microfinance institutions 2. Microloans that are not money based (for example, supply of fertilizer in return for crop share)
General tools	1. Education of people about their rights in land 2. Women's land groups 3. District dispute resolution tribunals and processes

Many other small initiatives are less formal, such as

◆ Providing credit in the name of the woman following the Grameen Bank model — the woman's name goes on the receipt and she manages repayment

◆ Allowing women to join farmers unions and organizations in their own right

◆ Rewarding labor contracts on land and tracking the contributions

◆ Inheritance tracking — including women in the path of formal recognition

◆ Allowing women vegetable plots for production of family food (Indonesia)

◆ Adopting a memory book (Uganda) as a method of making families aware of both assets and HIV status

4. HUMAN-RIGHTS TOOLS

The expanded project literature covers the variety of tenures used, their inherently local nature, new spatial identification systems, and the transitional processes experienced by many social groups. The "one size fits all" private title approach has been replaced in land administration theory with more resilient assumptions that the changing needs of a given population will drive changes in tenures at the same time as these needs drive normative shifts and behavioral modifications through very complex processes that are unique to the situation. These assumptions lead to bottom-up, not top-down project design. Sometimes, land administration and land reform projects seek specific formal legal changes even at a national level to complement these processes. These more complex projects are specifically designed to deliver political and social outcomes, not merely technical outcomes such as changed laws, registered titles, or private ownership. They do not involve changing institutional frameworks and laws in the hope that the changes will penetrate into the local practice according to a remotely devised, bureaucratic plan.

These complex projects rely on essential social components and processes to engage beneficiaries in change management, in addition to the technical components of legal change. These components build on the tools described in the previous sections but focus heavily on social components and processes and include

◆ A thorough, on-the-ground social evaluation of extant conditions to inform the design of the suite of changes

◆ A thorough appraisal of the tenure-related processes actually used to describe, allocate, distribute, conserve, and transfer land

- A vocabulary of the conceptual frameworks that give social, spiritual, and economic meaning to these processes

- A projection of the possible positive and negative impacts the project activities will have

- A monitoring and evaluation program to ensure results achieved are related to desired outcomes

Projects are then continuously evaluated against these information sets, with consequential and systematic adjustment to the overall design from time to time. With any luck, during the project and specifically at its completion, designers and financial contributors will be able to say with certainty that the intended beneficiaries received the desired benefit by virtue of project activities. Even with this armory of components, in most land administration experience, project designers still cannot determine why they failed to achieve desired results and how to do better the next time. Understanding how land works in relation to people is no easy task. But no self-respecting land administrator would jump in first with a proposal that tenure be fundamentally changed without a well-grounded preliminary evaluation and well-conceived, clear, follow-up measures. Moreover, the latest research suggests that tenure changes are insufficient in themselves (though sometimes necessary) to create desirable and sustainable outcomes, especially regarding land conservation. They must be accompanied by infrastructure improvements (including garbage treatment, roads, and utility supply), investment in education, social planning, and many other composite efforts to be able to attack poverty (Kabubo-Mariara 2006).

The complexity of titling programs compounded by the appalling human needs of the world's poor, especially those crowded into urban slums, has excited an urge to even more basic approaches to stabilizing people's relationships with land. Nothing is more basic than provision of clean water and sanitation (figure 12.34). Giving people secure water supplies can become a first step in building up land security, especially if the water system is organized. Unconventional approaches involve attempts to make the right to water and sanitation a fundamental human right and public good (Tipping, Adom, and Tibaijuka 2005). The provision of sanitation, water, and other amenities to socially tenured land remains problematic (Du Plessis and Leckie 2006) and must be done through welfare systems when the land generates no income stream suitable for cost-sharing.

Similar concepts rely on building evidence chains by tracking people's use of land through organization of electricity supplies (using regular receipts as land occupancy evidence) and payment of occupation taxes (again, using receipts or government records as proof of land-use patterns).

Figure 12.34 A land administration project in the Philippines considers access to water a fundamental right.

Thus far, land has been the central focus of LAS. But changes in global weather patterns have seen another, perhaps more fundamental aspect of security of tenure—the availability of water. There is no doubt that security of land tenure contributes to the willingness of people to move from vendor-provided water (which is notoriously more expensive per liter than reticulated water) to other kinds of service provision. A secure land and water relationship is also a vital component in the decision of commercial water system builders to provide water infrastructure to neighborhoods. In the cycle of land administration theory and practice, the future may demand that we focus on provision of water to land in addition to all the other demands made of modern systems.

Chapter 13

Project management and evaluation

13.1 Project context

13.2 Designing and building land administration systems

13.3 Evaluating and monitoring land administration systems

13

13.1 Project context

In this overview of project management and evaluation, specialized aspects of land administration projects are considered, as well as how to design and evaluate entire systems. It is not a manual, or even a detailed consideration of the topics. Useful management tools are introduced. The benefits, weaknesses, issues, and problems associated with project management are highlighted. In practice, the overview needs to be expanded by the specialized information available in the many books written on project management and evaluation, as well as the many project descriptions available on the Internet from organizations like the World Bank and the many development assistance organizations and NGOs involved in LAPs in developing countries. A key aspect of LAPs is to improve the quality of life in the affected region (figure 13.1). This requires LAPs to be carefully designed and managed.

However, without political support and leadership from government, land administration systems will not be successful, nor can they even commence. For example, governments may decide that a country needs a more efficient land market. A project with this objective has a chance of success if strong economic, social, and environmental justifications are recognized and political will is present. For most developing countries, organizations like the World Bank provide substantial loans and advice, while high-quality and appropriate technical assistance is usually provided by foreign countries through their development assistance programs. An appropriate project design might be produced, but without real political enablement, the project will fail. At the early stages, confirmation of commitment to the project and leadership at the highest levels of government, ideally from a prime minister or president, is essential. Additionally, every project needs a "champion" who has a clear management responsibility. This is particularly important when the international donors disagree with the host country or recipient on priorities and strategies. Difficulties can also arise when the international consultants to the project, who are paid by an international donor, and the recipient country have different priorities. In this case, strong leadership is required to balance the ambitions of each party and negotiate a way forward.

13.2 Designing and building land administration systems

A cleverly formulated strategy for designing, building, and managing a LAP is crucial for effective implementation. However, caution must be taken against taking a too-rigorous approach. While a vision and plan are essential for project funding and implementation, often the secret to government engagement and project success is "incremental opportunism"—that is, making a move when the move is right. Flexibility is essential, though it doesn't always sit well with donors, who may expect more rigid controls.

The management of LAPs and the evaluation of the LAS used to administer them require considerable expertise and experience that draw on generic project management and evaluation techniques, which are then modified to suit the project. For this discussion, the context is designing and building stand-alone land titling, land administration, or cadastral projects in developing countries. These have been major initiatives for organizations like the World Bank since the 1980s. These projects are generically referred to as LAS projects, although they are often officially called "land titling," "land administration," "land management," or "cadastral" projects. Many of these projects are stand-alone LAS projects, although they often form part of a larger structural reform project.

A focus on large LAS projects in developing countries permits consideration of all aspects of project management, including cultural, technical, financial, operations and management,

Figure 13.1 Life in the rural Philippines is improved through effectively designed LAPs.

institutional, legal, education and training, and political factors. Many of these considerations are also relevant to projects in developed countries, where realistically, a complete reengineering of LAS will rarely occur. Developed countries are more likely to undertake partial restructuring—for instance, when a country automates its land registry, introduces a coordinated cadastre, or converts its paper-based conveyancing to electronic or digital systems. Nonetheless, complete evaluation of LAS is often required in developed countries, and a capacity to evaluate total LAS performance is equally relevant.

PROFESSIONAL SERVICES AND TOOLS

In reality, all projects are different and will inevitably have their own characteristics. Identifying, preparing, implementing, evaluating, and monitoring LAS projects are therefore complex processes that require professional management expertise and experience. This expertise ranges from skills in community consultation and aspects of anthropology, case study analysis, strategic planning, and technical expertise in surveying and mapping to administration and legal expertise in land policy and associated laws.

Management expertise is especially important, including use of management tools such as

- ◆ SWOT analyses—a strategic planning tool used to assess the strengths, weaknesses, opportunities, and threats (SWOT) to a future project
- ◆ Fishbone charts to assist in problem identification

- ◆ Logical framework analysis (LFA), or "LogFrame" — a project definition and design methodology that clearly identifies project objectives

- ◆ Gantt charts — simple bar charts that show project schedules, financial management, procurement, and contract management arrangements

These tools are used with the most important tool of all — the ability to manage the project strategically from beginning to end. Often termed the "project cycle," this management workflow delivers a systematic approach to project management for the life of the project and beyond.

DEVELOPING A REENGINEERING FRAMEWORK

Outlining a broad framework of pressures and processes associated with land administration reform helps participants understand what is involved in project design (figure 13.2). Two key components of the framework are the vision of new people-to-land relationships and a sound concept for the land administration system being developed. Such global drivers for change as the need for sustainable development help dictate a new vision of people-to-land relationships. This vision may encompass poverty reduction, social equity, environmental management, or economic development. However, every country has some form of existing land administration system that, together with the new vision and strategic planning processes, result in a conceptual model for new LAS.

Usually, a project is designed to build and implement new LAS. Given the complexity and time involved, the project as implemented inevitably does not deliver the "ideal" conceptual system. However, successful projects deliver realistic and operational systems, which are then reevaluated and benchmarked against key performance indicators (KPIs). Over time, this feedback, together with new or modified external drivers, inevitably results in new or modified people-to-land relationships that require new conceptual LAS. And so, the process repeats itself.

The four key steps in this framework are

1. The development of a new vision for people-to-land relationships
2. The process of evaluating existing LAS to develop a new concept for LAS
3. The process of implementing the conceptual model using LAS project methodology
4. The processes of reviewing, evaluating, and benchmarking new LAS over time, which leads to repeating the earlier steps

The multistage approach ensures that LAS reform and LAPs are not done ad hoc, based on an inadequate understanding of local conditions. It offsets the temptation to base the design of a project on intuition, relying instead on experience and common sense informed by a thorough understanding of the local situation.

STEP 1: DEVELOPING A LAS VISION AND OBJECTIVES

For the project to be a success, it must address a real need in society. In most, if not all, countries, land issues are major political issues. But purely political outcomes are not enough. LAS reform or reengineering requires enormous commitment that can only be justified by aiming for significant results, such as poverty reduction, social justice and security of all tenures, equitable access to land, or economic development. The LAS project needs to be a key part of the government's broad objectives, or part of a national plan. Without broad justification, the project will lack an articulated vision and objectives, making it difficult, if not impossible, to chart a clear development path.

The development of a country's land administration vision is also an essential component of any reengineering process. Land administration reform by its very nature is long term, and a clear road map is essential to ensure that all developments and changes contribute to the overall LAS vision. The complexity of LAS suggests that projects should comprise "bite size" subprojects that have a clear focus—for example, administering one category of people-to-land relationships (i.e., individual private rights or traditional rights). These subprojects need to be

Williamson, I. P. and L. Ting 2001. Land administration and cadastral trends: A framework for reengineering. Computers, Environment, and Urban Systems 25:339–66.

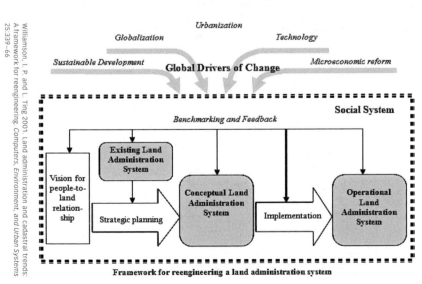

Figure 13.2 Creating new LAS may mean reengineering existing LAS to afford a new vision of people-to-land relationships.

undertaken as part of an agreed vision and within the broad land administration strategy for the jurisdiction.

The land administration model—design criteria

Important first steps in the design process are to consider the country's overarching development priorities alongside the existing environment and to prioritize the activities required to achieve policy goals. For example, while poverty reduction may be the prime policy objective, a decision needs to be made whether to focus on perhaps the urban or rural poor or indigenous communities. If growth in economic activity from a more efficient land market is the top government priority, then the focus may be on the urban sector, with a particular focus on improving security of tenure as a basis of collateral for greater bank lending. Essentially, there needs to be a clear link between the overriding government priorities and policies and the actual land-related policy initiatives.

Geography

Designers of a LAP must clearly identify where the project or subprojects will be implemented. In reality, even if a LAS design were to comprise a whole country, operational coverage of an entire national landscape is usually impossible. The recipient government and project donors will focus on priority areas identified by political pressures, such as urban areas with a large informal sector, rural areas with a large indigenous population operating in the informal sector, high-value lands

Figure 13.3 Fishbone charts are typically used to identify the overriding issue a LAP is designed to address.

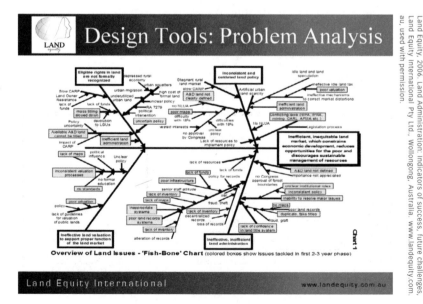

Land Equity, 2006. Land Administration: Indicators of success, future challenges. Land Equity International Pty Ltd., Wollongong, Australia. www.landequity.com. au, used with permission.

in the rural area already in the formal sector, or forest or state lands. Whatever the national design, each area will present unique characteristics and require different strategies.

Understanding the "problem" in a development context

An important initial step in designing a LAP is to clearly identify the main "problem" that the system needs to address or, in other words, the project objective. A LAP goal may or may not be compatible with existing government priorities and development objectives. However, usually these land-related issues are included in the objectives of a country's national development plan. A key objective may be promotion of economic development through efficient land markets, but poverty reduction is often equally important. Promotion of environmental sustainability, good governance, and gender equity are also key national objectives. These objectives are ones that tend to be an important focus of funding agencies, including the World Bank and development assistance agencies such as USAID, the U.S. Agency for International Development; AusAID, the Australian government's overseas aid agency; Germany's GTZ; CIDA, the Canadian International Development Agency; and SIDA, the Swedish International Development Cooperation Agency.

The process of identifying the problem to be resolved is usually undertaken through stakeholder workshops, which discuss issues and potential bottlenecks. Often, a SWOT analysis is used to focus on the strengths, weaknesses, opportunities, and threats within the current land administration environment. Problem identification is often presented and analyzed in a Fishbone chart (or Ishikawa diagram) commonly used in standard quality management processes (figure 13.3). These help in problem analysis and in understanding the cause-and-effect relationships that will influence project design.

STEP 2: UNDERSTANDING THE EXISTING SYSTEM

At the time of project preparation, the existing land-related systems and associated processes of landownership, transfer, inheritance, development, leasing, and so forth must be clearly documented in detail, whether the systems are formal or informal. There is a wealth of information on how to form an understanding of existing LAS, much of it derived from anthropology. People and land are inseparably related, so an operational land administration system is intimately tied to a country's social structure and culture. Therefore, a key step in the LAS reform process—or reengineering—is understanding local conditions and the current land administration system from the legal, technical, institutional, social, economic, and political perspectives. These observation processes have much in common with the approach taken by anthropologists.

Detailed case studies must be undertaken that include political, legal, social, anthropological, environmental, economic, and technical dimensions. The focus must look at different groups in society, who live in urban, rural, coastal, mountainous, and forest areas, to diagnose whether they use formal or informal processes. How people conceptualize and value land needs to be documented. A full understanding can often take months, if not years, but will ensure that the next steps in the design process are easy to implement. Again, a SWOT analysis can be helpful in understanding the existing system and identifying issues.

An alternative, three-stage approach to LAS reform, originally designed for cadastral systems, is shown in figure 13.4. The first stage uses case studies to form a general picture of the system. The case studies should not be tokenistic or overly donor-centric, since such issues may not be relevant to the situation. After the big picture and some major areas of interest are understood, the next stage of comparisons can be undertaken. In this example, the land administration system in the jurisdiction being studied is compared with similar systems to identify differences and commonalities, as well as intrajurisdictional and interjurisdictional linkages.

Describing each local component within the land management paradigm
It is useful to consider how a country relates to each element of the land management paradigm. The case study approach can again be used in developing LAS strategy. Project preparation involves investigating, understanding, and documenting these components as background information. All too often, both governments and lending institutions repeatedly try to reduce the time allocated to understanding the existing land-related environment, to the detriment of the long-term project. The aspects that need to be investigated include

- ◆ **The country context**, including the legal system, government institutions, the political structure, and the role of the judiciary. Population, key demographic indicators, the number of land parcels, and tenure types also need to be examined. Understanding the underlying social issues, such as the type and number of land disputes and how they are or are not resolved, is also essential. The role of local chiefs or village elders or courts in resolving land disputes, and court delays in getting land disputes heard and resolved, are also included.

- ◆ **The existing land policy framework**, including how a country currently deals with state lands, communal and indigenous lands, private property, leasehold land, resource tenures, and lands in both informal and formal systems. Foreign ownership, the environment, mining, water rights, use of state forests, and the marine environment are among the land-related laws and policies that need to be documented.

◆ **The existing land information infrastructure**, including current mapping and land records, either in digital or hard copy.

◆ **A careful documentation of the different tenures and land uses** across the country. This should be related to all land parcels (including buildings) irrespective of ownership and tenure. Relating tenures to land parcels is much more significant, because it indicates the effort required to undertake the project.

◆ **A thorough understanding of land values**, valuation practices, and the operation of banks and lending institutions, including their policies for mortgaging, lending, and debt recovery. This includes understanding the operation of land markets in both urban and rural areas.

Williamson, I. P., and C. Fourie. 1998. Using the Case Study Methodology for Cadastral Reform. *Geomatica*, used with permission.

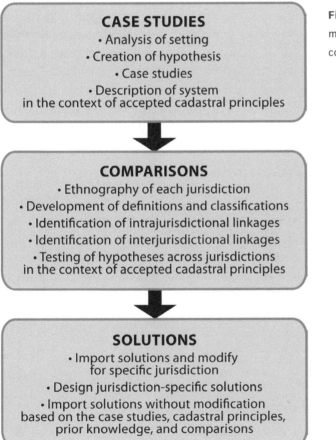

Figure 13.4 In the cadastral reform methodology, case studies are used for comparison to arrive at solutions.

CASE STUDIES
• Analysis of setting
• Creation of hypothesis
• Case studies
• Description of system
in the context of accepted cadastral principles

COMPARISONS
• Ethnography of each jurisdiction
• Development of definitions and classifications
• Identification of intrajurisdictional linkages
• Identification of interjurisdictional linkages
• Testing of hypotheses across jurisdictions
in the context of accepted cadastral principles

SOLUTIONS
• Import solutions and modify
for specific jurisdiction
• Design jurisdiction-specific solutions
• Import solutions without modification
based on the case studies, cadastral principles,
prior knowledge, and comparisons

◆ **A thorough understanding of the land-use control and development practices**, whether it is a subdivision of land for housing development or high-rise, commercial, or industrial development, or whether development is controlled at all.

◆ Importantly, all these activities must be investigated in **a social and economic context**. Understanding the social and economic environment requires an understanding of the capacity of both people and institutions within a country and especially the cognitive understanding people have of land uses in different parts of the country. Simply, it is essential to align any new or improved land administration system with how people actually think about and relate to land.

Documenting key LAS processes and practices

Documentation of the common land processes of titling, land transfer, mortgaging, leasing and subdivision, inheritance, and social transitions is at the heart of understanding and reforming LAS. Detailed "time and motion" studies of each process are required (see chapter 4, "Land administration processes"). It is essential to carefully track a land transfer or subdivision of a land parcel from beginning to end—in other words, to follow the "paper trail" and, in some countries, the "money trail." This detailed understanding of LAS processes makes reengineering possible. Without it, LAS reform will fail.

STEP 3: IMPLEMENTING THE CONCEPTUAL MODEL

Components of a LAS project

On completion of the first two stages, it is possible to start developing solutions based on strategies that appear to work in other jurisdictions and adapting them to the jurisdiction under study. Once the land-related policy initiatives have been identified and prioritized, then actual project design can commence. From a project management perspective, a number of critical components is usually required or is present in successful LAPs.

For example, once a project is conceived, it may then be divided into subprojects such as

◆ Development of land policy.

◆ Survey and mapping of land parcels.

◆ Adjudication of rights, and creation and issuing of land documents.

◆ Physical infrastructure building, such as the building of new regional land offices or a new building headquarters. It may also include improved filing systems or development of computerized indices.

◆ Education, training, and research that in turn consist of a range of smaller projects at the university, technical college, and "in-house" levels, as well as a commitment to ongoing research on land-related issues. Often, a government training center is included.

◆ Management functions, which may include finance, institutional reform, computerization, leadership, and project management.

◆ Improvement of valuation capacity of the country, land-use control and land development processes, environmental management, risk management, law reform, and the flow of private, bank-sourced finance for mortgages.

◆ Community and stakeholder engagement as well as media and public exposure and building the capacity of NGOs and professional societies to support the project.

Using best practices and applying the toolbox

Once the subprojects have been determined, a process has to be followed to determine which tools from the land administration toolbox are most appropriate (see chapter 12, "The land administration toolbox"). For example, will a systematic or sporadic approach be used for land titling? What form of boundary or title system will be used? Will an incremental approach to issuing full land titles be used such as qualified titles or titles where the boundaries are qualified? Should the project be part of another government initiative, such as provision of large-scale orthophoto mapping as part of a national SDI? Should the title system be judicial or administrative, and centralized or decentralized? How can the system ensure capture of all transactions and social transitions? And the list goes on.

The role of pilot projects

The choice of the most appropriate tools is difficult and often influenced by many factors such as cost, long-term sustainability, capacity of human resources, and the legal or social environment, among others. Without doubt, the best approach is to use pilot projects to try out different approaches and test their applicability. The biggest threat to success is imposition of off-the-shelf or "one size fits all" solutions. Another threat is when governments start to believe that building or reforming LAS is too difficult, so they consider a build, own, operate, transfer (BOOT) model created and managed by the private sector. Few, if any, BOOT approaches have been successful in land administration, especially at the early stages of building land markets in developing countries. They tend to institutionalize undesirable silo arrangements and prevent open access to information.

Social and economic analysis (baseline and longitudinal studies)
An important component in the design, implementation, and evaluation of LAPs is to undertake baseline studies before the project begins in order to monitor the project's future impact. This form of analysis can be either a baseline study or longitudinal study, which usually focuses on social and economic indicators. For example, investigation of the effectiveness of the Vietnamese LAS in rural areas involved a number of baseline studies to estimate the impact of the issuance of land-use certificates. These studies estimated the effect on the growth of a land market and the use of land-use certificates for collateral for bank loans or mortgages (Smith et al. 2007). A good example of a longitudinal study is the one undertaken for the Thailand Land Titling Project (Feder et al. 1988; Feder and Nishio 1998) to assess land titling in rural areas, and specifically whether land titling improved the value of the land over a number of years. Baseline and longitudinal socioeconomic studies provide important indicators that can be used for ongoing monitoring and evaluation of a project.

Community and stakeholder engagement (participatory development)
Community and stakeholder engagement is mandatory to the success of LAS. If a society does not want or need the system or see the benefit of it, then LAS have little chance of success. It is critically important that community and stakeholder engagement be at the local, village, or community level (figure 13.5), so that people understand the processes and results, and the concepts of tenure are aligned with those in day-to-day use. Additionally, a similar effort is required to ensure the engagement of all stakeholders. For example, an enormous effort may be required to introduce the new system if the professional private-sector surveyors and lawyers are not supportive and do not see a benefit. Even with the best community and stakeholder engagement, the project will fail without political commitment and leadership and appropriate tools from the land administration toolbox. And even with all parties satisfied, failure will occur if national governance capacity is inadequate.

Critical success factors
To meet the threat of project failure, repeated efforts have been made over several decades to list the critical success factors for LAPs. A summary of these follows:

◆ A long-term political commitment and a champion, especially in the early stages. This is important to ensure that the inevitable government infighting, jealousies, silo mentalities, and interdepartmental competition are managed to the benefit of the project.

◆ Equally important, strong political leadership is needed to ensure the project is not hijacked by professional elites. The most common instances are imposition of

Figure 13.5 Cambodian villagers participate in a LAP at the local level.

expensive and unrealistic surveying and mapping standards by land surveyors, as well as lawyers or a judiciary who block the introduction of simplified tenure, transaction, and administrative systems, thereby dooming the project to failure. Conversely, it is essential that NGOs and professional organizations are involved from the very beginning.

◆ A desire by the society for a land market and, in turn, a land administration system.

◆ A clear vision and objectives.

◆ An evolutionary and incremental approach that recognizes existing rights and an institutional base, whether the rights are formal or informal.

◆ Clear, transparent, and accepted institutional arrangements that identify which agencies are responsible for which activities in LAS. Making an agency responsible for the project that is accepted across government is highly desirable.

◆ Design of a system in sympathy with the capacity of the country. It is important that appropriate local counterparts are appointed to the project. Often, these are not forthcoming. If local disengagement is then compounded by a lack of government funding by the country where the project is being undertaken (a common occurrence), then delays and stress are placed on all parties.

◆ A commitment to building the capacity of both people and institutions, especially the capacity to manage and implement. This includes commitment to long-term education, training, and research. While education and training are often given lip service, in reality, many government bureaucrats in host countries see money spent on education and training as "wasted." At the same time, lending authorities, including even the World Bank, sometimes are too focused on measurable outputs, such as the number of land parcels surveyed and titled, than on developing the capacity of a country to sustainably manage a project. The key is that each country, and each project, must have the capacity to engage the intended beneficiaries and to transfer working knowledge and responsibility to them.

◆ A system where all stakeholders are adequately rewarded within the operations of LAS, whether in government or the private sector. In other words, a system where negative behaviors such as corruption and rent seeking are minimized and do not undermine public confidence in the system.

◆ A detailed understanding of the existing system and land administration processes.

◆ A simple and clear land code that provides clarity, transparency, and security. A project often requires substantial law reform to reengineer a complex legal environment where there are many laws and regulations, often overlapping and contradictory, leading to ambiguity, confusion, and an inability to adopt clear strategies or processes.

◆ A holistic approach to project design that includes all the components of the land administration toolbox within the context of the land management paradigm. A design based on a technical solution, particularly one that assumes building LAS is equivalent to buying GIS software and then building the system, is doomed to failure.

◆ Local implementation within a national vision. The public must be attracted to participate in both the initial and later stages, especially by bringing all changes in landownership into the system.

◆ A commitment to long-term sustainable funding, resourcing, and maintenance of LAS.

◆ A simple system that is low cost. Sometimes, projects are designed to be expensive and complex, with the result that the informal sector continues to flourish over the formal sector. In developing countries, a pro-poor approach is essential for social justice and equity as well as economic development.

Logical framework analysis

A LogFrame is a tool for planning and managing development projects. Ironically, while LogFrames are still in favor for LAS in developing countries, they are rarely used in the developed world. This simply emphasizes the need to adopt the best management approach under the circumstances. In many development agencies, a LFA is used to link the various project outputs and inputs needed to achieve a particular goal, along the lines of the generic LogFrame in table 13.1.

LFA is used by most development agencies to strengthen project design, implementation, and evaluation. It can be used in almost any context to identify what is to be achieved and to

TABLE 13.1 A LOGFRAME ANALYSIS FOR TITLING PROJECTS
(COURTESY LAND EQUITY INTERNATIONAL PT. LTD. 2006)

GENERIC LAND TITLING PROJECT LOGFRAME			
Project structure	Objectively verifiable indicator: Measures to verify achievement of the goal in terms of quality, quantity, and time	Means of verification: Sources of data needed to verify status of the goal level indicators	Important assumptions/risks: External factors necessary for sustaining objectives in the long run
GOAL			
Reduce poverty and enhance national economic growth	Increased economic growth and social stability	Government reports Socioeconomic impact assessments Land office records Project progress report	Political will and stability to engage in poverty reduction strategy
PURPOSE			
To improve security of tenure To provide a framework for an active formal land market in support of social and economic growth of a country	Increased number of registered titles Reduced number of land disputes Increased revenue and subsequent registrations Increased formal lending, equal access for women	Government reports Socioeconomic impact assessments Land office records Project progress report	Domestic market conditions in regard to demand and prices remain stable Financial sector is strengthened Social and political stability

Continued on next page

Continued from previous page

TABLE 13.1 A LOGFRAME ANALYSIS FOR TITLING PROJECTS			
Project structure	**Indicators of achievement**	**Means of verification**	**Important assumptions/risks**
OUTPUTS: MAJOR COMPONENTS			
A platform for long-term development in the land sector policies and regulatory framework	Established policy committees and studies; enhanced public awareness; regulations developed	Supervision mission reports	Government commitment; integration of institutions to form policy; increased civil society participation
A sustainable institutional environment for land administration	Laws and regulations passed on government functions; institutional decentralization capacity established; strengthened human capacity; M&E strategies developed	Supervision mission reports Progress reports Land office reports and statistics	High level of government commitment for reform
An equitable and efficient system of land registration and valuation	Registration standards developed and monitored; decentralized access to registration service; streamlined registration procedures. Registration is complete < 5 days; more than 80% subsequent transactions registered; valuation system established; valuation assessment obtainable < 5 days	Land office reports and statistics Progress reports Supervision mission reports	Timely availability of funds; government commitment to reform and strengthen human resources; community awareness of land registration benefits
Secure landownership for all landowners	Community relations program implemented; social awareness and safeguard monitoring; reduced number of court land disputes; % of registered rights; all individuals adequately represented in title distribution	Land office reports and statistics Socioeconomic impact study Supervision mission reports Progress reports and annual plan	Availability of funds; adherence to safeguard policies; availability of skilled professionals for adjudication and surveying
Improved capacity to manage project	Project management capacity improved; M&E established; socioeconomic impact study conducted	Progress and supervision reports	Availability of funds; staffing commitment

Continued on facing page

Continued from previous page

TABLE 13.1 A LOGFRAME ANALYSIS FOR TITLING PROJECTS

Project structure	Indicators of achievement	Means of verification	Important assumptions/risks
ACTIVITIES: SUBCOMPONENTS	**INPUT: TYPE OF INPUTS REQUIRED AND EXPECTED COST, ETC.**		
Review and develop land policy and regulatory framework	Develop capacity; formulate land policy; establish land information mechanism; > 10% of budget	Progress reports (quarterly) Disbursement reports (quarterly) Supervision mission reports	Vested interests opposing reform contained and political support for reform maintained
Institutional development	Develop modern office; human resources and training development; education strategy; < 10% of budget		
Modern land registration and valuation system	Strengthen service standards and procedures; strengthen valuation system; < 10% of budget		
Accelerated land titling through systematic registration	Community education and services strategy; systematic land titling; < 35% of budget		
Project management support and implementation	Technical assistance; support management and implementation; M&E; > 35 percent of budget		

determine to what degree the planned activity fits into broader or higher-level strategies (DFID 2002). The LogFrame matrix as shown in table 13.2 is a simple method of showing the relationships among goals and objectives, and inputs, processes, and outputs. It is useful to show the hierarchy of objectives, indicators for each, and major risks and assumptions. While a LogFrame is usually undertaken at the start of a project before detailed project design, it is useful in monitoring and evaluation throughout a project since it identifies key indicators and means of verification. Developing a LogFrame is a participatory process that requires some skill and experience. There is a great deal written on LogFrames and the process to develop

TABLE 13.2 THE LOGFRAME MATRIX
(DFID 2002)

Project structure	Indicators of achievement	Means of verification	Important risks and assumptions
GOAL			
What are the wider objectives that the activity will help achieve? Longer term program impact	What are the quantitative measures or qualitative judgments on whether these broad objectives have been achieved?	What sources of information exist or can be provided to allow the goal to be measured?	What external factors are necessary to sustain the objectives in the long run?
PURPOSE			
What are the intended immediate effects of the program or project, what are the benefits, and to whom? What improvements or changes will the program or project bring about? The essential motivation for undertaking the program or project	What are the quantitative measures or qualitative judgments by which achievement of the purpose can be judged?	What sources of information exist or can be provided to allow the achievement of the purpose to be measured?	What external factors are necessary if the purpose is to contribute to achievement of the goal?
OUTPUTS			
What outputs (deliverables) are to be produced in order to achieve the purpose?	What kind and quality are the outputs and by when will they be produced? (QQT: Quantity, quality, time)	What are the sources of information to verify the achievement of the outputs?	What are the factors not in the project's control that are liable to restrict the outputs from achieving the purpose?
ACTIVITIES			
What activities must be achieved to accomplish the outputs?	What kind and quality are the activities and by when will they be produced?	What are the sources of information to verify the achievement of the activities?	What factors will restrict the activities from creating the outputs?

them, with the UK Department for International Development (DFID) (2002) and BOND (British Overseas NGOs for Development) (2003) two good examples.

STEP 4: MANAGING AND MONITORING PROJECTS

Project management

Most LAS reform is done by creating a defined project. Sometimes, these are relatively small projects that are part of the broader land administration system, but sometimes the whole system is reformed and a project established to manage long-term change. In most developed countries, only a part of the land administration system is reformed, such as automation of the land registry, digital lodgment of cadastral survey plans, electronic conveyancing, or streamlining part of the supporting legislation. While there are always incremental reforms occurring in the more developed countries, in less developed countries, projects often focus on a major reform of the land administration system. Projects that involve large aerial mapping and surveying supported by new technologies are typically national in design (for example, Mongolia's project on cadastral surveying funded by the Asia Development Bank). Among a nation's many tenure types, formalization tends to involve a national approach, hopefully flexible enough to handle all land arrangements. A national legal focus, for example, should reflect all the tenures that function throughout the country in tenure recognition processes and support local land management and dispute management systems. A national initiative is the most complex and challenging reform and requires a carefully designed project to implement. These types of reforms are the focus of this chapter.

Once the scope is determined, questions need to focus on how to design the land administration system itself. A project-based development path is the most typical. Projects can have a variety of goals. Among them are land law reform (Bruce et al. 2006); land resettlement (ADB 1998); land redistribution and land administration (UNECE Guidelines 2004; Land Administration in the UNECE Region: Development trends and main principles, 2005a), and land markets; agricultural productivity improvement; and many others. Underpinning all these is the overall requirement that land must be identifiable. Management of the spatial component (i.e., the identification of boundaries, the representation of boundaries in a paper- or computer-based system, and the identification of interests in the land within and around these boundaries) is the overarching requirement. Without the spatial components being handled consistently, land distribution or improvement projects cannot effectively deliver land management capacity. Hence, the cadastral component is a core component of land project work. The building of this component is therefore the typical focus of the LAP cycle.

"The Project Cycle"

The best introduction on how to build a large-scale project remains the model provided by the World Bank (figure 13.6), even though each project has its unique characteristics.

Discrete phases of this generic but adaptable project design include

- The preproject identification phase
- The project identification phase
- The preparation, appraisal, and approval phases
- The implementation phases
- The evaluation phase

While each funding agency will have its own project standards and processes, and each nation (or even a department of government within a nation) will have its own method of proceeding, all projects generally follow this kind of generic pattern. Projects will also involve a mixture of public and private participation. This broad project design should be applied to LAPs, in addition to the specific technical factors. LAPs need to allow for different kinds of participation, ranging from projects undertaken entirely within government to projects paid for by government but undertaken by private-sector enterprises, including charitable and nongovernment sectors.

Feasibility and risk assessment

Key considerations when preparing a project are its feasibility and the risks involved. Usually, a project proposed to a funding agency must address these issues and often document them in a risk matrix that addresses the risks for each stage or activity within the project. A risk management matrix will consider such aspects as the actual risk, potential damage to the project, likelihood of the risk, potential impact, rating or importance of the risk, risk treatment, responsibility for addressing the risk, and timing of risk occurrence within the project. The broad risks would have been identified in a LogFrame. There are many dimensions to feasibility and risk: for example, whether the overall technical design is feasible or whether each stage of the project is feasible. This is usually a straightforward assessment. Another example is whether the project requires institutional or legal reform. If so, changes can be problematic. Legal change is hard to initiate and control, and frequently, even when the law is changed, social and institutional behavior continues as before. Other risks include funding, political support, capacity of the host country, capacity for appropriate external advisers, provision of education and training, maintenance of stakeholder support, and more.

Sustainability

Assuming the project is feasible and that an adequate risk management strategy can be developed, a key issue is whether the project is sustainable, both institutionally and financially over the long term—that is, when the external donor funding or bank funding ceases. This issue needs to be considered during project preparation. Key issues affecting sustainability include ongoing funding and political support, appropriate infrastructure, the cognitive capacity of society to integrate the reforms into daily life, and ongoing capacity building including education, training, and research. The system's capacity to draw derivative or post-titling transactions and changes to ownership through social transitions, especially inheritance, is particularly important.

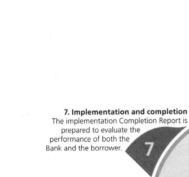

8. Evaluation
The Bank's independent Operations Evaluation Department prepares an audit report and evaluates the project. Analysis is used for future project design.

7. Implementation and completion
The implementation Completion Report is prepared to evaluate the performance of both the Bank and the borrower.

6. Implementation and supervision
The Borrower implements the project. The Bank ensures that the loan proceeds are used for the loan purposes with due regard for economy, efficiency, and effectiveness.

5. Negotiations and board approval
The Bank and borrower agree on loan or credit agreement and the project is presented to the Board for approval.

1. Country assistance strategy
The Bank prepares lending and advisory services, based on the selectivity framework and areas of comparative advantage, targeted to country poverty reduction efforts.

2. Identification
Projects are identified that support strategies and that are financially, economically, and environmentally sound. Development strategies are analyzed.

3. Preparation
The Bank provides policy and project advice along with financial assistance. Clients conduct studies and prepare final project documentation.

4. Appraisal
The Bank assesses the economic, technical, institutional, financial, environmental, and social aspects of the project. The project appraisal document and draft legal documents are prepared.

Figure 13.6 The World Bank has developed the Project Cycle as a model of how to conduct large-scale projects.

Figure 13.7 LAPs can be set in urban areas like India.

Project design, appraisal, inception, and mobilization

The Project Cycle used by the World Bank provides an insight into the bank's lending and project policies. However, it does not fully explore some of the practical steps in developing and implementing a project. For example, usually up to four or five different missions are undertaken before a project actually begins. Inception missions follow standard steps although they are always designed to suit the project (figures 13.7 and 13.8). Such steps can include the following:

- First, land-related issues are identified, and broad project assumptions are developed. This is often done by organizations such as the World Bank, usually undertaken with input from one of the development assistance agencies that specialize in land-related projects, or increasingly, done in response to pressures from NGOs.

- Second, the project is designed, which usually involves both international and local experts. Unfortunately, this is often a step that is underresourced, given the

Figure 13.8 LAPs can also be set in rural areas like Tibet.

complexity of large LAPs. It is not unreasonable for a large LAS project to have a team working on the project design and preparation for well over three months. In general, the more effort that is put into project design, the better its chances of success.

◆ Third, the appraisal of the project design usually involves all parties, such as the recipient country, the lending agency, sometimes the World Bank, and a development assistance agency, such as SIDA, USAID, or AusAID, that co-finances the project. The appraisal team usually includes representatives of these organizations as well as local and international experts. Once the project design has been amended and approved, the funding is organized.

◆ The next step usually involves the international development assistance agency that has agreed to co-finance the project appointing either a government or private-sector organization to manage and implement the project. This usually requires bids or tenders by interested parties to undertake the work; selection of the successful team, often through an extensive and rigorous evaluation process; contract negotiations; and finally, awarding and signing of the contract.

◆ Lastly, the successful contractor visits the site to undertake an inception study for project implementation. This usually involves all parties involved in the project. Once the inception study is approved (particularly by the recipient country), the project contractor commences the final mobilization mission to start the project.

Project management office

Establishment of an effective project management office (PMO) is critical to project success. The office should have an area where project overview, highlights, achievements, and progress can be exhibited. The PMO should have a meeting room and offices for both local staff and visiting experts. A well-appointed project office can make a significant difference to project morale and is a good indicator of the government support given to a project. Unfortunately, many PMOs are less than satisfactory with staff appointed late or not at all. Just as problematic is when local staffers are appointed in a part-time capacity in addition to their "permanent" position, with the result that they may be rarely available.

Project director, manager, and coordinator

Each project needs a management hierarchy. Usually, the recipient government provides a senior person as project director. It is important that this person has some knowledge of land-related issues and is well connected and relatively senior in the government hierarchy. Local operational staff reports to the project director. The project manager is usually a full-time position appointed by the managing contractor. The project manager typically has a dual role reporting to the project director and the project coordinator, who is a senior part-time employee representing the contracting organization. The project manager and project coordinator often have important liaison roles with the development assistance agency supporting the project and the main funding agency such as the World Bank. Usually, local and international experts report to the project manager. The PMO usually has a mix of administrative and technical staff that assists both the project director and project manager. The importance of experienced project directors and project managers cannot be overemphasized.

Managing changes and contract amendments

All projects change and evolve. In the case of LAPs, there are regular reviews that inevitably result in changes to the objectives, scope, tasks, and outputs. Often, the funding arrangements change or political priorities change. Whatever the reason, the best-designed LAP always seems to change and evolve. The result is that the management contract often needs to be amended. This sometimes requires lengthy contract negotiations and highlights the need for flexible contracts. It is essential that these contract changes are carefully documented.

Stakeholder and community consultation

One of the biggest changes over the past couple of decades to the way LAPs are designed is the emphasis on extensive stakeholder and community consultation. Today, such consultation is an integral and usually lengthy component of project design. Previously, government authorities were often the only source of information. Today, government still provides a great deal, if not most, of the data and information to support project design; however, now, NGOs, professional organizations representing surveyors, lawyers and other interested parties, the judiciary, and academics are all consulted through stakeholder forums and other mechanisms. At the local level, village meetings and forums are held to discuss the project and seek advice from landholders and tenants alike on the issues and necessary steps required to improve or introduce the new system. Organizing stakeholder and community consultation is a professional activity that requires professional skills. Importantly, it is an ongoing process that occurs throughout the project.

Community relations
Community relations include more than stakeholder and community consultations. Regular contact with all interested parties as well as the wider community is essential to a LAS project. This can include a wide range of media, including radio, television, posters, leaflets, newspaper articles, and advertisements. It includes seminars, conferences, workshops, and town and village meetings, as well as a permanent display within the PMO explaining the project, its objectives, progress, and achievements. The skills of an experienced media expert are useful in designing the community relations strategy and program.

Scheduling and project management
An important component of managing a project is the development of charts showing all activities as a schedule of outputs (description, duration, planned completion, planned deliverables, planned inputs), and often the main deliverables as set out in the contract. Several techniques are used such as Gantt charts, showing tasks, duration, start, and finish. Gantt charts are usually derived from the more strategic LogFrame discussed previously. For more complex projects, critical path analysis may be used, but this is not usually the case in LAPs. Scheduling is important to develop the required resources and staff for each task and translate these into financial requirements and cash flows. Again, scheduling and associated project management requires professional expertise. Today, a variety of computer programs can assist scheduling as well as managing finances.

Project monitoring and evaluation (project focus)
Monitoring and evaluation are used in two ways in land administration—for an overall systems focus or a specific project focus. Taking an overall systems focus can be more extensive and is discussed more fully in section 13.3. At the project level, monitoring and evaluation is an essential tool in managing the performance of specific LAP activities. Each activity is measured and assessed to check whether it starts and finishes on time, uses the allocated resources, and is within budget. At the same time, key performance indicators (KPIs) are used to measure overall project performance: for example, the number of photomaps produced in square kilometers per

year; the number of control points surveyed per day, per month, or per year; and the number of land parcels adjudicated and surveyed per day, per month, or per year. Similarly, KPIs can be the number of land titles registered and issued; the number of land offices built; the number of personnel trained to undertake a particular activity or educated in a discipline at a technical college or university; or the number of study tours successfully completed. Depending on the project, about ten or more KPIs are monitored and evaluated. Evaluation is important to identify problems or bottlenecks that need to be rectified. Usually, the results of the KPIs will be evaluated by the project director and project manager on a monthly basis, although some KPIs may be evaluated more often and some less often. Most project management offices have posters or charts showing the outputs of each KPI.

Financial management

Good financial management is central to good project management. Every LAP will have a financial manager or controller and often supported by a small team. When there is a contracted project manager and team, the contractor also has someone to monitor the finances associated with the project management contract such as salaries of long-term and short-term experts, living expenses and per diems, travel costs, etc. However, the main focus of financial management is on the actual project and such items as salaries, travel and vehicle costs, procurement including preparing and managing contracts for project equipment and resources, stakeholder and community interaction (seminars, conferences, workshops, etc), publicity, and establishing field offices.

Quality assurance

Historically, quality assurance was undertaken by appointing a team of experts (covering surveying and mapping, land registration, land valuation, education and training, adjudication, community interaction, law reform, strategic management, and so on, depending on the nature of the project) under the direction of the funding agencies that visited and reviewed the project on an annual basis (and sometimes more often). Each specialist would review their area of interest and make appropriate recommendations. In recent times, some projects have adopted a more sophisticated quality assurance (QA) approach by formally appointing a QA panel at the beginning of the project that is part of the project management contract team but operates at arm's length from project administration. AusAID and the World Bank have used this approach successfully in a LAP in the Philippines. While the concept offers many advantages, it may appear expensive (although in reality, it may be a low-cost option) and is still evolving as a project management tool.

Resources

All projects demand resources, and one of the primary roles of the PMO is to ensure there are adequate resources to meet the project schedule. The key categories of resources are people, equipment, accommodations, and transport. The management of staff (human resources management and development) requires a dedicated section to ensure the appointment of appropriate project personnel and adequate training. The other categories also require dedicated attention.

Procurement

Large comprehensive LAPs require significant equipment, vehicles, and resources. Most of the acquisition is in surveying equipment, such as digital theodolites (Total Stations) and satellite positioning equipment (GPS), and computing equipment. Usually, mapping is a central component of these projects, requiring the acquisition of either aerial photography or high-resolution satellite imagery. Producing appropriate maps from this imagery is also a major task. All these activities require professional expertise to evaluate the needs of the project and prepare procurement contracts. The whole process of evaluating user needs, developing contracts, seeking tenders, and awarding contracts, especially in the case of aerial photography, satellite imagery, and mapping, can involve significant delays. While an initial estimation suggests these activities can be accomplished within a few months or in less than a year, the reality is that they can often take two to three years before the product is eventually delivered. This time frame can have a major impact on a project and must be carefully considered during project design.

The four steps mentioned here and their related activities represent most of the key aspects for designing, building, and managing the daily operations of LAS.

13.3 Evaluating and monitoring land administration systems

THE NEED FOR BENCHMARKING AND EVALUATION

Project design and evaluation can encompass the whole program of development assistance, not just individual LAPs. LAS evaluation and monitoring can be relevant to entire systems, particularly the more developed ones. This approach tries to evaluate and benchmark existing systems to answer the question, How good is the current system performing compared with other systems? Or, is there justification for reengineering or reform? A recent version of this approach was developed by Daniel Steudler (2004) and is described in articles by Steudler, Abbas Rajabifard, and Ian Williamson (2004), and Steudler and Williamson (2005).

This is not to say that there has not been a great deal of effort over the past couple of decades to attempt to evaluate and compare LAS. The work of the World Bank in documenting LAS in developing countries is widely acknowledged, as is the work of UNECE for European countries (also see Steudler 2004). FIG has also been active for many years in benchmarking cadastral systems as described in the Cadastral Template Project in chapter 10 as well as previous initiatives in benchmarking cadastral systems (Steudler et al. 1997).

AN EVALUATION FRAMEWORK

Steudler (2004) developed a practical framework that evaluates LAS activities or outcomes at the policy, management, and operational level. He also includes external factors and review processes within the evaluation matrix (figure 13.9). For each activity or outcome, he identifies evaluation aspects, indicators (of the existing situation), and best practices. By combining the existing situation with best practices, he identifies performance gaps that are summarized in a SWOT matrix (table 13.3). This rather detailed figure contains all the relevant principles for this evaluation framework. Steudler tested the evaluation framework on several case studies (Steudler 2004; Steudler and Williamson 2005). The benefit of Steudler's approach is that it gives a rigorous framework for evaluation that recognizes the different levels of activities and associated indicators from the policy perspective (such as delivering economic development or addressing poverty reduction) to the operational level (such as the cost or time to transfer a property or survey a land parcel).

Figure 13.9 The evaluation methodology shows how LAS activities are weighed against best practices to identify both efficiencies and performance gaps.

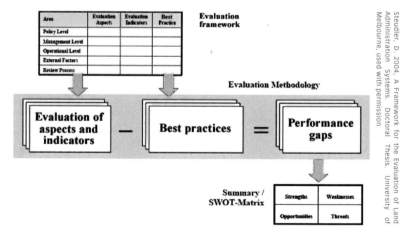

Steudler, D. 2004. A Framework for the Evaluation of Land Administration Systems. Doctoral Thesis. University of Melbourne, used with permission

TABLE 13.3 – SUMMARY OF EVALUATION FRAMEWORK FOR LAS (STEUDLER 2004)		
Evaluation areas	**Evaluation aspects**	**Best practices**
POLICY LEVEL		
Stakeholders: parliament, government (long-term implications, 5–20 years) Tasks: definition of the objectives, legal framework, long-term financial aspects, economic-social-environmental aspects (equitable, sustainable)	Land policy aspects, land tenure stabilization, land market improvement	Mentioned in constitution, laws, and regulations, and suitable to circumstances
	Are objectives defined?	Clearly defined and continuously acknowledged
	Historical background	Awareness
	Social background	Society benefits from LAS
	Political and administrative structures	Suitable to circumstances
	Good governance and civic participation	Efficient and transparent access to land information, supported by strategic and political decisions
	Land tenure aspects	Formal acknowledgment and suitable to circumstances
	People-to-land relationships	Recognized within government and suitable to circumstances
	Legal aspects	Suitable to circumstances
	Land market aspects (number of land sales, value of real estate markets, total value of mortgages, etc.)	
	Funding aspects (funding system, administrative levels involved)	Supportive of efficient LAS establishment
	Direct revenue (fees, stamp duties, land taxes)	Reasonable revenue, suitable to circumstances
	Cost recovery	Clear policy
	Environmental sustainability	LAS includes such duties and is supportive of environmental issues
	Strategic aspects and targets	Clearly defined and publicized

Continued on next page

Continued from previous page

TABLE 13.3 – SUMMARY OF EVALUATION FRAMEWORK FOR LAS

Evaluation areas	Evaluation aspects	Best practices
MANAGEMENT LEVEL		
Stakeholder: administration (medium-term implications, 1–5 years) Tasks: definition of strategic targets, setup of institutional and organizational structures	Institutional aspects: departments, agencies, centralized vs. decentralized	Suitable to circumstances
	Organizational aspects: how agencies themselves are organized	
	Private-sector involvement	
	Reform activities	Reform projects are carried out within a coordinated context
	Human resources and personnel aspects (number of personnel, salaries)	Suitable to circumstances
	Cadastral principles	Only one complete and comprehensive cadastral system, which is effective, efficient, and trustworthy
	Complete legal situation of land	Inclusion of all rights, restrictions, and responsibilities
	Cadastral surveying data as basis for LAS	Cadastral surveying data is updated at all times, standardized, and suitable for as many purposes as possible
	Cadastral transaction processes	Efficient and secure
	Users, products, services	Awareness of users, products, and services; suitable to circumstances
OPERATIONAL LEVEL		
Stakeholders: operational units (short-term implications) Tasks: to provide products, services, and interfaces (interfaces between units and user interface) in an efficient, reliable, and secure manner	Reliability (number of errors, number of title and boundary disputes)	Low number of errors and disputes
	Security	Well-defined notification process; established backup procedures
	Accuracy of information	Accurate registration
	Efficiency of transactions (time and money)	Transactions carried out in reasonably short time and at reasonable cost

Continued on facing page

Continued from previous page

TABLE 13.3 – SUMMARY OF EVALUATION FRAMEWORK FOR LAS		
Evaluation areas	**Evaluation aspects**	**Best practices**
OPERATIONAL LEVEL (CONTINUED)		
Stakeholders: operational units (short-term implications) Tasks: to provide products, services, and interfaces (interfaces between units and user interface) in an efficient, reliable, and secure manner	Transparency, clarity, simplicity	Transparent, clear, and simple system
	Accessibility	Open, transparent, and simple access to land information
	SDI aspects (digital data format, data modeling techniques)	Data in digital format; interoperable sharing of data
	IT aspects (IT and Web-enabled solutions)	Level of computerization suitable to country's capacity
	Data standards and integration	Unique parcel identifiers; linkage of data
	Mapping standards	Coordinated use of unique geodetic reference framework
	Complete coverage	100% coverage
	Completeness of records	Record of each parcel complete by itself
EXTERNAL FACTORS		
Stakeholders: industry, academia, etc. Tasks: capacity building, technological supply, human resources	Capacity building, education (number of universities, students)	Suitable to circumstances, good cooperation between academic, and public and private sectors
	Technological supply by local existing industry	Cost-efficient, appropriate, and suitable to circumstances
	Professional association aspects	Suitable to circumstances
REVIEW PROCESS		
Stakeholders: independent land board, for example Tasks: to review objectives and strategies, monitor user satisfaction, manage visions and reforms	Review process of objective and strategic targets	Regular review takes place, and objectives and strategic targets are either met or adapted
	User satisfaction	Regular review takes place, and customers are satisfied
	Visions and reforms	Closely monitored and acknowledged

Part 5

The future of land administration

Part 5 looks to the future, highlighting the key issues and concepts presented in the book that contribute to the future vision of land administration. Chapter 14 reviews the land administration journey and underscores the key role land administration plays in supporting sustainable development. The important emerging concept of the need to spatially enable society and how LAS fits into that vision is emphasized. The chapter and the book conclude by considering the key issues and challenges facing LAS in the next decade.

Chapter 14

Future trends

14.1 The land administration journey

14.2 LAS supporting sustainable development

14.3 LAS to support spatially enabled society

14.4 LAS issues in the next decade

14.5 The challenges ahead

14

14.1 The land administration journey

Any jurisdiction or country that seeks to establish or reengineer land administration systems needs a vision, strategies, and tools. The ten principles of land administration in chapter 1, section 1.6, can guide these processes of development. The principles highlight key themes—the need for an overall land management vision consistent with the land management paradigm, a focus on processes when reforming systems, the adoption of the toolbox approach, and, most importantly, a focus on supporting sustainable development as the primary objective of LAS. A number of crosscutting themes influence all well-designed reform projects—all systems are dynamic and evolutionary; spatial (place and location) technologies and the cadastre play central roles; and land administration is about people, politics, and places. The reform vision, strategies, and tools can be used in both the most advanced LAS and to support rudimentary initiatives in the poorest countries. LAS need to be resilient in the face of inevitable change.

The LAS journey is long term—it can take ten, twenty, thirty, or forty years or more to achieve major reform. There are few short-term fixes, if any. This is simple reality, even though reasonably short-term targets can be set for particular segments of the overall reform mission. Some technical changes can be done relatively quickly, but others can take decades. Some of the most developed countries have taken three or four decades to reorganize their systems to take advantage of technology. Sweden is a good example. Acknowledged as a world leader in the use of IT within LAS, Sweden acknowledges that, after four decades, the long journey of reform and reengineering continues.

Fundamentally, LAS are about formalizing tenure, irrespective of its local form and content, whether it's short-term occupation rights or full ownership. Simply, land administration is about formal systems. We don't apologize for this. We accept that informal systems are essential parts of any system or society, but without organizing a coherent, formal system for administering land as part of its incremental path toward reform, a country or society will be doomed to poverty.

This does not mean that the formal system needs to be complicated, national in scale, or expensive. On the other hand, any successful reform effort requires flexibility, innovation, and a focus on people. There are no single answers or guaranteed reforms since land administration is an evolving discipline. Just as the people-to-land relationship evolves, so does the land administration response.

The focus on people and society at large is particularly central to any reform project. Modern delivery systems must focus on training and understanding people's existing cognitive capacity about land, especially through opportunities to participate in land decisions. New tools need to be developed that include institutional tools aimed at capacity building and competencies, good governance, and the social tools of engagement and participation. Importantly, the land administration journey must move away from a government focus to a focus on partnerships with the private sector, NGOs, academia, and society at large.

The reality is, formal land titling systems of the type used in modern democratic countries will never be able to fully serve our planet. A global report card on current LAS strategies shows formal systems only serve the needs of about 20 percent of the countries in the world. By any measure, this is not a good endorsement of current LAS tools and strategies. For most countries, titling alone simply does not work (Payne, Durand-Lasserve, and Rakodi 2007). Titling on its own will not solve poverty. Formal titling has a role, but it is only part of the solution. There are many tools and tenure options in the land administration toolbox. Security of tenure is not about landownership for all. On the other hand, not having secure tenures is a recipe for national poverty on both the personal and government scale. Land is potentially wealth, and land administration supports wealth generation. What is needed for the developing world is more flexibility and more people-oriented land administration tools: new tenure tools, new gendering tools, and new local or village tools.

While there are many committed organizations around the world that have achieved some success with LAS, there are many more that are failures. The complex issues surrounding the establishment and maintenance of LAS need further research by the international community of scholars and development aid experts. There is a particular need to focus on a new range of tenure, institutional, framework, and technical tools to support the land administration path of all countries.

14.2 LAS supporting sustainable development

For more than 100 years, and in some cases, nearly 200 years, countries have relied on cadastral and land administration systems to deliver security of tenure, land markets, and land tax objectives. Even today, only a few such systems focus on the broader objective of supporting sustainable development to meet the demands of our changing world. LAS goals need to change focus. Future LAS strategies must support the well-being of countries such as Brazil, Russia, India, and China, which potentially stand to be the major economic powers of the future and will inevitably play key roles as the gatekeepers of sustainable development. LAS strategies also need to support initiatives aimed at meeting global warming and changing weather patterns, be sensitive to globalization of the world's economy and clashes between the haves and have-nots, and contribute to reduction of hunger and world poverty.

Though market reform is the primary motivation for most LAS improvement, the demands of population increases and environmental failure especially require nations to concentrate their systems on land management for sustainable development. These overarching priorities move the design focus away from technical systems and toward humanitarian concerns. Historically, the main output of LAS was delivery of title by building traditional "institutions" of land registries and cadastres. Now, modern delivery systems must focus on training and understanding people's existing cognitive capacity toward land, especially by providing opportunities to participate. The demand for change is not enough. One of the realities of land administration is that institutions (and the people who run them) are unable to change direction quickly. They need time to adapt. So, the processes that governments use to initiate change must be focused on people more than on land.

In the future, LAS will move more broadly into land governance, and the land administration discipline will add comprehensive "people" tools to its toolbox. People-oriented tools will be of two broad kinds: the institutional tools aimed at capacity building and competencies and the social tools of engagement and participation. People tools require project designers to understand existing cognitive capacity and to develop participation models for millions of people in relation to land. The new technologies will be crucial to these tools: They are already replacing static maps with accurate and up-to-date pictures, increasing opportunities for public engagement, and providing more useful information for implementing policy. The outcome of LAS will move the technical focus away from internal institutional systems to information sharing—and the use of place as the fundamental information-sorting tool, available not only to land institutions and agencies, but to the broader government, business, and society.

LAS can no longer afford the luxury of maintaining its existing, narrow focus. It will need to help solve broad societal issues highlighted by the Millennium Development Goals by contributing to poverty eradication, wealth distribution, the management of cities, and sustainable development in the broadest sense. It will be accomplished through e-governance and e-democracy, as well as knowledge management. As a result, LAS must be built to serve the sustainable development objectives of economic development, environmental management, social justice, and good governance. Given the emerging new technologies, it is not as difficult as it sounds.

14.3 LAS to support spatially enabled society

Spatial technologies are evolving quickly, particularly with regard to land-related data. IT companies such as those that produce Google Earth and Microsoft Bing Maps for Enterprise are partnering with countries to deliver their national cadastres linked to a geocoded national address database and high-resolution imagery over the Internet. In some countries, an Internet user can, in an instant, display any known postal address on screen with a combination of satellite images of on-the-ground reality, authoritative land boundaries, and address information. Seamless streaming of the images, and attachment of vital text information, are but two new technical facilities for presentation of land information. Yet remarkable though these developments are, this ability to "find, see, and describe" is only the beginning of spatial enablement.

Spatial enablement comes when countries capitalize on the power that is generated from land information within their land administration and related systems. For complex, developed economies, the land parcel is one of many possible property objects. Other objects of value include planning zones, heritage areas, recreational parks, and the hundreds of organizational arrangements made for better land management. The design of the land information system needs to be sufficiently comprehensive to take all land objects into account and manage them through an SDI. Given technological trends, the most effective management is likely to lie in spatial enablement through various sets of data information. Spatial enablement is ultimately a transformational technology that supports and benefits from efficient organization of government and its administrative systems.

Popular uses of spatial technology involve displaying imagery, then tracking assets and inventory through an increasing array of devices, the most common being the ubiquitous mobile phone. Remarkable as these applications are, spatial technology can be used in even more dynamic ways. Transformational use of spatial technology occurs when it is used to improve business processes of government, including equitable taxation, allocation of services, conservation of natural resources, and planning for rational growth. Use of this transformational capacity of spatial technology in government creates a spatially enabled government (SEG). SEG is achieved when location and spatial information are available to citizens and businesses to use in creative ways and when governments use place as a means of organizing their activities. The majority of countries fail to take advantage of this transformational capacity, thereby limiting their ability to capitalize on this new technology and constraining the future of spatial information professionals and businesses.

By combining the new concepts in sustainable development of land and resources with the energy and potential of spatial technology, countries can reengineer the work processes of government agencies and businesses, more than merely managing information. This is now a clear priority of governments in the most developed countries.

The central role of spatial technology is now moving beyond traditional land administration to nonspatial government functions. These technologies are now used throughout the world to visualize information and facilitate e-government and spatially enabled accounting systems. The transformational use of spatial technology expands the popular view of spatial enablement from the GPS "finders" in digital instruments (vehicle tracking systems, mobile phone systems, the assets and management of common electrical appliances) into a more comprehensive view of what can be done with spatially enabled systems. This expansion is probably as significant as the evolution from paper to digital systems.

Spatial tools are no longer sequestered in mapping agencies where they were created. The broader attraction of spatial technology lies in how it presents information, whether users rely on computers and the Internet or on communications technologies. The adage of "a picture tells a thousand words" has been superseded by "a map tells what's in a thousand spread-sheets." Spatial systems convert queries into results that are more people-friendly. The power of the visual over the verbal both reduces the amount of textual description necessary and organizes it into easy-to-digest information. Combined with the Web environment, opportuni-ties for communicating information across various levels in the managing agency and between the agency and its stakeholders are vastly improved. Data is easily converted into knowledge, so that managers and policy makers can make more informed decisions. Potential improve-ment in manageability of business processes is encouraging government agencies to take up spatial enablement, even if maps and visualization systems are not part of their normal IT repertoire.

Most agencies and businesses start spatial enablement by taking up the geocoded address file as a means of introducing spatial tools in their suite of technical supports and IT. A spatially enabled business organizes its activities and processes around "place"-based technologies, as distinct from simply using maps and visuals on the Web. Geocoding and other spatial informa-tion related to place and location is being used to organize business management and process-ing systems. This adds to or substitutes for the unique business file numbers, identification numbers, dates, and so on that now populate standard relational databases, and object-ori-ented and service-oriented architectures. The next innovation, and one that involves novel uses of spatial technology, will bring this tool into agencies that do not traditionally use it—tax offices, human services, health services, census, immigration, and other service agencies.

This transformation involves organizing social, employment, economic, and environmental data in relation to reliable and authoritative coordinate identification of significant places. Such systems facilitate the integration, not merely presentation, of information throughout an agency or board. At first, new spatial systems became popular because they delivered the ben-efits of business processes to traditional users of land information, including emergency man-agement, resource and water management, land management, and marine management operations. Now, spatial enablement supports identification of where nonspatial datasets apply and potentially allows for seamless interrogation and integration of that information, even by agencies that do not traditionally use spatial information.

The spatial enablement of society is only possible with forward planning and a shared vision of what is possible. This vision is at the heart of the next generation of LAS.

14.4 LAS issues in the next decade

Among the many issues that LAS will need to address in the future, a number stand out as key challenges or limitations that must be addressed if LAS is to achieve its full potential, among them:

- **Land governance:** The spatial dimension of governance relates to land, property, and natural resources. It is the governmental side of land management. The control and management of physical space is the basis for the distribution of power, wealth, opportunities, and human well-being. The key challenges of the new millennium are already clear in the international public arena: climate change, food shortages, energy scarcity, environmental degradation, and natural disasters. These issues all relate to governance and management of land. Land governance is a crosscutting activity that must be addressed holistically, posing a challenge to traditional LAS based on silo-organized information.

- **Urban growth:** According to UN–HABITAT, 2007 marked the year when the majority of the world's population resided in urban areas as distinct from rural areas, with urbanization ever increasing. In parallel, the number of people living in poverty in urban slums in very dangerous health and environmental conditions is also on the increase. Current LAS strategies have not been able to stabilize rural land sufficiently enough to slow this trend. Solutions must be found to reduce, or at least control, the divide between the haves and have-nots and all the consequences that flow from this scenario. Simply, new LAS tools must be developed to accommodate urban growth.

- **Tools to administer the continuum of tenures:** While individual private rights will continue to be an important component of future LAS, the focus must change to new tools to administer the wide range of tenures that are coming to be recognized through a continuum from simply short-term occupation to full ownership. Administering the continuum of tenures and having tools that allow tenures to evolve over time are central to the next generation of LAS.

- **Tools to manage RRRs:** The concept of what constitutes land is evolving. The unbundling of rights is occurring in formal systems more in accord with how informal or traditional systems work. At the same time, governments worldwide are accelerating evolution of their legislative and regulatory framework by creating legal restrictions and responsibilities related to land to support sustainable development. In most developed countries, the number of statutes that have a spatial footprint and impose some restriction or responsibility on land has grown to unmanageable proportions. A common approach to managing RRRs used to rely on land title offices or land registries. However, today, less than 1 percent of RRRs is usually managed through this approach. New and innovative LAS tools are required if these statutes and regulations are to have any chance of achieving their objectives.

♦ **LAS to capitalize on technology:** One of the major challenges of LAS worldwide is to catch up with technology, or capitalize on the promise of technology. There have been rapid developments in spatial and GIS technologies over the past decade. However, LAS generally has failed to capitalize on these opportunities. While the inability of current LAS tools and strategies to address urgent global issues is obvious, the power and promise of spatial technology offers hope to the global poor. Modern LAS can play a key role in e-government and e-democracy. Spatial technology can break down historic institutional silos through data sharing and interoperability within an SDI environment. Virtual jurisdictions, cities, and societies offer exciting options and challenges. The power of location and place to revolutionize the way governments do business through spatial enablement is also opening up. Spatial technology is at the heart of this new LAS evolution and the range of LAS tools now being developed.

♦ **Institutional catch-up:** LAS needs to evolve to reflect changes in the people-to-land relationship. Unfortunately, one of the biggest limitations to capitalizing on the new and innovative tools offered by modern LAS to support sustainable development is the historic institutional arrangement of key agencies into separate silos. Still, in both the developed and developing world, the historic cadastral and LAS silos, and topographic and geographic information silos, continue to compete and stall innovation and development. Only when the parcel layer common to the cadastre is available can land information layers provided by LAS be efficient and effective. Efforts to rebuild road, property, and ownership layers to make use of GIS outside registry and cadastral systems are both fallible and expensive. If sustainable development is to be a reality, countries need to model and measure the impact of human activity on the natural environment—that is, use the measurements evidenced in the cadastre against the information provided in the national geographic information database or by national mapping.

14.5 The challenges ahead

Land administration is about land management, but this cannot be achieved successfully unless the major focus is on building the capacity of people and institutions. This is far from simple, because our understanding of the nature of land has dramatically changed. What was thought of as a simple physical thing is now understood as interrelated bundles of opportunities used by ever-evolving groups of people for different purposes, coupled with a complicated array of interdependent responsibilities and restrictions. Land administration functions now need to include the "unbundling" of land from RRRs, the separation of resources from land, and the creation of complex commodities in land. These arrangements depend on sound and predictable administration and additionally on the cognitive capacity of the public to understand and make use of these arrangements. Building and maintaining these capacities is at the heart of modern land administration.

The sensitivity of policy makers and development aid experts to the cognitive reality of the intended beneficiaries of LAS has vastly improved project and system design in recent years, but there is still a long way to go. Sustainable LAS is owned by, and responsive to, its intended beneficiaries. Simply, the land administration strategies of the past have only been marginally successful and then only for about a fifth of the world's countries. New approaches and strategies are in demand.

Land administration in most countries involves a systematic approach to providing infrastructure to manage the normal processes related to delivering land and managing land markets. An analysis of which countries in the world are capable of providing this infrastructure suggests that only about forty or fewer of the approximately 200 countries in the world can do so satisfactorily.

In most countries, normalized and reliable infrastructure is a luxury that is not available. Paradoxically, land administration in these countries must seek to deliver even broader objectives—managing entrenched problems of population growth and movement, burgeoning urban slums, depletion of land quality, accelerating poverty, and maintaining postconflict peace. Countries with the least ability to manage land have the most desperate need of a basic LAS infrastructure to achieve governance goals and sustainable development, yet they are the least able to build it. This paradox is not easy to resolve, particularly because a world view of the land administration discipline indicates that each country must approach its land issues within a local historical and institutional framework and rely on its own capacity to deliver good governance.

Recent positive improvements in delivering LAS are heartening. Globalization, the growth of spatial technology, and the attractions of formal land markets are improving consistency and effectiveness in building infrastructure and transferring know-how. The influence of the simple numerical and quantitative comparisons in the World Bank's Doing Business Reports on Registering Property cannot be underestimated. Moreover, the maturity in land administration theory and practice has encouraged new ideas and approaches about how to build essential infrastructure to suit local contexts. Thus, land administration has evolved from merely a science of land measurement to a broad approach to land management. The idea of land as a mere physical object has been replaced by better appreciation of the cultural values and cognitive meanings of land. Basic competence in administering land and resource management is now seen as a foundation for leveraging wealth generation by overseeing an open-ended series of opportunities to build, develop, transfer, mortgage, unbundle interests and rights, and manage social transitions among owners. Systems that respond to these new demands will look very different from the technical frameworks built by heritage systems. LAS increasingly is called on to manage transition to peace after conflict and to repair damage done by natural disasters such as earthquakes and tsunamis. Stabilizing land is not only about measurement: It is about the sustainable use of social, institutional, technical, and governance tools.

Even in the developed world where LAS is well established, the focus for most systems is still on supporting simple land trading with little attention, if any, to supporting sustainable development. While LAS in the developed world has the best chance of capitalizing on established infrastructure, surprisingly few countries are grasping the opportunity to use their rich land information resources to spatially enable society at large. For most countries still, LAS' biggest constraint remains its institutional silos.

No country can leave land management to ad hoc and unplanned responses to land needs. The toolbox approach described in this book allows strategies to be coordinated in an inherently flexible way. While the array of tools described provides a structural framework for decision makers, the nature of these tools is constantly changing. One of the most significant changes is the move away from focusing solely on technical tools. Future LAS design will focus on land governance, capacity building, and reflecting the broad cognitive understanding about the roles land plays in society and the economy. These drivers of change are in addition to the forces of technology, irrespective of the level of a country's development. LAS in the future will be integrated with associated government functions, deliver well-organized information for policy makers and business investment, and guarantee security of all tenures, not just those based on the traditional land market.

Glossary

This glossary explains the way words are used in this book. The definitions are not necessarily technically correct in any particular jurisdiction, because meanings are specific to area and grow out of local history and usage. Rather, our aim is to reflect ever-changing global meanings based on the international discourse in land administration. The starting points are the United Nations Economic Commission for Europe, Working Party on Land Administration (WPLA) glossary (`http://www.unece.org/hlm/wpla/publications/laglossary.html`) and the Bathurst Declaration glossary (`http://www.fig.net/pub/figpub/pub21/figpub21.htm`).

Another useful glossary is the Land Tenure Lexicon: Glossary of terms from English- and French-speaking West Africa (`http://www.iied.org/pubs/pdfs/7411IIED.pdf`).

adjudication A process whereby the ownership and rights in land are officially determined.

adverse possession The occupation of land inconsistent with the rights of the true owner.

alienation The power of an owner to dispose of an interest in land or property. In particular, land may be alienated from the state and granted to private individuals.

allodial title A title that is authoritative and absolute and not held through the state, in contrast with titles that reflect feudal tenures derived from a lord or king. Allodial titles are generally not taxable and cannot be resumed by the state.

appraisal An estimation of the market value of real property.

approximate boundary A boundary of a property that has not been determined other than

approximately. The boundary is less accurate than a fixed or general boundary.

assessment A determination of the tax level of a property based on its relative market value.

basic property unit (BPU) The extent of land that is recorded in the register as one homogeneous unit.

boundary Either the physical objects marking the limits of a property or an imaginary line or surface marking the division between two legal estates. Boundary is also used to describe the division between features with different administrative, legal, land-use, and topographic characteristics.

build, own, operate, and transfer (BOOT) A management term referring to development projects in which the contracts between the landowner (usually the government) and the developer include arrangements for building, owning, operating, and transferring the assets. BOOT contracts are popular with countries that need to develop basic infrastructure, such as roads, electricity grids, industrial estates, and so on.

cadastral index map A map showing the legal property framework of all land within an area, including property boundaries, administrative boundaries, parcel identifiers, and sometimes the estimated area of each parcel, road reserves, and administrative names.

cadastral map An official map showing the boundaries of land parcels, often buildings on land, the parcel identifier, and sometimes references to boundary corner monumentation. Cadastral maps may also show limited topographic features.

cadastral mapping The process of producing a cadastral map, usually as a result of cadastral surveying.

cadastral surveying The surveying and documenting of land parcel boundaries in support of a country's land administration or land registration system. The survey often results in a cadastral survey plan that may or may not be used to create or update a cadastral map.

cadastre A register of land information. According to the International Federation of Surveyors (FIG) definition, a cadastre is normally a parcel-based and up-to-date land information system containing a record of interests in land (i.e., rights, restrictions, and responsibilities). It usually includes a geometric description of land parcels linked to other records describing the nature of the interests,

the ownership or control of those interests, and often the value of the parcel and its improvements. It may be established for fiscal purposes (e.g., valuation and equitable taxation), legal purposes (e.g., conveyancing), to assist in the management of land and land use (e.g., for planning and other administrative purposes), and to facilitate sustainable development and environmental protection.

capacity The ability of individuals and organizations or organizational units to perform functions effectively, efficiently, and sustainably.

capacity assessment A structured and analytical process whereby the various dimensions of capacity are assessed within a broader systems context, as well as being evaluated for specific entities and individuals within the system. Capacity assessment may be carried out in relation to donor projects (e.g., in land administration), or it may be carried out as an in-country activity of self-assessment.

capacity building The creation of an enabling environment with appropriate policy and legal frameworks, institutional development, including community participation, human resources development, and strengthening of managerial systems in a long-term, continuing process, in which all stakeholders participate. It is a comprehensive methodology aimed at providing sustainable outcomes through assessing and addressing a wide range of relevant issues and their interrelationships.

capacity development The process by which individuals, groups, organizations, institutions, and societies increase their abilities to perform core functions, solve problems, define and achieve objectives and understand and deal with their development needs in a broader context and in a sustainable manner.

chain of title The set of deeds and other legal instruments evidencing the changes of ownership of a parcel in a deeds-based conveyancing system. The term also refers to the title itself as it passes through these deeds and instruments. Most systems set a time beyond which no searches of the chain need be made—sometimes sixty years.

chattel Goods like cars, stoves, carpets, and so on. In property law, chattels are goods sold or mortgaged with land that are not part of or attached to the land so that they become fixtures. They must be specifically described in the legal document that transfers the title to the land.

civil law Internationally, there are two large families of legal systems: civil-law systems and common-law systems. Civil-law systems are more extensively used. Their features are a heritage of ancient Roman law, the use of codes rather than statutes as basic legislative instruments, and inquisitorial (rather than adversarial) court systems. Concepts of ownership, mortgage, usufructs, servitudes, and good faith are related to their historical sources in Roman law.

collateral Security for a loan, additional to the principal security.

commoditization The treatment of rights in land as marketable commodities. Sometimes, this is called commodification.

common law The second-largest family of legal systems, based on English law. Countries using common law are associated with colonization by the British, which applied the customs and precedents of the English system to aspects of colonial management. The major features include large bodies of specific legislation (not simple, short codes), extensive jurisdiction in the courts to interpret the legislation and make new law, and the power of a decision to operate as a precedent, binding on lower courts and influencing decisions in courts at the same level. Basic property concepts of ownership, adverse possession, mortgage, covenant, easement, trust, and collateral are related to English principles.

condominium An apartment block or development in which individuals own specific apartments and share responsibility for and ownership of property used in common, such as staircases, elevators, driveways, roofs and walls, and other common facilities. Condominium schemes vary from country to country.

consideration The price paid or value given by a purchaser for land or a right in land.

consolidation The amalgamation of land parcels into units of a different size, shape, and location. In some jurisdictions, consolidation refers to the planning and redistribution of land into units of more economic and rational size, shape, and location.

conveyance A method or a document whereby rights in land are transferred from one owner to another. The rights may be full ownership or a mortgage, charge, or lease.

covenant An agreement, either expressed or implied, contained in a deed that creates an obligation between parties. A covenantor gives rights to the covenantee who obtains the benefit. Some covenants operate as proprietary interests and bind people who acquire the land after the covenants were made.

customary law Unwritten law established by long usage. Sometimes, this law is called traditional law, or indigenous law.

customary tenure The holding of land in accordance with customary law.

data custodian The entity charged with ensuring appropriate care and maintenance of information.

deed A legal document evidencing legal rights and obligations. The most important deeds contain the conditions upon which land is transferred, mortgaged, or leased.

deeds registration A system of tracking changes of ownership of land in a public registration program that involves deposit of the deed (or a copy) that makes the change into the registry.

demarcation The marking of the boundaries of each land parcel on the ground.

digital cadastral database (DCDB) A term used to describe a statewide or jurisdiction-wide digital cadastral map.

digitizing The process of converting analog data such as graphic maps into digital form using scaling or other graphic means.

easement A right enjoyed by one landowner (the dominant tenement) over that of another (the servient tenement) — for instance, a right of access or for the passage of water or electricity. The right is regarded as existing for the benefit of the land, not its owner, and accordingly will not be extinguished if there is a change in ownership.

e-land Conducting land administration processes through the use of information and communications technology.

eminent domain The right of the state to take private property for public use, in well-organized systems, upon the payment of just compensation to the property owner. In civil law, eminent domain is not used. The principle is referred to as expropriation and can only be done when warranted by the public interest.

encroachment An unauthorized intrusion on the land of another.

encumbrance A right to or an interest in land that belongs to someone other than the person having the benefit of the right or interest, which represents a burden on the land. The encumbrance will not prevent a transfer of title by the owner of the land, but may reduce its value.

equity In common-law systems, equity is another system of rules based on principles of fairness, formulated and administered by the courts, that supplements the rules of law. Equity historically was administered by the courts of chancery. The rules of law were administered by the king's courts. The result is that common-law systems can simultaneously recognize two kinds of owners of land—an owner at law and an owner in equity.

estate In common-law systems, a proprietary right in land originally granted for a defined period and subject to the observance of tenurial duties. Thus, in English law, people do not own the land, but rather they own estates in the land. There are two kinds of estates: freehold and leasehold. The basic freehold estates run for life (life estate) or as long as the owner has heirs or descendants in perpetuity (a fee simple). The fee simple is now close to absolute ownership. Leasehold estates are for specific periods or periods that can be made specific (year to year).

expropriation The compulsory depriving of an owner of property, in systems that apply the rule of law, in return for compensation.

fixed boundary The legal boundary of a property where the precise line has been agreed and recorded. Usually evidenced or described mathematically.

fixture A chattel that has become so affixed to land as to become part of it, so that ownership of the chattel attaches to the owner of the land.

forfeiture A right to regain possession of leased or mortgaged premise, if the tenant or borrower breaches conditions contained in the agreement.

forgery A document that tells a lie about itself. A forgery is null and void and cannot affect title to land.

fragmentation The division of land units too small for rational exploitation, usually as a result of the system of inheritance. The process may lead to a multiplicity of parcels for one owner or a multiplicity of owners of one parcel.

fraud A deliberate misstatement made to influence another to act. A fraudulent statement gives the person affected a right to set aside the contract.

freehold A free tenure, distinct from leasehold, in which the owner has the maximum rights permissible within the tenure system for indefinite duration.

general boundary A legal boundary of a property where the precise line on the ground has not been determined. A general boundary is usually evidenced by physical monuments.

geodetic network A scientifically measured network of monuments laid over the Earth's surface identified by surveying equipment or by satellite geodesy.

geographic information system (GIS) A system for capturing, storing, checking, integrating, analyzing, and displaying data about the Earth that is spatially referenced. It is normally taken to include a spatially referenced database and appropriate applications software.

GPS A global positioning system using satellites.

grant A general word to describe the transfer of property whereby rights pass from the "grantor" to the "grantee."

i-land Information about land. It is the term for a new vision in a spatially enabled society in which land administration makes comprehensive use of ICT but awaits an effective land information system, based on the SDI that facilitates development.

interest in land A general term to describe proprietary rights in respect to land and its use, entitlement to rent or income derived from land and its use, and entitlement to the whole or part of the proceeds of sale of an estate in land.

interoperability The capability to communicate, execute programs, or transfer data among various functional units in a manner that requires the user to have little or no knowledge of the unique characteristics of those units.

land In most systems of law, land is the surface of the Earth, the materials beneath, the air above, and all things fixed to the soil. Notable exceptions are found in communist countries and countries, such as Indonesia, where land is controlled by the nation.

land administration The processes run by government using public- or private-sector agencies related to land tenure, land value, land use, and land development.

land administration projects (LAPs) Projects to build, reengineer, or improve land administration systems. Includes institutionalization of land administration systems capable of both reflecting and improving existing people-to-land relationships as the focus of many international aid and antipoverty initiatives.

land administration system: An infrastructure for implementation of land policies and land management strategies in support of sustainable development. The infrastructure includes institutional arrangements, a legal framework, processes, standards, land information, management and dissemination systems, and technologies required to support allocation, land markets, valuation, control of use, and development of interests in land.

land governance The activities associated with determining and implementing sustainable land policies.

land information management The managing of information about land.

land information system A system for acquiring, processing, storing, and distributing information about land.

land management The activities associated with the management of land as a resource to achieve social, environmental, and economic sustainable development.

land management paradigm A conceptual framework for understanding and innovation in land administration systems (LAS). The paradigm is the set of principles and practices that define land management as a discipline. The principles and practices relate to the four functions of LAS—namely, land tenure, land value, land use, and land development—and their interactions.

land reform The various processes involved in altering the pattern of land tenure and land use of a specified area. Some of the processes involve land administration, but most of the processes are intensely political.

land register A register, usually public, used to record the existence of deeds or title documents, thereby protecting rights in land and facilitating the transfer of those rights.

land registration The process of recording rights in land either in the form of registration of deeds or else through the registration of title to land, so that any person acquiring a property in good faith can trust in the information published by the registry. Land registration programs range from well-run, deeds-based registration systems, which virtually guarantee title, to Torrens-style systems, which guarantee the title. Registration is positive in nature, and confers and protects the title. Contrast with deeds registration systems, which provide a degree of confidence through registration but do not positively confer the title.

land tenure The manner of holding rights in and occupying land.

land title The title can be the evidence of a person's rights in land or ownership (the deeds or certificate of ownership) or the ownership itself, depending on the context.

land transfer The transfer of rights in land.

land use The manner in which land is used.

land value The worth of a property, determined by one of a variety of ways, each of which can give rise to a specific estimate.

leasehold The property right created by a lease, which is a contract by a landlord (the lessor) giving exclusive possession to a tenant (the lessee) for an agreed amount of money for an agreed period of time.

lessee A tenant holding land or buildings under leasehold.

marine cadastre A management tool that spatially describes, visualizes, and realizes formally and informally defined boundaries as associated rights, restrictions, and responsibilities in the marine environment.

marine SDI A spatial and temporal data infrastructure comprising a system of data and enabling technologies that are critical to sustainable development, management, and control of national marine, coastal, and freshwater areas.

market value The most probable sales price of a real estate property in terms of money, assuming a competitive and open market.

metadata A structured summary of information that describes the data (data about data).

metes and bounds A property description by reference to the bearings and lengths of the boundary lines (metes) together with the names of adjoining properties (bounds), often including features such as walls, river banks, and so on.

monumentation A generic term used to describe the processes and marks used to identify land parcel boundaries.

mortgage An interest in land created by a written instrument providing security for the mortgagee (the lender) for performance of a duty or the payment of a debt by the mortgagor (the borrower). In some legal systems, the mortgagee has the power to sell or forfeit the property when interest is not paid in time or the loan is not paid off in accordance with the contract.

multipurpose cadastre A cadastre that records interests about land parcels relating to tenure, value, use, and development.

mutation The division of land parcels into smaller units, for instance, as a result of inheritance or commercial development.

orthophoto map A map that looks like an aerial photograph or satellite image but which is geometrically accurate.

overriding interest A legal interest in land that has legal force even though not recorded in the public land registers; also called a statutory interest.

ownership The most comprehensive right a person can have with respect to a thing (in this context, land). Full ownership usually includes the exclusive right to use and dispose of the thing (land), but the exact rights vary from country to country.

parcel An area of land with defined boundaries, under unique ownership for specific real property rights.

parcel identifier A unique reference that identifies a parcel in a cadastre or cadastral map.

passport Official title for an asset.

photogrammetry The science and art of taking accurate measurements from photographs.

plot An area of land identifiable on a map.

policy The stated goals for determining how land should be used, managed, and conserved in order to meet social, environmental, and economic objectives.

possession Actual occupation of land.

preemption A right to be offered a property if the owner decides to sell, which does not impose any obligation to buy.

prescription The gaining of a right through a lapse of time. Systems generally work by barring the original owner's right to take action to stop the behavior, rather than by conferring the positive right.

private property Ownership of assets by individuals or legal entities (e.g., companies, cooperatives, and so on).

property Something that is capable of being owned, either in the form of real property (land) or personal property (chattels). The interest can involve physical aspects, such as use of the land, or conceptual rights, such as a right to use the land in the future. However, property has many other meanings, too.

real property Land and any things attached to the land, including buildings, apartments, and other construction, and natural objects such as trees, and in some jurisdictions, minerals. Communist and some ex-communist countries only recognize property in buildings, not land.

rectification The legal process whereby errors on a land register may be corrected.

rent seeking Occurs when an individual, organization, or firm seeks to make money through economic rent. Rent seeking generally implies the extraction of uncompensated value from others without making any contribution to productivity, such as by gaining control of land and other preexisting natural resources or by imposing burdensome regulations or making other government decisions that may affect consumers or businesses.

security An interest in an asset given to secure repayment of a debt.

security of tenure At the most basic level, security exists when "the fact that other people believe the land you occupy and use is the land that you are allowed to live on and use" (UN–HABITAT 2004, 13). Legal security exists insofar as the law of a country protects the continuing use.

silo An agency in a land administration system that operates according to its internal norms and functions and does not interact with other agencies. Historically, most agencies in land administration were established as silos. Modern policy imperatives and technology demand the reconstruction of silos into cooperative and interactive agencies.

spatial data/information Data/information relating to the land, sea, or air that can be referenced to a position on the Earth's surface. It is also the key to planning, sustainable management, and development of our natural resources at the local, national, regional, and global level.

spatial data infrastructure (SDI) A term that describes the fundamental spatial datasets, the standards that enable them to be integrated, the distribution network to provide access to them, the policies and administrative principles that ensure compatibility among jurisdictions and agencies, and the people including user, provider, and value adder who are interested at a certain area level, starting at the local level and proceeding through the state, national, and regional levels to the global level. This has resulted in the development of the SDI concept at these levels.

spatially enabled government Achieved when location and spatial information are available to citizens and businesses to encourage creative uses and when governments use place as a means of organizing their activities and information.

sporadic registration A method of bringing land into a registration program through ad hoc methods, usually on transfer of the land.

squatter A person who uses land without title. Many countries are unable to provide titles to citizens who are inevitably squatters, especially those who live in urban slums. Here, imperatives of housing and livelihood demand that even squatters are protected against arbitrary eviction.

strata title An entitlement to a three-dimensional space within a larger space (usually a defined parcel) in a cadastre, often in high-rise buildings, but also in freestanding villas, commercial developments, and car spaces. Strata titles usually include a share in common property that is managed by an owners corporation. The boundaries of the spaces are usually defined by walls, floors, and ceilings using the general boundary concept.

subdivision The process of dividing a land parcel into smaller parcels.

sustainable development Development that meets the needs of the present without compromising the ability of future generations to meet their own needs. The field of sustainable development can be conceptually broken into three constituent parts: environmental sustainability, economic sustainability, and sociopolitical sustainability.

systematic registration A method of bringing all parcels of land in a defined region into the system through a single process of public education, adjudication of titles, surveying or other means of identifying the parcels, creating unique parcel numbers, and issuing titles.

tenure The way in which the rights, restrictions, and responsibilities that people have with respect to the land are held. The cadastre may record different forms of land tenure such as ownership, leasehold, and different types of common, communal, or customary land tenure.

title insurance A system for compensating people who suffer losses through a title system. The insurance can be provided by private insurers, for example, in the United States, or by the government through title guarantees and insurance schemes related to registration programs.

topography The physical features of the Earth's surface.

transfer Either the act by which title to property is conveyed from one person to another or the document used to pass registered land to the transferee.

trust In common law, an arrangement by which legal title to property is held by one person on behalf of and for the benefit of another.

usufruct The restricted right by which a person is entitled to use and to enjoy the fruits of a property that is owned by another person.

value Either the market value (sales price paid), the rental value (based on how much it can be rented out for), the use value (the potential of the land—for example, for agriculture), the investment value (what income it should generate), or the assessed value (the official value for tax purposes).

Reference list

Alberta Land Surveyors Association. 2007. The subdivision process. http://www.alsa.ab.ca (accessed August 1, 2009).

Aldrich, D., J. C. Bertot, and C. R. McClure. 2002. E-government: Initiatives, developments, and issues. *Government Information Quarterly* 19:349–55.

Angus-Leppan, P. V., and I. P. Williamson. 1985. A project for upgrading the cadastral system in Thailand. *Survey Review* 28:215–16.

Arizona Department of Revenue. 2001. Rectangular survey system. Land Manual. Appendix A. http://www.revenue.state.az.us/Forms/Property/LandManual/AppendixA.pdf (accessed August 1, 2009).

Asia Development Bank (ADB). 1998. Handbook on resettlement: A guide to good practice. http://www.adb.org/Documents/Handbooks/Resettlement/default.asp (accessed August 1, 2009).

Astke, H., G. Mulholland, and R. Nyarady. 2004. Profile definition for a standardized cadastral model. Proceedings of Joint FIG–Commission 7 and COST Action G9 Workshop on Standardization in the Cadastral Domain (December 9–10). Bamberg, Germany. http://www.fig.net/commission7/bamberg_2004/index.htm (accessed August 1, 2009).

Atwood, D. A. 1990. Registration in Africa: The impact on agricultural production. *World Development* 18 (5): 659–71.

Augustinus, C., C. Lemmen, and P. van Oosterom. 2006. Social tenure domain model: Requirements from the perspective of pro-poor land management. Proceedings of 5th FIG Regional Conference on Promoting Land Administration and Good Government. (March 8–11). Accra, Ghana. http://www.fig.net/pub/accra/papers/ps03/ps03_02_lemmen.pdf (accessed August 1, 2009).

Auzins, A. 2004. Institutional Arrangements: A gateway towards sustainable land use. *Nordic Journal of Surveying and Real Estate Research* 1:57–71. http://mts.fgi.fi/njsr/issues/2004/ njsrvln12004_auzins.pdf (accessed August 1, 2009).

Barry, M., I. Elema, and P. van der Molen. 2003. Ocean governance in the Netherlands North Sea. FIG Working Week. Paris, France. http://www.fig.net/pub/fig_2003/index.htm (accessed August 1, 2009).

Beardsall, T. 2004. E-conveyancing: A challenge and a prize. Proceedings of International FIG Seminar (June 2–4): 104–11. Innsbruck, Austria. http://www.fig.net/commission7/innsbruck_2004/index.htm (accessed August 1, 2009).

Bell, K. 2006. World Bank support for land administration and management: Responding to the challenges of the Millennium Development Goals. 23rd International FIG Conference (October 8-13). Munich, Germany. http://www.fig.net/pub/fig2006/ (accessed August 1, 2009).

Bennett, R., J. Wallace, and I. P. Williamson. 2006. Achieving sustainability objectives through better management of property rights, restrictions, and responsibilities. In *Sustainability and Land Administration Systems*. Ed. I. Williamson, S. Enemark, and J. Wallace, 197–212. Department of Geomatics, University of Melbourne, Australia. http://www.geom.unimelb.edu.au/research/SDI_ research/EGM%20BOOK.pdf (accessed August 1, 2009).

Bennett, R., J. Wallace, and I. P. Williamson. 2008. Organizing property information for sustainable land information. *Land Use Policy* 25:126–38.

Biau, D. 2005. New legal framework and tools. Proceedings of Expert Group Meeting on Secure Land Tenure (opening remarks, November 11–12, 2004). UN Gigiri, Nairobi, Kenya. Denmark: International Federation of Surveyors. http://www.fig.net/commission7/nairobi_2004/program.htm (accessed August 1, 2009).

Binns, A. 2004. Defining a marine cadastre: Legal and institutional aspects. Master's thesis, University of Melbourne, Australia.

Binns, A., A. Rajabifard, P. A. Collier, and I. P. Williamson. 2004. Developing the concept of a marine cadastre: An Australian case study. *Trans-Tasman Surveyor Journal* (Australia), No. 6 (August 2004).

Binns, Sir B. O. 1953. *Cadastral surveys and records of rights in land.* FAO. Rome.

Binns, Sir B. O., and P. F. Dale, P. 1995. *Cadastral surveys and records of rights in land*, based on the 1953 study by Sir Bernard O. Binns, revised by P. F. Dale. Land Tenure Studies 1, FAO Rome. http://www.fao.org/docrep/006/V4860E/V4860E00.htm (accessed August 1, 2009).

Binswanger, H., K. Deininger, and G. Feder. 1993. Power, distortions, revolt, and reform in agricultural land relations. Policy Research Working Paper Series, No. 1164. Washington, D.C.: World Bank.

Bishop, I., F. J. Escobar, S. Karuppannan, I. P. Williamson, P. Yates, K. Suwarnarat, and H. W. Yaqub. 2000. Spatial data infrastructures for cities in developing countries: Lessons from the Bangkok experience. *Cities* 17:85–96.

Bledsoe, D. 2006. Can land titling and registration reduce poverty? In *Land Law Reform: Achieving Development Policy Objectives*. Ed. J. Bruce, G. Giovarelli, L. Rolfes Jr., D. Bledsoe, and R. Mitchell. Washington, D.C.: World Bank.

Bogaerts, T., I. P. Williamson, and E. M. Fendel. 2002. The role of land administration in the accession of Central European countries to the European Union. *Journal of Land Use Policy* 19:29–46.

BOND. 2003. Logical framework analysis. Guidance note, No. 4. http://www.bond.org.uk (accessed August 1, 2009).

Britton, W., K. Davies, and T. Johnson. 1980. *Modern Methods of Valuation of Land, Houses, and Buildings*. 7th ed. London: Estates Gazette.

Bromley, D. 2006. Land and economic development: New institutional arrangements for the 21st century. Toward a 2015 Vision of Land, conference sponsored by the International Center for Land Policy and Training in Taiwan and the Lincoln Institute of Land Policy (October 24–25). Taipei, Taiwan.

Bruce, J. W. 1998. Learning from comparative experience with agrarian reform. Proceedings of International Conference on Land Tenure in the Developing World (January 27–29). University of Cape Town, South Africa.

Bruce, J. W., R. Giovarelli, L. Rolfes Jr., D. Bledsoe, and R. Mitchell. 2006. *Land Law Reform: Achieving Development Policy Objectives*. Washington, D.C.: World Bank.

Bruggemann, H. 2004. The German GDI: A public-private cooperation project. Proceedings of International FIG Seminar (June 2–4): 88–97. Innsbruck, Austria. http://www.fig.net/commission7/innsbruck_2004/index.htm (accessed August 1, 2009).

Brundtland Report. 1987. Our common future. Brundtland Commission. Oxford: Oxford University Press.

Brunner, H. 2004. CYBERDOC: Archives for e-government. Proceedings of International FIG Seminar (June 2–4): 130–40. Innsbruck, Austria.

Burns, A. F., R. Eddington, C. Grant, and I. Lloyd. 1996. Land titling experience in Asia. Proceedings of International Conference on Land Tenure and Administration in Developing Countries (November 23–26, 1996). Orlando, Florida. http://www.surv.ufl.edu/publications/land_conf96/Barnstoc.htm (accessed August 1, 2009).

Burns, A. F. 2006. Land administration reform: Indicators of success and future challenges. Land Equity International. Wollongong, Australia. http://www.landequity.com.au (accessed August 1, 2009).

Burns, A. F. 2007. Land administration reform: Indicators of success and future challenges. Working Paper No. 37, Report No. 41893 (January 1). Washington, D.C.: World Bank.

Byamugisha, F. K. 1999. The effects of land registration on financial development and economic growth: A theoretical and conceptual framework. Policy Research Working Paper, No. 2240. Washington, D.C.: World Bank.

Cadman, D., and L. Austin-Crowe. 1993. *Property Development*. London: Chapman and Hall.

Chauveau, J. P., J. P. Colin, J. P. Jacob, P. L. Deville, and P. Y. Le Meur. 2006. Land tenure and resource access in West Africa. In *Changes in Land Access, Institutions, and Markets in West Africa*. London: IIED Publications.

Colby, B.G. 2000. Cap-and-trade policy challenges: A tale of three markets. *Land Economics* 76 (4): 638–48.

Courtney, J. M. 1983. Intervention through land-use regulation. In *Urban Land Policy: Issues and Opportunities*. Ed. H. B. Dunkerley, 153–70. Washington, D.C.: Oxford University Press.

Cowen, D. J., and W. J. Craig. 2003. A retrospective look at the need for a multipurpose cadastre. *Surveying and Land Information Science* 63 (4): 205–14. http://www. nationalcad.org/data/documents/Cowen_Craig.pdf (accessed August 1, 2009). Also see ArcNews Summer 2004. http://www.esri.com/news/arcnews/ summer04articles/a-retrospective-look.html (accessed August 1, 2009).

Dale, P. F. 1976. *Cadastral Surveys within the Commonwealth.* London: HM Stationery Office.

Dale, P. F., and R. Baldwin. 1998. Lessons learned from the emerging land markets in Central and Eastern Europe. Working Paper under the Action for Cooperation in the Field of Economics, EU program.

Dale, P. F., and R. Baldwin. 2000. Emerging land markets in Central and Eastern Europe. Structural change in the farming sectors in Central and Eastern Europe: Lessons for the EU accession. Ed. C. Csaki and Z. Lerman. Technical Paper No. 465:81–109. Washington, D.C.: World Bank.

Dale, P. F., and J. D. McLaughlin. 1988. *Land Information Management: An Introduction with Special Reference to Cadastral Problems in Third World Countries.* Oxford: Clarendon Press.

Dale, P. F., and J. D. McLaughlin. 1999. *Land Administration.* Oxford: Oxford University Press.

Dalrymple, K. 2005. Expanding rural land tenures to alleviate poverty. Doctorate thesis, University of Melbourne, Australia.

Dalrymple, K., I. P. Williamson, and J. Wallace. 2003. Cadastral systems within Australia. Australian Surveyor 48 (1): 37–49.

Davies, W., and P. Fouracre. 1995. *Property and Power in the Early Middle Ages.* Cambridge: Cambridge University Press.

Deininger, K., ed. 2003. *Land Policies for Growth and Poverty Reduction.* Published for the World Bank by Oxford University Press in New York.

Deininger, K., and H. Binswanger. 1999. The evolution of the World Bank's land policy: Principles, experience, and future challenges. *World Bank Research Observer* 14:247–76.

Deininger, K., and G. Feder. 1999. Land policy in developing countries. Rural Development Note 3, Report No. 20876: 4. Washington, D.C.: World Bank.

Denman, D. R. 1978. *The Place of Property: A New Recognition of the Function and Form of Property Rights in Land.* Hertfordshire, UK: Geographical Publications.

De Soto, H. 2000. *The Mystery of Capital: Why Capitalism Triumphs in the West and Fails Everywhere Else.* London: Black Swan.

DFID. 2002. Tools for development. http://www.dfid.gov. uk (accessed August 1, 2009).

DFID. 2003. Literacy, gender, and social agency. Research report.

DFID. 2004. Decentralization and governance. Key Sheets for Sustainable Livelihoods. http://www. keysheets.org/red_11_decentra_gov.html (accessed August 1, 2009).

Dowson, Sir E., and VLO Sheppard. 1952. *Land Registration.* London: HM Stationery Office.

Drucker, Peter. 1946. *The Concept of the Corporation.* New York: John Day Co.

Dunkerley, H. B. 1983. *Urban Land Policy, Issues and Opportunities.* New York: Oxford University Press.

Dunkerley, H. B., and CME Whitehead. 1983. *Urban Land Policy, Issues, and Opportunities.* Published for the World Bank by Oxford University Press in New York.

Du Plessis, J., and S. Leckie. 2006. The need for more inclusive concepts. In *Realizing Property Rights.* Ed. H. de Soto and F. Cheneval. Swiss Human Rights Books 1:194–203. Bern, Switzerland: Rüffer and Rug.

Effenberg, W. W., S. Enemark, and I. P. Williamson. 1999. Framework for discussion of digital spatial dataflow within cadastral systems. *Australian Surveyor* 44 (1): 35–43.

Elfick, M., T. Hodson, and C. Wilkinson. 2005. Managing a cadastral SDI framework built from boundary dimensions. Proceedings of FIG Working Week 2005 and GSDI-8 (April 16–19). Cairo, Egypt. http://www. fig.net/pub/cairo/ (accessed August 1, 2009).

Enemark, S. 1999. *Denmark: The EU Compendium of Spatial Planning Systems and Policies.* Brussels: European Union, Office for Official Publications of the European Communities.

Enemark, S. 2001. Merging the efforts of CLGE (Council of European Geodetic Surveyors) and FIG to enhance professional competence. Ed. S. Enemark and WP Prendergast. Enhancing professional competence of surveying in Europe (November 3, 2000): 1–75. Copenhagen, Denmark. http://www.fig. net/pub/CLGE-FIG-delft/report-1.htm (accessed August 1, 2009).

Enemark, S. 2004. Building land information policies. Proceedings of Special Forum on Building Land Information Policies in the Americas (October 26–27). Aguascalientes, Mexico. http://www.fig.net/pub/ mexico/papers_eng/ts2_enemark_eng.pdf (accessed August 1, 2009).

Enemark, S. 2006a. Responding to the Millennium Development Goals. 23rd International FIG Conference (October 8–13). Munich, Germany. http://www.fig.net/pub/fig2006/ (accessed August 1, 2009).

Enemark, S. 2006b. The land management paradigm for sustainable development. In *Sustainability and Land Administration Systems*. Ed. I. Williamson, S. Enemark, and J. Wallace. Department of Geomatics, University of Melbourne, Australia. http://www.geom. unimelb.edu.au/research/SDI_research/EGM%20BOOK. pdf (accessed August 1, 2009).

Enemark, S., and R. Ahene. 2003. Capacity building in land management: Implementing land policy reform in Malawi. *Survey Review* 37 (287): 20–30.

Enemark, S., and I. P. Williamson. 2004. Capacity building in land administration: A conceptual approach. *Survey Review* 39 (294): 639–50.

Enemark, S., I. Williamson, and J. Wallace. 2005, Building Modern Land Administration Systems in Developed Economies, *Journal of Spatial Science* 50 (2): 51–68.

Enemark, S., and R. McLaren. 2008. Preventing informal urban developments through means of sustainable land-use control. Proceedings of FIG Working Week (June 14–19). Stockholm, Sweden. http://www.fig.net/ pub/fig2008/ (accessed August 1, 2009).

Enterprise Research Institute for Latin America. 1997. An exploration of issues related to land titling programs. Special Report.

Epstein, R. 1993. Possession and title. *Georgia Law Review* 13, No. 1221.

Erba, D. A. 2004. Latin American cadastres: Successes and remaining problems. *Land Lines Newsletter* 16, No. 2. Lincoln Institute of Land Policy. http://www. lincolninst.edu/pubs/pub-detail.asp?id=883 (accessed August 1, 2009).

Ereata, T. 2003. Kiribati Country Report. The Cadastral Template Project. http://www.cadastraltemplate.org/ (accessed August 1, 2009).

ESRI. 2006, OGC support. http://www.esri.com/software/ standards (accessed August 1, 2009).

European Commission. 1997. *The EU Compendium of Spatial Planning Systems and Policies*. Brussels: European Union, Office for Official Publications of the European Communities.

European Commission. 2002. Guidelines for best practice in user interface for GIS. http://www.gisig.it/ best-gis/Guides/chapter8/ch8.pdf (accessed August 1, 2009).

European Union (EU). 2006. *Declaration of the cadastre in Latin America*. http://www.eurocadastre.org/ pdf/2006-02_declaration_of_cadastre.pdf (accessed August 1, 2009).

FAO. 1953. Cadastral surveys and records of rights in land. (Republished 1995). http://www.fao.org/ docrep/006/V4860E/V4860E00.htm (accessed August 1, 2009).

FAO. 2006. Access to rural land and land administration after violent conflict. FAO Land Tenure Series, No. 8. http://www.fao.org/sd/dim_in1/in1_060501_en.htm (accessed August 1, 2009).

FAO. 2007. Good governance in land tenure and administration. FAO Land Tenure Series, No. 9. http://www.fao.org/docrep/010/a1179e/a1179e00.htm (accessed August 1, 2009).

Feder, G., and D. Feeney. 1991. Land tenure and property rights: Theory and implications for development policy. *World Bank Economic Review* 5, No. 1.

Feder, G., and A. Nishio. 1998. The benefits of land registration and titling: Economic and social perspectives. *Land Use Policy* 15 (1): 24–44.

Feder, G., T. Onchan, and Y. Chalamwong. 1988. Land policies and farm performance in Thailand's forest reserve areas. *Economic Development and Cultural Change* 36 (3): 483-501.

Feder, G., T. Onchan, Y. Chalamwong, and C. Honglada-
rom. 1988. Land policies and farm productivity in
Thailand. World Bank research publication.
Baltimore: John Hopkins University Press.

Feeney, M., and I. P. Williamson. 2000. Researching
frameworks for evolving spatial data infrastructures.
Proceedings of SIRC. 12th Annual Colloquium of the
Spatial Information Research Center, University of
Otago (December 10-13): 93–105. Dunedin, New
Zealand.

FIG. 1995. Statement on the cadastre. Fig publication No
11. FIG Office, Copenhagen. http://www.fig.net/pub/
figpub/pubindex.htm (accessed August 1, 2009).

FIG. 1996. Continuing professional development. FIG
policy statement. FIG publication No. 15. http://www.
fig.net/pub/figpub/pub15/figpub15.htm (accessed
August 1, 2009).

FIG. 1998. Cadastre 2014. A vision for a future cadastral
system. http://www.fig.net/cadastre2014/ (accessed
August 1, 2009).

FIG. 2001. Women's access to land. FIG guidelines. FIG
publication No. 24. http://www.fig.net/pub/figpub/
pub24/figpub24.htm (accessed August 1, 2009).

FIG. 2002. Mutual recognition of professional qualifica-
tion. FIG publication No. 27. http://www.fig.net/pub/
figpub/pub27/figpub27.htm (accessed August 1, 2009).

FIG. 2004. Aguascalientes Statement. FIG publication No.
34. http://www.fig.net/pub/figpub/pub34/figpub34.pdf
(accessed August 1, 2009).

FIG. 2006. Administering marine spaces: International
issues. Commissions 4 and 7. http://www.fig.net/pub/
figpub/pub36/figpub36.htm (accessed August 1, 2009).

FIG. 2008a. Capacity assessment in land administration.
FIG publication No. 34. http://www.fig.net/pub/
figpub/pub41/figpub41.htm (accessed August 1, 2009).

FIG 2008b, Costa Rica declaration on pro-poor coastal
zone management. FIG publication No. 43. http://
www.fig.net/pub/figpub/pub43/figpub43.htm (accessed
August 1, 2009).

FIG/CLGE. 2001. Enhancing professional competence of
surveyors in Europe. http://www.fig.net/pub/
CLGE-FIG-delft/report-1.htm (accessed August 1, 2009).

Fitzpatrick, D. 1997. Disputes and pluralism in modern
Indonesian land law. Yale Journal International Law
Review 22 (1): 171–212.

Fitzpatrick, D. 2005. Best-practice options for the legal
recognition of customary tenure. Development and
Change 36 (3): 449–75.

Fitzpatrick, D. 2006. Evolution and chaos in property
systems: The Third World tragedy of contested
access. Yale Law Journal 11 (5): 996–1048.

Fonseca, F. 2005. System Heterogeneities: Analyses of
Interoperable Geospatial Information Systems.

Foste, I., and C. Kesselman. 1999. The Grid: Blueprint for
a Future Computing Infrastructure. San Francisco:
Morgan Kaufman Publishers.

Foste, I., and C. Kesselman. 2004. The Grid 2: Blueprint
for a Future Computing Infrastructure. San Francisco:
Morgan Kaufman Publishers.

Fowler, C., and E. Treml. 2001. Building a marine
cadastral information system for the United
States—a case study. Special issue, International
Journal on Computers, Environment & Urban Systems
25 (4–5): 493–507.

Frumkin, H. 2002. Urban sprawl and public health.
Public Health Report 117:210–17. Washington.

Fukuzaki, Y. 2003. Japan Country Report. The Cadastral
Template Project. http://www.cadastraltemplate.org/
(accessed August 1, 2009).

Galal, A., and O. Razzaz. 2001. Reforming real estate
markets. Policy Research Working Paper 2616.
Washington, D.C.: World Bank.

Gilbert, A. 2002. On the mystery of capital and the myths
of Hernando de Soto: What difference does legal title
make? International Development and Planning
Review 24 (1): 1–19.

Giovarelli, R. 2006. Overcoming gender biases in
established and transitional property rights systems.
In Land Law Reform: Achieving Development Policy
Objective. Ed. J. Bruce, R. Giovarelli, L. Rolfes Jr., D.
Bledsoe, and R. Mitchell. Washington, D.C.: World
Bank.

Glenn, P. 2004. Legal Traditions of the World: Sustainable
Diversity in Law, 2nd ed. New York: Oxford University Press.

GLTN/FIG. 2008. Workshop report. Land Professionals
Workshop on Gendering Land Tools (March 10–11).
Bagomoyo, Tanzania. http://www.fig.net/news/
news_2008/bagamoyo_march_2008.htm (accessed
August 1, 2009).

Gore, A. 1998. *The Digital Earth: Understanding Our Planet in the 21st Century*. Los Angeles: California Science Center.

Grant, D. 1997. Territoriality: Concept and delimitation. First Trans-Tasman Surveyors Conference (April 12–18). Newcastle, New South Wales, Australia.

Grant, D. 1999. Principles for a seabed cadastre. New Zealand Institute of Surveyors Conference and AGM FIG–Commission 7 Conference (October 9–15). Bay of Islands, New Zealand.

Grant, D. 2004. Cadastral automation and related e-government initiatives in New Zealand. Proceedings of International FIG Seminar (June 2–4). Innsbruck, Austria.

Green, D., and T. Bosomair. 2001. *Online GIS and Spatial Metadata*. New York: Francis and Taylor.

Greenland, A., and P. van der Molen. 2006. Administering marine spaces: International issues. FIG–Commissions 4 and 7, Working Group 4.3. FIG publication No. 36. **http://www.fig.net/pub/figpub/pub36/figpub36.htm** (accessed August 1, 2009).

Greunz, M., B. Schopp, and J. Haes. 2001. Integrating e-government infrastructure through secure XML document containers. Proceedings of 34th Hawaii International Conference on System Sciences. IEEE, Hawaii.

Groot, R., and J. McLaughlin. 2000. *Geospatial Data Infrastructure: Concepts, Cases, and Good Practice*. New York: Oxford University Press.

GTZ. 1998. Land tenure in development cooperation guiding principles. Gesellschaft fur Technische Zusammenarbeit (GTZ), Eschborn. Wiesbaden, Germany: Universum Verlagsanstalt. **http://www.gtz.de/** (accessed August 1, 2009).

Hakimpour, F. 2003. Using ontologies to resolve semantic heterogeneity for integrating spatial database schemata. Doctorate thesis, Zurich University, PhD Hall 1895.

Haldrup, K. 2002. Mainstreaming gender issues in land administration: Awareness, attention, and action. International Federation of Surveyors 22nd annual Congress (April 19–26). Washington, D.C. **http://www.fig.net/pub/fig_2002/fig_index.htm** (accessed August 1, 2009).

Hall, A. F. 1895. The survey system of New South Wales. *Australian Surveyor* 8 (7): 149.

Harcombe, P., and I. P. Williamson. 1998. A cadastral model for low-value lands: The NSW western lands experience. FIG–Commission 7 21st International Congress: Developing the Profession in a Developing World (July 19–25): 569–79. Brighton, UK.

Harrison, L. E., and S. P. Huntington, eds. 2000. *Culture Matters: How Values Shape Human Progress*. New York: Basic Books.

Hecht, L. 2004. Observations on the proposed standardized cadastre domain model: Where do we go from here? Proceedings of Joint FIG–Commission 7 and COST Action G9 Workshop on Standardization in the Cadastral Domain (December 9–10). Bamberg, Germany. **http://www.fig.net/commission7/bamberg_2004/index.htm** (accessed August 1, 2009).

HMT. 2000. Public services productivity: Meeting the challenge. **http://www.hm-treasury.gov.uk** (accessed August 1, 2009).

Holstein, L. 1996a. What are the roles of the public and private sectors in land titling and registration systems? International Conference on Land Tenure and Administration (November). Orlando, Florida. **http://sfrc.ifas.ufl.edu/geomatics/publications/land_conf96/Barnstoc.htm** (accessed August 1, 2009).

Holstein, L. 1996b. Toward best practice from World Bank experience in land titling and registration. International Conference on Land Tenure and Administration (November). Orlando, Florida. **http://sfrc.ifas.ufl.edu/geomatics/publications/land_conf96/Barnstoc.htm** (accessed August 1, 2009).

ILC. 2004. EU land policy guidelines (draft). EU Task Force on Land Tenure. Rome: International Land Coalition.

Inproteo. 2005. Definition for Interoperability. **http://www.inproteomics.com** (accessed August 1, 2009).

International Valuation Standards. 2001. A submission to the International Accounting Standards Board in respect to an issues paper issued for comment by the IASC Steering Committee on Extractive Industries (June). London. **http://www.ivsc.org/pubs/** (accessed August 1, 2009).

ISO and OGC. 2004. *Geography Markup Language*. International Standard Organization Open Geospatial Consortium. **http://www.opengeospatial.org/** (accessed August 1, 2009).

Jacobs, H. M. 2007. Social conflict over property rights. *Land Lines* (April): 14–19. Lincoln Institute of Land Policy in Cambridge, Massachusetts.

Jaeger, P. T., and K. M. Thompsom. 2003. E-government around the world: Lessons, challenges, and future directions. *Government Information Quarterly* 20:389–94.

Jones, M., and G. Taylor. 2004. Data integration issues for a farm decision support system. *Transactions in GIS* 8 (4): 459-77.

Kabubo-Mariara, J. W. 2006. Land conservation in Kenya: The role of property rights. Paper No. RP 153. African Economic Research Consortium. Nairobi, Kenya. http://www.aercafrica.org/publications/item. asp?itemid=242&category= (accessed August 1, 2009).

Kain, RJP, and E. Baigent. 1992. *The Cadastral Map in the Service of the State: A History of Property Mapping.* Chicago: University of Chicago Press.

Kalantari, M. 2003. Design and implementation of an agent-based distributed GIService. Master's thesis, KN Toosi University of Technology, Tehran, Iran.

Kalantari, M. 2004. Agent technology as a solution for network-enabled GIS. Proceedings of International Society for Photogrammetry and Remote Sensing (July 13–23). Istanbul, Turkey.

Kalantari, M. 2008. Cadastral data modeling: A tool for e-land administration. Doctorate thesis, University of Melbourne, Australia.

Kalantari, M., A. Rajabifard, J. Wallace, and I. P. Williamson. 2005. Toward e-land administration: Evaluating Australian online land administration services. Proceedings of Spatial Science Conference (September 14–16). Melbourne, Australia.

Kalantari, M., A. Rajabifard, J. Wallace, and I. P. Williamson. 2006. A new vision on cadastral data model. Proceedings of 23rd FIG Congress on Shaping the Change (October 8–13). Munich, Germany. http:// www.fig.net/pub/fig2006/ (accessed August 1, 2009).

Kalantari, M., A. Rajabifard, J. Wallace, and I. P. Williamson. 2008. Spatially referenced legal property objects. *Land Use Policy* 25:173–83.

Kanji, N., L. Cotula, T. Hilhorst, C. Toulmin, and W. Witten. 2005. Can land registration serve poor and marginalized groups? Summary report. IIED Publications. http://www.iied.org/pubs/ pdfs/12518IIED.pdf (accessed August 1, 2009).

Kaufmann, J., and D. Steudler. 1998. *Cadastre 2014: A Vision for a Future Cadastral System.* http://www.fig. net/cadastre2014/ (accessed August 1, 2009).

Knetsch, J., and M. Trebilcock. 1981. Land policy and economic development in Papua New Guinea. Institute of National Affairs Discussion Paper No. 6. Institute of National Affairs, Port Moresby.

Land Equity. 2006. Land administration: Indicators of success, future challenges. Rural Development Discussion Paper No. 37. Land Equity International in Wollongong, Australia.

Larsson, G. 1991. *Land Registration and Cadastral Systems: Tools for Land Information and Management.* New York: Wiley.

Larsson, G. 1996. *Land Registration and Cadastral Systems.* Essex, UK: Addison Wesley Longman.

Lastaria-Cornhiel, S. 1997. Impact of privatization on gender and property rights in Africa. *World Development 35* (8): 1317–41.

Lavigne Delville, P. 2000. Harmonizing formal law and customary land rights in French-speaking West Africa. In *Evolving Land Rights, Policy, and Tenure in Africa.* Ed. C. Toulmin and J. Quan, 97–122. London: DFIF/IIED/NRI.

Lavigne Delville, P. 2002a. Customary to modern transition. Proceedings of World Bank regional meeting on land issues (April 29 to May 2). Kampala, Uganda.

Lavigne Delville, P. 2002b. Towards an articulation of land regulation modes, recent progress, and issues at stake. Proceedings of World Bank regional meeting on land issues (April 29 to May 2). Kampala, Uganda.

Le Moule, C. 2004. Impact of decoupling and modulation in the enlarged union: A sectoral and farm-level assessment (IDEMA Project). Agricultural land markets, main issues in the recent literature, Deliverable 2. A European research project supported by the European Union. http://www.sli. lu.se/IDEMA/WPs/IDEMA_deliverable_2.pdf (accessed August 1, 2009).

Lemmen, C., P. van Oosterom, J. Zevenbergen, W. Quak, and P. van der Molen. 2005. Further progress in the development of the core cadastral domain model. Proceedings of FIG Working Week and GSDI-8 (April 16–21). Cairo, Egypt. http://www.fig.net/pub/ cairo/ (accessed August 1, 2009).

Leonard, R., and K. Ayutthaya. 2003. Thailand Land Titling Program. Monitoring paper. Northern Development Foundation. Chiang Mai, Thailand. http://www.landaction.org/spip/?lang=en (accessed August 1, 2009).

Lindsay, J. 2002. Land in law and sustainable development since Rio: Legal trends in agriculture and natural resource management. FAO Legislative Study No. 73, chap. 8. http://www.fao.org/DOCREP/005/Y3872E/Y3872E00.HTM (accessed August 1, 2009).

Ljunggren, T. 2004. Moving focus from organization to information. Proceedings of Joint FIG–Commission 7 and COST Action G9 Workshop on Standardization in the Cadastral Domain (December 9–10). Bamberg, Germany.

Lohmann, L. 2002. Polanyi along Mekong: New tensions and resolutions over land. www.thecornerhouse.org.uk/item.shtml?x+52214 (accessed August 1, 2009).

Longley, P. A., and M. C. Batty. 2003. *Advanced Spatial Analysis: The CASA book of GIS*. Redlands, California: ESRI Press.

Louwman, W. 2004. Legal consequences of the electronic transfer of immovable property in the Netherlands. Proceedings of International FIG Seminar (June 2–4): 112–15. Innsbruck, Austria. http://www.fig.net/commission7/innsbruck_2004/index.htm (accessed August 1, 2009).

Lyons, K., E. Cotterell, and K. Davies. 2002. On the efficiency of property rights in Queensland. Research report. Queensland government, Department of Natural Resources and Mines, Brisbane, Australia. http://www.anzlic.org.au/pubinfo/2393092707.html (accessed August 1, 2009).

Manes, A. T. 2003. *Web Services: A Manager's Guide*. Boston: Addison Wesley.

Matsukawa, T., and O. Habeck. 2007. *Review of Risk Management Instruments for Infrastructure: Financing and Recent Trends and Developments*. Washington, D.C.: World Bank.

McAuslan, P. 1998. Making the law work: Restructuring land relations in Africa. *Development and Change* 29 (3): 525–52.

McAuslan, P. 2003. The International Development Act, 2002: Benign imperialism or a missed opportunity. *Modern Law Review* 66:563–603.

McGrath, G., T. MacNeill, and I. Ford. 1996. Issues and key principles related to the implementation of cadastral and land registration systems: A perspective from Eastern Europe and the former Soviet Union. Proceedings of International Conference on Land Tenure and Administration in Developing Countries (November 23–26). Orlando, Florida. http://www.surv.ufl.edu/publications/land_conf96/Barnstoc.htm (accessed August 1, 2009).

McLaughlin, J. D. 1975. The nature, design, and development of multipurpose cadastres. Doctorate thesis, University of Wisconsin–Madison.

McLaughlin, J. D. 1998. Land administration guest lecture, Department of Geomatics, University of Melbourne, Australia (March 9).

Meyer, N. V. 2004. Cadastral core data. Draft report, version 5 (October).

Mhlanga, A., and I. Greenway. 1999. Recent experiences in developing Swaziland's Surveyor General's Department. *Survey Review* 35 (274): 231–42.

Mohammadi, H., A. Rajabifard, A. Binns, and I. P. Williamson. 2006. Development of a framework and associated tools for the integration of multisource spatial datasets. 17th UNRCC–AP (September 18–22). Bangkok, Thailand.

Moyer, D. D., and K. P. Fisher. 1973. *Land Parcel Identifiers for Information Systems*. Chicago: American Bar Foundation.

Mulolwa, A. 2002. Integrated land delivery: Towards improving land administration in Zambia. DUP Science. Delft, Netherlands: Delft University Press.

Nasoetion, L. I. 2003. Indonesia Country Report. The Cadastral Template Project. http://www.cadastraltemplate.org/ (accessed August 1, 2009).

National Research Council. 1980. *Need for a Multipurpose Cadastre*. Panel on Multipurpose Cadastre. Committee on Geodesy, Assembly of Mathematics and Physical Sciences. Washington, D.C.: National Academy Press. http://www.nap.edu/catalog.php?record_id=10989 (accessed August 1, 2009).

National Research Council. 2007. *Land Parcel Databases: A National Vision*. http://www8.nationalacademies.org/cp/projectview.aspx?key=219 (accessed August 1, 2009).

Nichols, S., D. Monahan, and M. Sutherland. 2000. Good governance of Canada's offshore and coastal zone: Towards an understanding of the marine boundary issues. *Geomatica* 54 (4): 415–24.

North, D. C. 1990. *Institutions, Institutional Change, and Economic Performance*. Cambridge: Cambridge University Press.

OECD. 2005. Fueling the future: Security, stability, and development. Convention on the Organization for Economic Cooperation and Development (May 2–3). Paris.

OGC. 2003. OpenGIS Reference Model. Open Geospatial Consortium. **http://www.opengeospatial.org/** (accessed August 1, 2009).

OGC. 2005. Specification for GIS services. Open Geospatial Consortium. **http://www.opengeospatial.org/** (accessed August 1, 2009).

Ogilvie, M., and G. Mulholland. 2004. Land information (LIN): Catalyst for integrated e- government. Proceedings of International FIG Seminar (June 2–4): 69–77. Innsbruck, Austria. **http://www.fig.net/commission7/innsbruck_2004/index.htm** (accessed August 1, 2009).

Onsrud, H. 1999. Ensuring success in land administration projects in countries in transition. Paper prepared for first session of UNECE Working Party on Land Administration (November 15–16). Geneva, Switzerland. **http://www.unece.org/hlm/wpla/welcome.html** (accessed August 1, 2009).

Oram, A. 2001. *Peer to Peer: Harnessing the Power of Disruptive Technology*. Sebastopol, UK: O'Reilly.

Österberg, T. 2003. Sweden Country Report. The Cadastral Template Project. **http://www.cadastraltemplate.org/** (accessed August 1, 2009).

Owolabi, K. 2003. Namibia Country Report. The Cadastral Template Project. **http://www.cadastraltemplate.org/** (accessed August 1, 2009).

Paasch, J. M. 2004. Modeling the cadastral domain. Proceedings of 10th European Commission, Geographic Information, and GIS Workshop on ESDI (European SDI) State of the Art (June 23–25). Warsaw, Poland.

Panayotou, T. 1994. Economic instruments for environmental management and sustainable development. International Environment Program. Harvard Institute for International Development, Harvard University, Massachusetts.

Park, M. 2003. The effect of adverse possession on part of a registered title land parcel. Doctorate thesis, University of Melbourne, Australia. **http://www.geom.unimelb.edu.au/research/publications/MMP_PhD.pdf** (accessed August 1, 2009).

Payne, G. 2001. Settling for more: Innovative approaches to tenure for the urban poor. Seminar on Securing Land for the Urban Poor (October 2–4). UN Center for Human Settlements (Habitat) and UN Economic and Social Council for Asia and Pacific (UNESCAP). Fukuoka, Japan.

Payne, G., ed. 2002. *Land Rights and Innovation: Improving Tenure Security for the Urban Poor*. UK: ITDG Publishing.

Payne, G. 2004. Land tenure and property rights: An introduction. *Habitat International* 28:167–79.

Payne, G., A. Durand-Lasserve, and C. Rakodi. 2007. Social and economic impacts of land titling programs in urban and peri-urban areas: A review of the literature. World Bank Urban Research Symposium (May 14–16). Washington, D.C. **http://www.worldbank.org/** (accessed August 1, 2009).

PCGIAP. 2000. Permanent Committee on GIS Infrastructure for Asia and the Pacific. **http://www.pcgiap.org/** (accessed August 1, 2009).

PCGIAP. 2004. International Workshop on Marine Administration—the Spatial Dimension, Working Group 3. Permanent Committee on GIS Infrastructure for Asia and the Pacific (May 4–6). Kuala Lumpur, Malaysia.

Peng, Z. R., and M. H. Tsou. 2003. *Internet GIS*. Hoboken, N.J.: Wiley.

Platteau, J. 1996. The evolutionary theory of land rights as applied to sub-Saharan Africa: A critical assessment. *Development and Change* 27:29–86.

Platteau, J. P. 2000. Does Africa need land reform? In *Evolving Land Rights, Policy, and Tenure in Africa*. Ed. C. Toulmin and J. Quan. London: DFIF/IIED/NRI.

Potsiou, C., and C. Ionnidis. 2006. Informal settlements in Greece: The mystery of missing information and the difficulty of their integration into the legal framework. Proceedings of 5th FIG Regional Conference (March 8–11). Accra, Ghana. http://www.fig.net/pub/accra/ (accessed August 1, 2009).

Proenza, F. J. 2006. Information systems and land administration. FAO Report No. 6. http://www.eforall.org/pdf/ICTs&LandAdmin_16Sep2006.pdf (accessed August 1, 2009).

PSMA. 2008. G-NAF, Australia's Geocoded National Address file. http://www.psma.com.au/ (accessed August 1, 2009).

Radwan, M. M., Y. Bishr, B. Emara, A. Saleh, and R. Sabrah. 2005. Online cadastre portal services in the framework of e-government to support real estate industry in Egypt. Proceedings of FIG Working Week and GSDI-8 (April 16–21). Cairo, Egypt. http://www.fig.net/pub/cairo/ (accessed August 1, 2009).

Raff, M. 2003. *Private Property and Environmental Responsibility: A Comparative Study of German Real Property Law.* The Hague: Kluwer Law International.

Rajabifard, A. 2002. Diffusion of regional spatial data infrastructure: With particular reference to Asia and the Pacific. Doctorate thesis, Department of Geomatics, University of Melbourne, Australia.

Rajabifard, A., A. Binns, and I. P. Williamson. 2006. Virtual Australia: An enabling platform to improve opportunities in the spatial information industry. Special issue, *Journal of Spatial Science* 51, No. 1.

Rajabifard, A., F. Escobar, and I. P. Williamson. 2000. Hierarchical spatial reasoning applied to spatial data infrastructures. *Cartography Journal* (Australia) 29, No. 2.

Rajabifard, A., M. Feeney, and I. P. Williamson. 2002a. Future directions for the development of spatial data infrastructure. *International Journal of Applied Earth Observation and Geoinformation* 4 (1): 11–22.

Rajabifard, A., M. Feeney, and I. P. Williamson. 2002b. The cultural aspects of sharing and dynamic partnerships within an SDI hierarchy, *Cartography Journal* (Australia) 31, No. 1.

Rajabifard, A., and I. P. Williamson. 2001. Spatial data infrastructures: Concept, SDI hierarchy, and future directions. GEOMATICS '80 Conference. Tehran, Iran.

Rajabifard, A., and I. P. Williamson. 2004. SDI development and capacity building. Proceedings of 7th GSDI Conference (February 2–6). Bangalore, India.

Rajabifard, A., I. P. Williamson, and A. Binns. 2006. Marine administration research activities within Asia and the Pacific region—towards a seamless land–sea interface. Administering marine spaces: International issues. FIG–Commissions 4 and 7 Working Group 4.3. FIG publication No. 36: 21–36. http://www.fig.net/pub/figpub/pub36/pub36.pdf (accessed August 1, 2009).

Rajabifard, A., I. P. Williamson, D. Steudler, A. Binns, and M. King, 2007. Assessing the worldwide comparison of cadastral systems. *Land Use Policy Journal* 24:275–88.

Ratcliffe, J., and M. Stubbs. 1996. *Urban Planning and Real Estate Development.* London: UCL Press.

Rawat, S. 2003. Interoperable geospatial data model in the context of the Indian NSDI. Master's thesis, International Institute for Geo-Information Science and Earth Observation. Holland.

Razzaz, O., and A. Galal. 2000. Reforming land and real estate markets. Draft research paper. Washington, D.C.: World Bank.

Remkes, J. W. 2000. Foreword to *Geospatial Data Infrastructure: Cases, Concepts, and Good Practice.* Ed. R. Groot and J. McLaughlin. New York: Oxford University Press.

Robertson, B., G. Benwell, and C. Hoogsteden. 1999. The marine resource: Administration infrastructure requirements. UN–FIG Conference on Land Tenure and Cadastral Infrastructures for Sustainable Development. Melbourne, Australia.

Roux, P. L. 2004. Extensible models and templates for sustainable land information management intent and purpose. Proceedings of Joint FIG–Commission 7 and COST Action G9 Workshop on Standardization in the Cadastral Domain (December 9–10). Bamberg, Germany. http://www.fig.net/commission7/bamberg_2004/index.htm (accessed August 1, 2009).

Ryttersgaard, J. 2001. Spatial data infrastructure: Developing trends and challenges. International Conference on Spatial Data Information for Sustainable Development. Nairobi, Kenya.

Sambura, A. 2004. E-land administration in accession countries — experience in Poland. Proceedings of International FIG Seminar (June 2–4, 2004): 78–87. Innsbruck, Austria. http://www.fig.net/commission7/innsbruck_2004/index.htm (accessed August 1, 2009).

Schech, S., and J. Haggis. 2002. Critical introduction to *Culture and Development*, chapter 6. Oxford: Blackwell.

SDI Cookbook. 2000. Developing spatial data infrastructures: The SDI cookbook, version 1.0. Prepared and released by the GSDI — Technical Working Group. http://www.gsdi.org/pubs/cookbook (accessed August 1, 2009).

SDI Cookbook. 2004. Developing spatial data infrastructures: The SDI cookbook, version 2.0. Prepared and released by the GSDI-Technical Working Group.

Simpson, R.W. 1976. *Land Registration*. Cambridge; Cambridge University Press.

Simsion, G. C., and G. C. Witt. 2005. *Data Modeling Essentials*, 3rd ed. San Francisco: Elsevier.

Smith, W., I. P. Williamson, A. Burns, T. K. Chung, NTV Ha, and H. X. Quyen. 2007. The impact of land market processes on the poor in rural Vietnam. *Survey Review* 39:303–20.

Steudler, D. 2004. A framework for the evaluation of land administration systems. Doctorate thesis, University of Melbourne, Australia. http://www.geom.unimelb.edu.au/research/publications/PhDThesisDanielS.pdf (accessed August 1, 2009).

Steudler, D., and J. Kaufmann, eds. 2002. Benchmarking cadastral systems. FIG–Commission 7 on Cadastre and Land Management. Working Group 1998–2002 on Reforming the Cadastre (April). Denmark. http://www.fig.net/pub/compub/FIG2002-BenchmarkingCadastralSystems.pdf (accessed August 1, 2009).

Steudler, D., A. Rajabifard, and I. P. Williamson. 2004. Evaluation of land administration systems. *Journal of Land Use Policy* 21:371–80.

Steudler, D., I. P. Williamson, J. Kaufmann, and D. M. Grant. 1997. Benchmarking cadastral systems. *Australian Surveyor* 42 (3): 87–106.

Steudler, D., I. P. Williamson, and A. Rajabifard. 2003. The development of a cadastral template. *Hong Kong Journal of Geospatial Engineering* 5 (1): 39–48.

Steudler, D., I. P. Williamson, A. Rajabifard, and S. Enemark. 2004. The Cadastral Template Project. FIG Working Week on Good Practices in Land Administration and Cadastre (May). http://www.fig.net/pub/athens/ (accessed August 1, 2009).

Steudler, D., and I. P. Williamson. 2005. Evaluation of the national land administration system in Switzerland — case study based on a management model. *Survey Review* 38 (298): 317–30.

Stolk, P. A. 2004. E-conveyancing using public key infrastructure in the Netherlands. Proceedings of International FIG Seminar (June 2–4). Innsbruck, Austria. http://www.fig.net/commission7/innsbruck_2004/index.htm (accessed August 1, 2009).

Stoter, J. E. 2004. 3D cadastre. Doctorate thesis, University of Technology, Delft, the Netherlands. Publications on Geodesy 57. Netherlands Geodetic Commission. www.ncg.knaw.nl (accessed August 1, 2009).

Stoter, J. E., and P. van Oosterom. 2002. Incorporating 3D geo-objects into a 2D Geo-DBMS. Proceedings of American Congress on Surveying and Mapping/American Society for Photogrammetry and Remote Sensing annual conference.

Survey Practice Handbook Victoria. 1997a. Surveyors Registration Board of Victoria. *Survey Practice Handbook*. Part 1, appendix A8. Plan layouts required for Subdivision Act 1988. http://www.vic.gov.au/ (accessed August 1, 2009).

Survey Practice Handbook Victoria. 1997b. Surveyors Registration Board of Victoria. *Survey Practice Handbook*. Part 1, appendix A8. Plan of survey. http://www.vic.gov.au/ (accessed August 1, 2009).

Survey Practice Handbook Victoria. 1997c. Surveyors Registration Board of Victoria. *Survey Practice Handbook*. Part 1, appendix A8. Abstract of field records of survey. http://www.vic.gov.au/ (accessed August 1, 2009).

Ting, L., and I. P. Williamson. 1999. Cadastral trends: A synthesis. *Australian Surveyor* 4 (1): 46–54.

Ting, L., I. P. Williamson, D. Grant, and J. R. Parker. 1998. Lessons from the evolution of Western land administration systems. Proceedings of International Conference on Land Tenure in the Developing World with a focus on South Africa (January 27–29). Cape Town, South Africa.

Ting, L., I. P. Williamson, D. Grant, and J. R. Parker. 1999. Understanding the evolution of land administration systems in some common-law countries. *Survey Review* 35 (272): 83–102.

Tipping, D. C., D. Adom, and A. K. Tibaijuka. 2005. Achieving healthy urban futures in the 21st century: New approaches to financing and governance of access to clean water and basic sanitation as a global public good. Helsinki Process Publication Series (February). Helsinki, Finland. http://www. helsinkiprocess.fi/netcomm/ImgLib/24/89/helsinki_ process_publication_series_2_2005.pdf (accessed August 1, 2009).

Todd, P. 2001. Marine cadastre: Opportunities and implications for Queensland; A Spatial Odyssey. 42nd Australian Surveyors Congress.

Torhonen, M. 2001. Developing land administration in Cambodia. *Computers, Environment, and Urban Systems* 25 (4–5): 407–28.

Tosta, N. 1997. Building national spatial data infrastructures: Roles and responsibilities. http://www.gisqatar. org.qa/conf97/links/g1.html (accessed August 1, 2009).

Toulmin, C. and J. Quan. 2000. Registering customary rights. In series Evolving Land Rights, Policy, and Tenure in Africa, 207–08. London: FID/IIED/NRI.

Toulmin, C., P. Lavigne-Delville, and S. Traore, eds. 2002. Introduction to *In the Dynamics of Resource Tenure in West Africa*. Portsmouth, New Jersey: Heinemann.

Toynbee, A. 1884. Lectures on the Industrial Revolution in England: England in 1760–Agriculture. http:// socserv.mcmaster.ca/econ/ugcm/3ll3/toynbee/indrev (accessed August 1, 2009).

Tuladhar, A. M., M. Radwan, F. A. Kader, and S. El-Ruby. 2005. Federated data model to improve accessibility of distributed cadastral databases in land administration. Proceedings of FIG Working Week and GSDI-8 (April 16–21). Cairo, Egypt. http://www.fig. net/pub/cairo/ (accessed August 1, 2009).

UNDP. 1998. Capacity assessment and development. Technical Advisory Paper No. 3. http://portals. wi.wur.nl/files/docs/ppme/capsystech3-98e.pdf (accessed August 1, 2009).

UNDP. 2002. *Developing Capacity through Technical Cooperation: Country Experiences*. Ed. Stephen Browne. United Nations Development Programme. New York.

UNDP. 2008. Making the law work for everyone. Commission on Legal Empowerment of the Poor. http://www.undp.org/legalempowerment/reports/ concept2action.html (accessed August 1, 2009).

UNECE. 1996. Land administration guidelines with special reference to countries in transition. United Nations Economic Commission for Europe. New York and Geneva, Switzerland. http://www.unece.org/hlm/ wpla/publications/laguidelines (accessed August 1, 2009).

UNECE. 2001. Land (real estate) mass valuation systems for taxation purposes in Europe. Working Party on Land Administration. Geneva, Switzerland.

UNECE. 2004. Guidelines on real property units and identifiers. United Nations Economic Commission for Europe. New York and Geneva, Switzerland. http:// www.unece.org/hlm/publications.htm#WPLA (accessed August 1, 2009).

UNECE. 2005a. Land administration in the UNECE Region: Development trends and main principles. Geneva, Switzerland. http://www.unece.org/env/ documents/2005/wpla/ECE-HBP-140-e.pdf (accessed August 1, 2009).

UNECE. 2005b. *Inventory of Land Administration Systems in Europe and North America*. 4th ed. London: HM Land Registry.

UNECE. 2005c. *Social and Economic Benefits of Good Land Administration*. 2nd ed. http://www.unece.org/hlm/ wpla/publications/UNECE%20Statement%20-%20Final%20 version.pdf (accessed August 1, 2009).

UNESCAP. 2007. UN Economic and Social Commission for Asia and the Pacific, chapter 10. http://www. unescap.org/stat/data/syb2007/index.asp (accessed August 1, 2009).

UN–FIG. 1996. The Bogor Declaration. United Nations Interregional Meeting of Experts on the Cadastre. FIG publication No. 13A (March 18–22). Bogor, Indonesia. http://www.fig.net/pub/figpub/pub13a/ figpub13a.htm (accessed August 1, 2009).

UN–FIG. 1999. The Bathurst Declaration on Land Administration for Sustainable Development. Report from UN-FIG Workshop on Land Tenure and Cadastral Infrastructures for Sustainable Development (October 18–22). Bathurst, NSW, Australia. A joint initiative of FIG and the United Nations. http:// www.fig.net/pub/figpub/pub21/figpub21.htm (accessed August 1, 2009).

UN–HABITAT. 1999. Implementing the Habitat Agenda: Adequate shelter for all. Global Campaign for Secure Tenure. UNCHC (Habitat). Nairobi, Kenya: United Nations.

UN–HABITAT. 2002. Global Campaign on Urban Governance. Nairobi, Kenya: United Nations. http://ww2.unhabitat.org/campaigns/governance/docs_pubs.asp (accessed August 1, 2009).

UN–HABITAT. 2003. *Handbook on Best Practices Security of Tenure and Access to Land: Implementation of the Habitat Agenda*. Ed. Clarissa Augustinus. Nairobi, Kenya: United Nations. http://ww2.unhabitat.org/publication/hs58899e/hs58899e.pdf (accessed August 1, 2009).

UN–HABITAT. 2004. *Pro-Poor Land Management: Integrating Slums into City Planning Approaches*. Nairobi, Kenya: United Nations. http://www.chs.ubc.ca/archives/?q=node/921 (accessed August 1, 2009).

UN–HABITAT. 2005. *Shared Tenure Options for Women: A Global Overview*. Nairobi, Kenya: United Nations.

UN–HABITAT. 2006a. *State of the World's Cities 2006–07*. Nairobi, Kenya: United Nations. http://www.unhabitat.org/pmss/getPage.asp?page=bookView&book=2101 (accessed August 1, 2009).

UN–HABITAT. 2006b. Report of 3rd session of World Urban Forum (June 19–23). Vancouver, Canada: United Nations. http://www.unhabitat.org/downloads/docs/3406_98924_WUF3-Report.pdf (accessed August 1, 2009).

UN–HABITAT. 2006c. Mechanism for gendering land tools: A framework for delivery of women's security of tenure. High-Status Round Table on Gendering Land Tools. Nairobi, Kenya: United Nations.

UN–HABITAT. 2007. How to develop pro-poor land policy: Process, guide, and lessons. United Nations Human Settlements Programme. http://www.unhcr.org/refworld/docid/4a69b5f72.html (accessed August 1, 2009).

UN–HABITAT. 2008. Secure land rights for all. UN–HABITAT at Global Land Tool Network. http://www.gltn.net/index.php?option=com_content&task=view&id=121&Itemid=17 (accessed August 1, 2009).

United Nations. 1948. The universal declaration of human rights. Adopted and proclaimed by General Assembly resolution 217 A (III) of December 10. http://www.un.org/Overview/rights.html (accessed August 1, 2009).

United Nations. 1973. Report of the Ad Hoc Group of Experts on Cadastral Surveying and Mapping. New York: United Nations.

United Nations. 1985. Conventional and digital cadastral mapping. Report of the Ad Hoc Group of Experts on Cadastral Surveying and Land Information Systems. Economic and Social Council E/CONF.77/L.1.

United Nations. 2003. Oceans and the law of the sea. Oceans and the Law of the Sea home page. http://www.un.org/Depts/los/index.htm (accessed August 1, 2009).

United Nations. 2004. Guidelines on real property units and identifiers. United Nations Economic Commission for Europe, New York, and Geneva.

UNRCC–AP. 2006. Resolution 3: Marine administration—the spatial dimension. 17th UNRCC–AP (September 18–22). Bangkok, Thailand. http://www.pcgiap.org/ (accessed August 1, 2009).

U.S. Department of the Interior. 2002. Principal meridians and base lines. Bureau of Land Management. http://www.blm.gov/wo/st/en/prog/more/cadastralsurvey/meridians.html (accessed August 1, 2009).

Usery, E. L., M. P. Finn, and M. Starbuck. 2005. Integrating data layers to support the national map of the United States. International Cartographic Conference. A Coruna, Spain.

Van der Molen, P. 2005. Authentic registers and good governance. FIG Working Week (April 16–21). Cairo, Egypt. http://www.fig.net/pub/cairo/ (accessed August 1, 2009).

Van der Molen, P. 2006. Unconventional approaches to land administration: The need for an international research agenda; Promoting land administration and good governance. Proceedings of 5th FIG Regional Conference (March 8–11). Accra, Ghana. http://www.fig.net/pub/accra/ (accessed August 1, 2009).

Van der Molen, P., and S. Mishra. 2006. Land administration and social development: Enhancing land registration and cadastre. *GIM International* (April): 15–17.

Van Oosterom, P., C. Lemmen, RAD By, and A. M. Tuladhar. 2004. *Geo-ICT Technology Push vs. Cadastral Market Pull* http://ictupdate.cta.int/en (accessed August 1, 2009).

Vckouski, A. 1998. *Interoperable and Distributed Processing*. Padstow, Cornwall: Taylor and Francis.

Wallace, J. 2005. Using remedies to secure access to land-regularizing occupation. FIG-Commission 7 Expert Group Meeting on Secure Land Tenure: New legal frameworks and tools in Asia and the Pacific (December 8–9). Bangkok, Thailand.

Wallace, J., and I. P. Williamson. 2004. Developing cadastres to service complex property markets. Proceedings of Joint FIG–Commission 7 and COST Action G9 Workshop on Standardization in the Cadastral Domain (December 9–10). Bamberg, Germany. http://www.fig.net/commission7/bamberg_2004/index.htm (accessed August 1, 2009).

Wallace, J., and I. P. Williamson. 2006. Building land markets. *Land Use Policy* 23 (2): 123–35.

Webster, C., and L. Wai-Chung Lai. 2003. *Property Rights, Planning, and Markets: Managing Spontaneous Cities.* Cheltenham, UK: Edward Elgar.

Williamson, I. P. 1983. A modern cadastre for New South Wales. Doctorate thesis, School of Surveying, University of New South Wales, Australia. http://www.sli.unimelb.edu.au/research/SDI_research/publications/files/ipw_83_PhD.pdf (accessed August 1, 2009).

Williamson, I. P. 1985. Cadastres and land information systems in common-law jurisdictions. *Survey Review,* pt. 1, vol. 28 (217): 114–29; pt. 2, vol. 28 (218): 186–95.

Williamson, I. P. 1986. Cadastral and land information systems in developing countries. *Australian Surveyor* 33 (1).

Williamson, I. P. 1996. Appropriate cadastral systems. *Australian Surveyor* 41:35–37.

Williamson, I. P., and S. Enemark. 1996. Understanding cadastral maps. *Australian Surveyor* 41 (1): 38–52.

Williamson, I. P., S. Enemark, and J. Wallace, eds. 2006. Sustainability and land administration systems. Proceedings of Expert Group Meeting (November 9–11, 2005). Department of Geomatics, University of Melbourne, Australia. http://www.geom.unimelb.edu.au/research/SDI_research/EGM%20BOOK.pdf (accessed August 1, 2009).

Williamson, I. P., S. Enemark, and J. Wallace. 2006. Incorporating sustainable development objectives into land administration. Proceedings of FIG 23rd Congress on Shaping the Change. Munich, Germany. http://www.fig.net/pub/fig2006/ (accessed August 1, 2009).

Williamson, I. P., and C. Fourie. 1998. Using the case study methodology for cadastral reform. Geomatica 52:283–95.

Williamson, I. P., and L. Ting. 2001. Land administration and cadastral trends: A framework for reengineering. *Computers, Environment, and Urban Systems* 25:339–66.

World Bank. 1996. Toolkit: Gender issues in agriculture. http://www.worldbank.org/ (accessed August 1, 2009).

World Bank. 2000. Entering the 21st century. World Development Report 1999/2000. New York: Oxford University Press.

World Bank. 2003a. Land policies for growth and poverty reduction. World Bank Research Report. Washington, D.C.

World Bank. 2003b. A comparative study of land administration systems. World Bank policy changes. Washington, D.C.

World Bank. 2004. Doing business in 2004: Understanding regulations. Washington, D.C.: World Bank. http://www.doingbusiness.org/Downloads/ (accessed August 1, 2009).

World Bank. 2005. Doing business in 2005: Removing obstacles to growth. Washington, D.C.: World Bank. http://www.doingbusiness.org/documents/DoingBusiness2005.PDF (accessed August 1, 2009).

World Bank. 2006. Doing business in 2006: Creating jobs. Washington, D.C.: World Bank. http://www.doingbusiness.org/documents/DoingBusines2006_fullreport.pdf (accessed August 1, 2009).

World Bank. 2007. Doing business in 2007: How to reform. Washington, D.C.: World Bank. http://www.doingbusiness.org/documents/DoingBusiness2007_FullReport.pdf (accessed August 1, 2009).

Young, A.J., and F. R. Baker. 2005. Digital mapping data currency through sharing: A practical study. Spatial Science Conference on Spatial Intelligence, Innovation, and Praxis (September). Melbourne, Australia.

Ziemann, H. 1976. *Land unit identification.* Ottawa: National Research Council of Canada.

Zlatanova, S., and J. Stoter. 2006. The role of DBMS in the new generation GIS architecture. In *Frontiers of Geographic Information Technology.* Ed. S. Rana and J. Sharma. Berlin: Springer.

Index

#

3D cadastre. See also cadastres: use of, 44; using with building tenures, 386

300 CE, Roman survey in, 46

1086, Domesday Book of, 46

1700s, enclosure movement in, 50, 53

1844, Danish cadastre, 53–54

1858, Torrens system, 50–51

1882, Settled Land Act of, 50

1925, UK Land Registration Act, 61

1980s, 74–76; land scarcity in, 52; National Research Council (NRC) study, 74

1987, Brundtland Report, 84

1989, fall of Berlin Wall, 76

1990s and beyond, 76–77

1992, UN Rio Earth Summit (Agenda 21), 84

1995, Indonesia's Land Administration Project, 86

1997, European Commission, 176–77

2014, cadastre projected for, 76–77

A

abbreviations, xiv–xv

absentee ownership, 107

academic research, international growth of, 81

access and sharing tools, 256

accountability, role in good governance, 31

Acquis Communautaire, 30

adjudication in TLTP (Thailand Land Titling Project), 101

Africa: land registration in, 147; need for pro-poor tools in, 390; UN–HABITAT's prediction about, 21

Agenda 21: 1992 UN Rio Earth Summit, 84

agent computing, 244–45

agricultural holdings, land consolidation of, 189–91

agricultural policies, 186

agricultural sectors, productivity in, 19

Aguascalientes Statement, 79

airspace, cubes of, 38

Alberta, Canada subdivision process (figure), 108

America. See United States

APR (Asia and the Pacific Region), problems in, 389–90

aquifer, Ogallala, 20

Aral Sea (Central Asia), 20

assets, recording in Domesday Book of 1086, 46

Atwood, David, 147

Australia: cadastral survey (figure), 380–81; cadastral systems, 55; DCDB (figure), 377; German-style system adopted in (South), 50–51; ICT in, 238; land transfer process (Victoria), 102; modern urban infrastructure (Melbourne), 43; plan of subdivision (figure), 375; schedule of coordinates (figure), 382; subdivision for strata titles (figure), 383

Austria, CYBERDOC archive in, 239

B

Baldwin, R., 158

Bangkok, Thailand (figure), 45

baselines and meridians in United States (figure), 62

basemaps in multipurpose cadastre, 55

Bathurst Declaration, 77

Belgium, planning system in, 177

benchmarking, need for, 78, 429–30

Benwell, George, 211

Berlin Wall, fall in 1989, 76

best practices: in evaluation framework, 430–33; interest in, 112

Binns, Bernard O., 73

bloodline inheritance system, 105–6

Bogor Declaration on Cadastral Reform, 77, 96

Bonaparte, Napoleon, 48

BOOT (build, own, operate, transfer) model, 413

boundaries: approximate, 361–62; creating and marking, 364; defining precisely, 361; fixed vs. general, 360; guaranteed, 364; identification of, 359; marking, 362; "movement" of, 363–64; natural, 360

boundary overlay in TLTP (Thailand Land Titling Project), 101

boundary points, accuracy in Danish DCDB, 379

boundary processes: in Denmark, 109; determining, 107–9; in Korea, 109–10; in Zambia, 109, 111

boundary systems, strengths and weaknesses, 364–65

boundary tools, 358–65

Britain, comparison of ownership in, 143

British Colonial Office, 72–73

Bruce, J. W., 73

Brundtland Report (1987), 84

building authorities, role in development process, 198

building contractors, role in development process, 198

building permit control, 180–81, 195

building tenures, 381–86. See also tenure tools

buildings, problems created for LAS, 356

business models, tools, 330–32

butterfly diagram: data dissemination in, 243–44; figure, 127; layers in, 128; process and information integration, 227; SDI as body of, 129

C

cadastral activities, collecting information, 269–71

cadastral challenges, analysis of, 289–90

cadastral concept (figure), 56

cadastral data, integrating with other data, 124

cadastral data modeling: approaches, 256–57; and coordination among subsystems, 258–61; and coordination of subsystems, 258–61; and data management, 257–58; figure, 258–59; flexible approach, 259; informal approach, 259; and land administration, 256–58

cadastral database, features of, 257–58

cadastral datasets, 234

cadastral elements, 57

"cadastral engines," approaches toward, 66

cadastral information, reliability of, 67

cadastral issues, analysis of, 289–90

cadastral maps, benefits of, 22

cadastral principles, proposal of, 57

cadastral process vs. land-use regulations, 54

cadastral reform, focus of, 96

cadastral reform methodology (figure), 411

cadastral surveying and mapping, 366–67; coordinated cadastral survey, 369–70; DCDB (digital cadastral database), 370; isolated survey system, 368; legal coordinates, 370; local plane coordinate system, 368–69; overview, 365–67; tools, 376–77, 380–81; true meridian, 368

cadastral surveying tools. See also surveying; tools: digitizing and scanning, 374; GPS (Global Positioning System), 374; graphic, 371–72; nonrectified photomaps, 372; numerical, 372–73; offset methods, 373; orthogonal method, 371;

orthophotography, 372; photogrammetry, 371–73; plane table, 371; polar method, 373; rectified photomaps, 372; satellite mapping, 372; stadia, 371; topographic mapping, 372

Cadastral Surveys and Records of Rights in Land, 73

Cadastral Surveys within the Commonwealth, 73

cadastral systems, 121–25; boundary identification in, 109–10; broadening focus to land administration, 270; in Denmark, 49; foundation for, 49; growing importance of, 74; influences on design of, 122; linking with land-valuation systems, 55; in North America, 55; overview, 121–25; reforms in 1980s, 75

cadastral template project: countries in, 277; descriptive analysis, 275; design, 274–275; indicator 1, 276–77; indicator 4, 283–85; indicator 8, 289–90; indicators 2 and 3, 277–83; indicators 5, 6, and 7, 286–89; outcomes, 274–75; overview, 271–73; populations in, 277; principles and associated indicators, 275; statistical analysis, 275; structure, 274–75

cadastral tools, examples of, 43. See also noncadastral approaches and tools

cadastral units: entities, 125–26; parcels, 125–26; properties, 125–26

"Cadastre 2014: A vision for a future cadastral system," 76–77

cadastres. See also 3D cadastre; marine cadastres; PCGIAP–FIG cadastral template project: applying to marine management, 210–11; combining with land registry functions, 268–69; components of, 65; creation by Torrens system, 51; Danish, 53–54; describing, 55–56; as engines of LAS, 127; European and German approach, 9–10; European-style, 47–52; as fiscal tool, 48; French/Latin approach, 9; generic definitions of, 52; in hierarchy of land issues, 132–33; importance of, 9; including land administration in, 78; influences on, 57–58; in land administration theory, 52; as land management tool, 52; as land market tool, 49; vs. land registries, 66; as LIS (land information system), 374; local design, 58; multipurpose, 52, 54–58, 67; as planning tool, 51; principle, 34; projection for 2014, 76–77; significance, 38; significance of, 65–67, 127; Torrens title approach, 9

Canada: development in Maritime Provinces in, 52; subdivision process (Alberta), 108

capacity: improving for global comparisons, 290–91; overview, 297–99

capacity assessment, 298, 304

capacity building. See also LAS capacity: approach, 299; barriers, 300–301; commitment to, 26; conceptual framework, 296–97, 301; defined, 298; maintaining continuity, 302–3; maintaining sustainability, 302–3; meanings and interpretations, 296; self-assessment of needs, 302; strategies for, 297; sustainability and continuity, 302–3; tools, 327–28

capacity development, 299, 304

capital gains tax, 167

capitalism, significance of land to, 156–57

CARIS Land Information Network, 240

CCDM (Core Cadastral Domain Model), 392

central planning approach, 174–76

centralist economies, control over land markets in, 145

centralized land organization, conversion of, 27

change management, 8

Chile: challenges related to, 8–9; mixed rural land (figure), 11

China: decline of water table in North Plain, 20; taxation system in 700 CE, 46

cities. See also modern cities; urban areas: building, 21–22; demands for space in, 43; estimated populations in, 20

citizenship, role in good governance, 31

city planning, use of cadastre in, 51

civic engagement, role in good governance, 31

claims to land, priorities among, 346–47

clean water. See also groundwater policy area: access to, 20; providing access to, 25

coastal waters, 208

coastal zones: conflicting interests in, 203; importance of, 207; management, 188–89; maritime activities in, 212; stakeholders, 209; sustainable management of, 213

cognitive approach: applying to land use, 38; and LAS evolution, 156–62

cognitive capacity: pillars of, 158–59; supporting, 158

cognitive framework, management of, 18

colonization: spread of legal systems by (figure), 59; variations in, 58. See also postcolonial countries

commodities: gaining support for, 159; popularization of, 38–39; waves of creativity, 157–58

commodities market: characteristics of, 152; infrastructure and tools (table), 154

commoditization systems, 155–56

common processes, 6

community, engaging in tenure processes, 99–100

"Comparative study of land administration systems," 271

complex commodities market: characteristics of, 152; infrastructure and tools (table), 154

complex markets: characteristics of, 159; components of (figure), 161

compulsory purchase, 199

computers, influence of, 82–83

computing technology tools: agent computing, 244–45; grid computing, 245; P2P computing, 245

The Concept of the Corporation, 88

conferences in 1990s, 76

conservation regulation, provisions of, 185–86

contiguous zone, 209

continental shelf, 209

continuum of secure access (figure), 395

control points in TLTP (Thailand Land Titling Project), 101

Copenhagen, Ørestad section of, 202

Core Cadastral Domain Model (CCDM), 392

corporatization process, 162

countries, dividing into zones, 185

country context, considering in LAS design, 410

CPD (continuing professional development), 309–10, 312

credit: availability of, 19; security for, 17

crises, managing, 20–21

cubes of airspace, 38

cultural origins, anthropological view of, 58

Czech Republic, ratio of parcels to people in, 280

D

Dale, P., 73–74, 158, 376

Dalrymple, K., 112

Danish boundary process, 109

Danish cadastre, evolution of, 53–54

data capture tools, using in ICT for LAS, 242

data catalog tools, using in ICT for LAS, 243

data conversion tools, using in ICT for LAS, 243

data dissemination, using in ICT for LAS, 243–44

data integration: components of, 235; success of, 234

data management: of LAS, 257; role of modeling in, 258; tools, 255

data modeling: approaches, 256–57; tools used in ICT for LAS, 242 data models, impact on DBMS, 254

data sharing: concerns about, 82; impediment to, 234

database system tools, using in ICT for LAS, 242–43

datasets: cadastral and topographic, 234; integrating for natural and built environments, 233–34

DBMS (database management systems), impact of data models on, 254

DCDBs (digital cadastral databases): boundary coordinates in, 375; case of Denmark, 378–379; as component of SDIs, 376; data in, 374; graphically accurate, 375; survey-accurate, 375; updating, 257, 376; upgrading, 376

de Soto, Hernando, viii, 60–61, 63, 112, 148–50

decentralization, outcome from, 194

deeds: evolution in Europe, 50; registration, 59; vs. title registration, 341–42; vs. title system, 59

Denmark: cadastral system in (figure), 49; DCDB (digital cadastral database), 378–79; forestry policies in, 186; planning system in, 177–78; subdivision process in, 196; zone development in, 185

developers, role in development process, 197–98

developing countries, 168; cadastral construction in, 56; estimated populations in, 20; land transactions in, 104; LAS in, 33; mortgage process, 103; reform efforts, 79; tax collection in, 168

development, illegal instances of, 182

development aid projects, influence of, 80–81

development control. See land development

development gains/betterment tax, 166

development projects, using LogFrame tool with, 417–21

digital cadastral databases (DCDBs): boundary coordinates in, 375; case of Denmark, 378–79; as component of SDIs, 376; data integration, 374; graphically accurate, 375; survey-accurate, 375; updating, 257, 376; upgrading, 376

digital environment, emergence in 1990s, 76–77

digital signature, 246

digital technologies, original intention of, 246

Domesday Book of (1086), 46

Dowson, E., 73

Drucker, Peter, 88

E

Eastern Europe, failure of economies in, 30

e-citizenship vs. e-government, 131

economic context, considering in LAS design, 412

economic development paradigm, 73

economic improvement, impact in Thailand, 146–47

economic transition, assistance efforts, 79

economies: application to land markets, 138; building, 19; distinctions in, 162–63; failures in Central and Eastern Europe, 30; free-market vs. centralist, 144–45

e-conveyancing system, development of, 260

education: challenge for surveyors, 310–11; lifelong learning vs. vocational training, 309–10; national, 307–8; project-organized vs. subject-based, 308–9; virtual academy vs. classroom lectures, 309

educational trends: professional competence, 311–12; for surveyors and engineers, 308–10

Effenberg, W. W., 112

efficiency, role in good governance, 31

EFT (electronic fund transfer), 245–46

e-government: vs. e-citizen ship, 131; factors contributing to, 83

Egypt: registration of apartments in 2001, 139; Royal Registry of, 46

e-land, 83

e-land administration: applying phase, 240–41; challenges of, 248; emerging phase, 240–41; figure, 259; five phases of, 247–248; infusing phase, 240–41; interoperability framework, 252–54; interoperability toolbox, 254–56; overview, 246–48; transforming phase, 240–41

e-LAS: analyzing user requirements, 250–51; evaluating, 248–51; functionality support, 249–50; monitoring, 249; performance, 249; popularity, 251; user interactivity support, 249–50

electronic banking, 245–46

electronic documents, 246

eminent domain legislation in U.S. (map), 175

enabling platform, creating for SDIs, 233

enclosure movement (1700s), 50, 53

Enemark, S., 112

engineering focus, importance to LAS, 26

engineers, educational trends, 308–10

England: common-law approach toward land, 49–50; deeds system, 61; feudal system extended to, 46; organization of land records in, 61; registries in, 61

English LAS, development of, 60–62

English system, 66

enterprise architecture design tools, 255

enterprise facilitators: digital signature, 246; electronic banking, 245–46; electronic documents, 246

entity-relationship (E-R) approach, 242

environment, management of, 18

environmental protection, 187–88

environmental sustainability, ensuring, 24

equity of access, role in good governance, 31

E-R (entity-relationship) approach, 242

EU (European Union): rules of, 30

EUROGI questionnaire (2002), 270

Europe. See also UNECE (UN Economic Commission for Europe): history of land surveying in, 71–72; land taxation model in, 167; land-use planning in, 175; spatial planning traditions in, 176–77

European Commission 1997, 176–77

European countries, classification of, 177–78

European inheritance system, 105–6

European LAS: development of, 60–62; evolution of, 47; vs. United States, 63–64

European planning systems, operation of, 177

Europeans, approach toward mapping, 48

European-style cadastre, development of, 47–52, 55

evaluation framework, 430–33; external factors, 433; management level, 432; operational level, 432–33; overview, 430; policy level, 431; review process, 433

evidence of transfer, 102

exclusive economic zone, 209

exploitation tools, 256

expropriation, 199

F

FAO (Food and Agriculture Organization): Land Tenure Series, 73; sustainable LAS, 32

farming, optimizing (figure), 190

"feoffment by livery of seisin," 100–101

feudal systems of tenure, organization of, 46

feudalistic society, transition to market-based society, 53

field cadastral map in TLTP (Thailand Land Titling Project), 101

Fifth Amendment, "takings" phrase in, 174

FIG (International Federation of Surveyors), 54–55, 78

FIG congresses, 78

FIG–Commission 7, 76, 78, 270

Fiji, ratio of parcels to people in, 280

financial incentives, 198–99

financial institutions, role in development process, 198

Finland, planning system in, 177

fiscal cadastres, use in France and Germany, 49

fiscal tool, cadastre as, 48

Fishbone chart (figure), 408

Fisher, Kenneth, 355

"flying freeholds," 382

flyovers in TLTP (Thailand Land Titling Project), 101

food production, impact of standards on, 32

foreign aid, role of land administration in, 80

forestry policies, 186–87

formal land markets. See also land markets: difficulties with establishment of, 145–146; vs. informal markets, 138–40; negative economic consequences of, 143–44; support for, 17

Fourie, C., 112

Fowler, Cindy, 210

fragmentation, solutions to, 106

France: planning system in, 177; use of fiscal cadastres in, 49

free-market approach, 144–45, 173–74

French/Latin cadastral systems, 66, 122

funding tools, 330–32

G

gender equity, 24; goal of, 20; tools, 396–98. See also women

geocodes, significance of, 356

German-style land registration system, 50–51, 60, 66, 123–24

Germany: cadastral administrations of, 239; land-use planning, 175; planning system in, 177; use of fiscal cadastres in, 49

GIS services, OGC specifications for, 244

Global Plan of Action, 77

GLTN (Global Land Tool Network), 43, 82, 148, 387

GML (Geography Markup Language), 243

good governance, 30–33

governance: delivering land information for, 22–23; good, 30–33; improving, 39; in land administration, 30–33; vs. land administration, 332; support for, 17 government, protection of property rights by, 81

government administration, role of land administration in, 80

government land, treatment of, 87

government organizations, 305–7

Great Britain, Land Registry proposal, 239

Greece, planning system in, 178

grid computing, 245

groundwater policy area, importance of, 187–88. See also clean water

"Grundbuch" (land book) approach, 49

H

Hall, A. F., 97

Hamburg-Hanseatic land title registration system, 61

Hanoi: informal land market of, 140; land-use certificates in 2005, 139

Hanseatic land title registration system, 61

Hawaii, inheritance system in, 106

high seas, 209

Honduras, cost of LAP in, 268

Hong Kong, ratio of parcels to people in, 280

Hoogsteden, Chris, 211

human-rights tools, 398–400

I

ICT (information and communications technology), 241; development phases in LAS, 240–41; distributed tools, 244–45; e-land administration, 246–48; enterprise facilitators, 245–46; evaluating e-LAS, 248–51; figure, 238; in land administration, 238–40, 259; tools, 327

ICT options for LAS: data capture tools, 242; data catalog tools, 243; data conversion tools, 243; data dissemination, 243–44; data management tools, 241; data modeling tools, 242; database system tools, 242–43; Web services tools, 244

IHO (International Hydrographic Organization), 207–8

indigenous groups, support for, 140

individual vs. mass valuation, 165

Indonesia: Land Administration Project (1995), 86; transfers of land in, 104

Industrial Revolution, 50

informal settlements, focusing on, 183–84

informal systems, variety of, 15

informal vs. formal land markets, 138–40. See also land markets

information: collection about national systems, 269–71; digital integration of, 82; management of, 18

information and communications technology (ICT). See ICT (information and communications technology)

information strategy, integration of, 79

information systems, interoperability in, 251

infrastructure: development, 200; development of, 18; and tools (figure), 154

inheritance systems, 104–6

institutional capacity: development of, 303–4; government organizations, 305–7; overview, 303–5

institutional development: commitment to, 26; comprehensive approach, 304–5; impact of, 305

institution-building tools, 327–28

institutions principle, 34

insurance, obtaining for risk management, 330–32

"integrated legal property systems," 60–61

intercommunity interoperability, 253–54

International Federation of Surveyors (FIG), 54–55, 78

International Hydrographic Organization (IHO), 207–8

International Office of Cadastre and Land Records (OICRF), 78

international support, institutionalizing, 81–82

Internet, arrival of, 83

interoperability: framework for e-land administration, 252–54; in information systems, 251; institutional, 234; intercommunity, 253–54; legal, 234, 253; national, 253; nontechnical issues, 252; policy, 234; semantic, 252–53; social, 234; technical, 234–35, 254

interoperability toolbox: access and sharing tools, 256; data management tools, 255; enterprise architecture design tools, 255; exploitation tools, 256; overview, 254–55

"Inventory of land administration systems in Europe and North America," 78

Iran: challenges related to, 8–9; urban sprawl (figure), 9

Ireland, planning system in, 177

Islamic inheritance system, 105–6

Italy, planning system in, 178

K

Kelo v. City of New London, 174

Kenya, challenges related to, 8

Kibera, Kenya informal settlement (figure), 184

Kiribati, ratio of parcels to people in, 280

KPIs (key performance indicators), 427–28

L

land: as capital, 40; changing concept of, 43; characteristics of, 152; commoditization of, 153; as commodity, 41; as community, 40; concepts of, 38–44; as consumption good, 41; cultural and spiritual meanings of, 39; as deity (spiritual), 40; as environment, 41; as factor of production, 40; human approach toward, 41; as human right, 41; infrastructure and tools (table), 154; investigation by buyer, 102–3; managing perception of, 18; as nature, 41; perception of, 150; physical and cognitive aspects, 38; as physical space, 40; as property institution, 40; relationship of people to, 153; as resource, 41; vs. resource registries, 339; retrieving value for government purposes, 163–64; significance to capitalism, 156–57; as terra firma, 40; values associated with, 150

"land," meaning in modern administration, 5

land acquisition, 198–99

Land Administration, 77

land administration, 118; achieving four functions of, 120; activities, 117; in Africa and Latin America, 72; applying theory, 97; basic reasons for, 15; and cadastral data modeling, 256–58; capacity building vs. technology, 261; consolidation in 1990, 116; core processes, 97–99; debates, 26–28; defined, 37; defining, 27–28; development assistance in 1980s, 75–76; discipline of, 10, 30; evolution, 38; evolution of, 47; evolution of concept, 78–80; formalization, 16–18; four functions of, 118–19; framework, 27–28; function, 226; global perspective, 119–21; growth of literature about, 80; historical approach, 91; including in cadastres, 78; integration of, 6; interoperability in, 251–52; vs. land management, 116–17; vs. land reform, 29–30; modern approach, 91; need for, 44; origin of, 71; processes, 87, 95–97; professionals, 10; vs. reengineering, 97; reforming via process management, 112; reforms in 1980s, 75; response, 92–93; role in foreign aid, 80; stabilizing and improving, 29; tasks in settled societies, 95; theoretical and practical approaches, 74; theory, 9–10, 15, 72; tools, 44–47; UNECE principles of, 30–31; UNECE view of, 116; United Nations' definition, 96; world leaders in, 62; after World War II, 72–74

"Land administration in the UNECE region," 78, 96

land administration issues, international investigation of, 81

land administration principles: 10; cadastre, 34; land management paradigm, 34; LAS (land administration systems), 34; LAS are dynamic, 34; measures for success, 35; people and institutions, 34; principles, 10; processes, 35; rights, restrictions, responsibilities, 34; spatial data infrastructure, 35; technology, 35

land administration projects (LAPs). See LAPs (land administration projects)

land administration systems (LAS). See LAS (land administration systems)

land administration toolbox: emerging tools, 318, 320; general tools, 318, 320; professional tools, 318, 320

land arrangements, recording of, 45–46

land banking, 199

land consolidation: programs, 106; and readjustment, 189–91

land development: acquisition and financial incentives, 198–200; control, 195–96; control of, 195–96; explained, 120; figure, 123; infrastructure, 200; integrated and land use, 203; as outcome of planning process, 195; practices, 412; processes and actors, 196–98; rural, 203; terminology, 194; tools, 327; urban, 200–201, 203

land disputes: impact of, 20; management of, 17

land distribution, equity in, 19

land governance, addressing in next decade, 443

land holdings, registers of, 46

land information: access during 1980s, 75; availability of, 22; capabilities of, 22; considering in LAS design, 411; delivering, 22–23; managing with flexibility, 44; organization of, 121; tools, 327

Land Information Management, 74, 376

land issues, hierarchy of, 132–33

Land Law and Registration, 73

land management: infrastructure, 317; institutional capacity in, 303–7; vs. land administration, 116–17; process of, 303; vision, 129–32

land management paradigm, 6, 115–18, 173, 266–67; (figure), 117; adoption of, 226; applying to LAS design, 410–12; applying to marine needs, 222; in hierarchy of land issues, 132; implementing, 127–29; LAS in, 241; neutrality of, 128; operational component, 118; point of, 316; principle of, 34

land management systems, requirements of, 214–15

land management tool: cadastre as, 52; vs. titling tool, 395

land market tool, cadastre as, 49

land markets. See also formal land markets; informal land markets: achievement of, 152; analysis of, 10; building, 19, 163; challenges of, 162–63; characteristics of, 152; controls on, 144–45; difficulties with formal type, 145–46; evolutionary stages of, 151–52; formal, 140–43; formal vs. informal, 138–40; infrastructure and tools (table), 154; and land projects, 146–48; vs. land trading, 155; management of, 144; managing, 137–38; and national land policy, 143–44; relationship to private ownership, 143; significance of property rights in, 155; support for, 17; tools, 326; variable operation of, 146

land parcel maps. See cadastres

land parcels: explained, 125; framework, 356; in hierarchy of land issues, 132–33; identification in United States, 62; marking, 362; population and number of strata titles, 278–79; vs. properties 125–26; in urban areas, 284–85

land planning, improvement of, 17

land policies, 86–87. See also land administration; approach of World Bank, 73; approaches toward, 29; effect of changes in, 80; goal of, 176; in hierarchy of land issues, 132; and integrated land-use management, 192; leadership in, 75; relative to land markets, 143–44; variations in, 117–18

land policy framework, considering in LAS design, 410

land policy instruments, applying to plan objectives, 199

land policy journals, international articles in, 81

land policy tools, 320–26, 330–32

land readjustment systems, 189–91

Land Reform and Settlement and Cooperatives, 73

land reform projects, assessing scope of, 268

land reform vs. land administration, 29–30

land register, creation by Torrens system, 51

Land Registration, 73

Land Registration and Cadastral Systems, 77

land registration systems: analysis of, 73; figure, 61; in Thailand, 100

land registries. See registries

land registry functions, combining cadastres with, 268–69

land registry operations, transparency of, 22

land rights: characteristics of, 152; continuum (figure), 392; infrastructure and tools (table), 154; management of, 27–28

land stabilization, 32

land surveying, history of, 71

land taxation: explained, 166; fairness of, 19; principle, 166; reintroduction of, 168; security for, 17

Land Tenure Center, establishment of, 73

Land Tenure in Development Cooperation—Guiding Principles, 76

Land Tenure Series (FAO), 73

land tenures. See also security of tenure: explained, 41, 119–20; figure, 42, 123; social, 42; types of, 41–42

land titling: processes, 98–101; systematic and sporadic, 350–54

land trading: characteristics of, 152; infrastructure and tools (table), 154; vs. land markets, 155

land transfer process (Victoria, Australia), 102

land transfers: by agreement, 100–104; through social events, 104–7

land units: best-practice principles, 357–58; unique identification of, 355–56

land use. See also rights: central planning approach, 174–76; control of, 182; documenting in LAS design, 411; explained, 120; figure, 123; free-market approach, 173–74; high-density, 44; integrated management, 192–94; management, 172, 192–94; planning, 173; planning and regulation, 171–72; planning and restrictions, 173; regulation in communities, 51; rights, 173; tools, 327

land-use patterns, changes in, 107

land valuation. See also valuation principles: tools, 327

land value: considering in LAS design, 411; explained, 120; figure, 123; increase due to development, 196–97

landownership: documentation of, 45–46; early records of, 46; role in development process, 197

land-sea interface, complexity of, 188–89

land-use control, considering in LAS design, 412

land-use management, integrated, 192–94

land-use regulations vs. cadastral process, 54

land-valuation systems, linking with cadastral systems, 55

LandXML standard, 243

LAPs (land administration projects): analysis of, 266–68; Central European projects, 265; centralist government reconstruction, 264; cost in Honduras, 268; cost in Nicaragua, 268; cost in Panama, 268; cost of, 266; defining main problem in, 409; delivery of security tenure, 265; disaster response, 264; East European projects, 265; expense of, 268–69; failures of, 266–67; funding tools for, 332; geographical component, 408–9; improving people's lives, 263–66; for improving people's lives, 263–69; large-scale national projects, 265; large-scale systematic projects, 265; Latin American land reform projects, 265; management of mass relocation, 265; pilot titling programs, 266; postcolonial nationalism, 264; postconflict management, 264; problems with, 302; recognition of indigenous titles, 264; reconstruction of government postevolution, 264; redesign of native title, 264; resource administration, 265; rural vs. urban land, 267; suggestions for improvements in, 303; titling for the poor, 265; urban slum management and reconstruction, 264

LAS (land administration systems): in 1970s and 1980s, 146; adoption of, 10; benchmarking, 429–30; benefits of, 121; best practices, 430–33; congruity with ways people think, 39; data management of, 257; in developing countries, 33; dynamic nature of, 26; English, 60–62; European, 60–62; evaluation framework, 430–33; evolution of, 84; future challenges, 445–46; future of, 440; goals of, 121; in hierarchy of land issues, 132–33; holistic approach, 33, 35; improvement efforts, 79; improving, 316; influences on, 26; infrastructure and tools (table), 154; modern benefits, 16; next generation of, 23–24; objectives of, 438; parcel-based (figure), 75; perception of, 19; principle of, 34; processes, 330; purpose of, 5; in silos, 64–65; spatial enablement of, 130; success of, 26; synchronizing with cognitive impact, 39; theoretical framework, 10; traditional benefits, 16; transforming to e-land administration, 240–41; in United States, 62–64; in United States vs. Europe, 63–64

LAS are dynamic, principle, 34

LAS capacity, development of, 153. See also capacity building

LAS design elements: changes in, 18; common processes, 6; considerations, 23; describing components of, 410–12; developing reengineering framework, 406–7; developing vision and objectives, 407–9; future focus of, 446; implementing conceptual model, 412–21; land management paradigm, 6; managing projects, 421–29; monitoring projects, 421–29; objectives of, 15–16; problems with, 7; professional services and tools, 405–6; sustainable development, 6; toolbox approach, 6; understanding existing system, 409–12

LAS designers: challenge for, 131–32; message for, 153

LAS discipline, broadening of, 80

LAS evolution: and cognitive capacity, 156–62; relationship to cognitive capacity, 156–62

LAS future trends: capitalizing on technology, 444; challenges ahead, 445–46; institutional catch-up, 444; land governance issues, 443; overview, 437–39; supporting spatially enabled society, 440–42; supporting sustainable development, 439–40; tools for continuum of tenures, 443; tools to manage RRRs, 443–44; urban growth issues, 443

LAS improvement, motivation for, 439

LAS institutions, relationships among, 79

LAS issues: institutional catch-up, 444; land governance, 443; LAS capitalization on technology, 444; tools for continuum of tenure, 443; tools for managing RRRs, 443–44; urban growth, 443

LAS model design criteria, 408

LAS problems, universal nature of, 8

LAS processes and practices, documenting, 412

LAS projects: applying toolbox, 413; appraisal, 424–25; baseline studies, 414; community consultation, 426–27; community engagement, 414; community relations, 427; components of, 412–13; critical success factors, 414–16; design, 424–25; economic analysis, 414; establishing PMO, 426; evaluating, 427–28; feasibility and risk assessment, 422; financial management, 428; focus of, 427–28; goals of, 421; inception, 424–25; KPIs (key performance indicators), 427–28; LogFrame tool, 417–21; logical framework analysis, 417–21; longitudinal studies, 414; management hierarchy, 426; managing, 421; managing changes, 426; managing changes and amendments, 426; managing spatial components, 421; mobilization, 424–25; monitoring, 427–28; participatory development, 414; procurement, 429; project coordinator, 426; "project cycle," 422; project director, 426; project manager, 426; quality assurance, 428; resources, 428–29; role of pilot projects, 413; scheduling and project management, 427; social analysis, 414; stakeholder consultation, 426–27; stakeholder engagement, 414; sustainability, 423; using best practices, 413

LAS records, digital conversion in 1970s, 82

LAS reform, approach to, 410–11, 421

LAS tools, projected transporting of, 26–27

Latin America: land administration in, 72; Permanent Committee on Cadastres in, 57

Lavigne Delville, Philippe, 147

lawyers in cadastral template project, 286–89

legal interoperability, 253

legal origin, influence of, 59

legal planning control, decision options (figure), 179

legal traditions, approaches to land issues, 58

LFA (Logical Framework Analysis), 417–21

Lindsay, Jon, 27–28

LinkPopularity Web site, 251

LIS (land information system): development of, 54–55; relationship to cadastre, 9

LogFrame tool, 417–21

Logical Framework Analysis (LFA), 417–21

Luxembourg, planning system in, 177

M

Macao, ratio of parcels to people in, 280

Malawi: capacity-building project, 298, 300; gender equity issue (figure), 21

management tools: need for, 388–93; use of, 405–6

Manila, the Philipines, formal and informal markets (figure), 141

map use, universality of, 44

mapping, European approaches toward, 48

mapping information, militarization of, 234

maps: early examples of, 45; regarding as military information, 22

marine administration: challenges of, 206–7; coastal zones, 208; components of, 221; contiguous zone, 209; continental shelf, 209; exclusive economic zone, 209; existing, 207–10; high seas, 209; improving, 205–6; vs. land administration, 207; territorial waters, 208; tools, 326

marine boundaries, importance of, 220

marine cadastres. See also cadastres: concept, 210–14; functionality of, 216, 219, 221

marine environments: diverse interests in, 212, 214; spatial dimension of, 221

marine needs, applying land management paradigm to, 222

marine registers, 210, 214–16

marine resource titling standards: compliance aspects, 219; information aspects, 219; property aspects, 217; third-party property aspects, 217; title aspects, 218

marine SDI. See also SDIs (spatial data infrastructure): demands of, 210; developing, 216, 219, 221–22

maritime boundaries, describing and visualizing, 214

Maritime Provinces, development in, 52

maritime zones (figure), 209

market development, stages of, 150–53

market-based approach: predominance of, 6; transportability of, 18

markets. See complex markets; land markets

mass vs. individual valuation, 165

McLaughlin, J., 74, 376

MDGs (Millennium Development Goals), delivering 24–25; for slum dwellers, 183–84

Measure 37 (2004 ballot initiative), 174

measures for success principle, 35

Meeting of Officials on Land Administration (MOLA), 78

megacities: number of, 201; projected spatial distribution in 2015 (figure), 201; unmanaged, 21–22

Melbourne, Australia, modern urban infrastructure (figure), 43

meridians and baselines in United States (figure), 62

metadata, use in cadastral data modeling, 258

Mexico, formal and informal markets (figure), 142

militarization of mapping information, 234

military information, regarding maps as, 22

modern cities. See also cities; urban areas: building, 21–22; high-rise buildings in, 44; impact on concept of land, 43

MOLA (Meeting of Officials on Land Administration), 78

momentum, maintaining, 258–61

Moyer, David, 355

Mozambique, challenges related to, 8

multipurpose cadastre: figure, 123; international adoption of, 52, 54–58; vision of, 74

Mystery of Capital, The, 148–49

N

Namibia, ratio of parcels to people in, 281

national lands, alienation of, 349

national systems, collecting information about, 269

nature protection and management, 185–86

nature resource management, 187

Netherlands: ICT in, 239; "key registers," 347–48; planning system in, 177

New Brunswick, ICT in, 240

New South Wales (NSW) strata title, 383

New Zealand: ICT in, 238; Landonline program in, 238–39

Nicaragua, cost of LAP in, 268

Nichols, Sue, 212

noncadastral approaches and tools, 394–96. See also cadastral tools

North America, cadastral systems in, 55

North China Plain, decline of water table beneath, 20

NRC (National Research Council) study and development, 52, 54

NSW (New South Wales) strata title, 383

O

OECD (Organization for Economic Cooperation and Development), 139

Ogallala aquifer, depletion of, 20

OGC (Open Geospatial Consortium Inc.), 244

OICRF (International Office of Cadastre and Land Records), 78

Ordnance Survey, 60–61

Ørestad urban neighborhood (sidebar), 202

organizational development, impact of, 305

organizational success, UK model for, 305

"overreaching," use in land markets, 106

ownership: concept of, 150; evidence of, 153

P

P2P computing, 245

Panama, cost of LAP in, 268

parcel identifiers, use of, 355–56

parcel maps vs. text information, 347

parcel numbering system, 356

parcel-based LAS (figure), 75

parcels. See land parcels

parcels registered, percentage of, 283–85. See also registration systems

parcels to people, ratios of, 280

"passporting" property, role of, 148–50

Payne, Geoffrey, 143

PCGIAP–FIG cadastral template project. See also cadastres: countries in, 277; descriptive analysis, 275; design, 274–75; indicator 1, 276–77; indicator 4, 283–85; indicator 8, 289–90; indicators 2 and 3, 277–83; indicators 5, 6, and 7, 286–89; outcomes, 274–75; overview, 271–73; populations in, 277; principles and associated indicators, 275; statistical analysis, 275; structure, 274–75

PCGIAP Working Groups, 272

PCGIAP–FIG (2003), 271

people: approach toward land, 41; and institutions principle, 34; relationship to land, 37, 153

"people" components, importance of, 39

people to parcels, ratios of, 280

people-to-land relationships, 43; changes in, 47, 52; land administration response to, 48; in reengineered framework, 406; stability in, 44; symbiosis in Europe, 47

personal privacy, protection in e-land administration, 248

personal wealth, growth of, 19

PGCIAP (Permanent Committee on GIS Infrastructure for Asia and the Pacific), 219, 221

Philippines land-use pattern (figure), 122

photogrammetric measurements in TLTP (Thailand Land Titling Project), 101

plan objectives, land policy instruments related to, 199

planning. See also spatial planning: comprehensive, 192–94; politics of, 183; regulated, 174–75

planning authorities, role in development process, 198

planning control systems: approaches, 176–77; legal means of, 178–79; operation of, 177–78

planning journals, international articles in, 81

planning regulations, establishment of, 178–79

planning responsibilities, decentralization of, 192, 194

planning tool, cadastre as, 51

PMO (project management office), establishing, 426

Poland, e-land administration efforts in, 239

policy, paying attention to in project design, 266–67

Polish cadastral parcels (figure), 105

pollution control, 187–88

poor and rich nations, reducing divide between, 24. See also poverty; pro-poor

population: distribution in cadastral template project, 277; estimate in world, 20

Portugal, planning system in, 178

positive registries, 346

postcolonial countries, dualities in, 64. See also colonization

poverty. See also poor and rich nations; pro-poor: alleviation of, 17, 27; perpetuating levels and scale of, 184

poverty-level statistic, 29

preemption rights, 199

principles of land administration: cadastre, 34; land management paradigm, 34; LAS (land administration systems), 34; LAS are dynamic, 34; measures for success, 35; people and institutions, 34; processes, 35; rights, restrictions, responsibilities, 34; spatial data infrastructure, 35; technology, 35

privacy, protection in e-land administration, 248

private ownership: cultural concept of, 42; relationship to land markets, 143

private property, promotion of, 175–76

private rights, registers of, 339

private vs. public sectors, 178

"privatization oligarchs," 145

process management, improving, 112

processes principle, 35

professional advisers, role in development process, 198

professional competence (figure), 311–12

professional research, international growth of, 81

professional tools. See also tools: adjudication tools, 348–50; boundary tools, 358–65; building title tools, 381–86; cadastral surveying tools, 365–77, 380–81; land unit tools, 350, 355–58; mapping tools, 365–77, 380–81; registration system tools, 339–48; tenure tools, 333–39; titling tools, 348–50

project context, 403–4

Project Cycle (figure), 423

project design, paying attention to policy in, 266–67

project management office (PMO), establishing, 426

project management tools, 328–29

properties: assessing potential of, 197; explained, 125; historical concept, 91; incorporating into LAS, 348; vs. parcels, 125–26

"property" concept, changing in LAS, 90–92

property development, process of, 197

property management framework, 89

property markets, complexity of, 158

property prices. See land value

property rights: activists, 173; converting into tradable assets, 142; focus on, 39; protection by government, 81; relationship to land markets, 155; in Western systems, 42

property taxation: explained, 166; introducing, 167; payment of, 164; security for, 17

property theory, application of, 42, 155–56

property transfer tax, 166

pro-poor: management, 149; tools, 388–93. See also poor and rich nations; poverty

public administration theory, transformation of, 80

public vs. private sectors, 178

Q

QA (quality assurance), 428

R

Rajabifard, A., 112

readjustment systems, 189–91

rectangular survey system (figure), 63

rectification in TLTP (Thailand Land Titling Project), 101

reengineering: framework, 406–407; in land administration, 97

reform projects, focus of, 438

Register Search Statement (figure), 344

registers: early use of, 45–46; making authentic, 347

registration, relationship to economic improvement, 146–47

registration systems. See also parcels registered: vs. establishment approach, 277–283; evolution of, 59–60; national approaches, 339; vs. registration methods, 276–77; tools, 339–48

registries: accuracy and completeness, 346; vs. cadastres, 66; deeds vs. title registration, 341–42; features of, 340, 342, 345–46; land vs. resource, 339; positive, 346; in Western Europe, 61

regulation theory and practice, changes in, 91

regulations, growth of, 88–90

regulatory environment, features of, 90

resource vs. land registries, 339

resources, management of, 18

responsibilities, growth of, 88–90

restrictions and responsibilities: management of, 89–93; records maintenance, 92

rich and poor nations, reducing divide between, 24

rights, explained, 88–89. See also land use

rights, restrictions, and responsibilities (RRRs): incorporation into cadastral data models, 261; principle, 34; vs. RORs (rights, obligations, and restrictions), 88

risk management tools, 330–332

Robertson, Bill, 211

Romans, survey in 300 CE, 46

Royal Registry of ancient Egypt, 46

RRRs (rights, restrictions, and responsibilities): incorporation into cadastral data models, 261; principle, 34; vs. RORs (rights, obligations, and restrictions), 88

rule of law, support for, 17

rural areas, conflicting interests in, 203

rural development, 203

rural environments, planning, 185

rural planning: agricultural policies, 186; coastal zone management, 188–89; environmental protection, 187–88; forestry policies, 186–87; natural resource management, 187; nature protection and management, 185–86; pollution control, 187–88

rural population movements, responses to, 21

rural resources, rights and access to, 32

rural: vs. urban land, 267; vs. urban populations, 277

rural zone development, 185

S

SAPs (Structural Adjustment Programs), 80

SDI framework, technical issues in, 235

SDI model (figure), 237

SDI strategy, adoption of, 226

SDIs (spatial data infrastructures). See also marine SDI: changing roles of, 231–32; components of, 230; concepts and hierarchy, 229–31; design and implementation of, 230; designers of, 227; development of, 74; emerging vision of, 228; as enabling platform, 232–33; figure, 260; global, 231; hierarchy, 230–31; in hierarchy of land issues, 132–33; justifying expense of, 230; local, 231; national, 231, 234; need for, 225–28; opportunities created by, 227; organizational, 231; regional, 231; state, 231; tools, 327; virtual jurisdictions, 231–32

sectoral land-use management, 185, 193

securitization process, 161–62

security, role in good governance, 31

security of tenure, 17, 19–20, 24–25. See also tenures: delivering, 338–39; delivery of, 24–25; importance to stability, 267; undermining, 336

SEG (spatially enabled government), 441

semantic interoperability, 252–53

separation process, 162

service tax, 166

service-oriented architecture (SOA), 244

services, framework for delivery of, 25

Settled Land Act of 1882, 50, 106

settlements, informal, 183–84

SGD (Surveyor General's Department, Swaziland), 306–7

Sheppard, V.L. O., 73

silo agencies, digital conversions in 1980s, 82

silos: LAS (land administration systems) in, 64–65; problem of, 7

Simpson, R. W., 73

skills, management vs. specialist, 308

slum dwellers: improving lives of, 183–84; number of, 183

SOA (service-oriented architecture), 244

social changes, tracking in LAS, 106

social context, considering in LAS design, 412

social events, transferring land through, 104–7

social goals, achieving, 20

social land tenure, reliance on, 42

social processes, accommodating, 107

Social Tenure Domain Model (STDM), 392

social tenures: examples of, 42; models, 391–92

South Australia, German-style system adopted in, 50–51

Spain, planning system in, 177

spatial data infrastructures (SDIs). See SDIs (spatial data infrastructures)

spatial data integration: challenges of, 235–36; institutional issues, 236; legal issues, 236; nontechnical issues, 235–36; policy issues, 235–36; and SDIS, 236–37; social impediments to, 237; social issues, 236; technical issues, 236–37

spatial datasets. See datasets

spatial enablement of LAS, 130–31, 440–42

spatial information: capabilities of, 22; forms of, 129–30; precision of, 232–33

spatial knowledge, power of, 228

spatial planning. See also planning control systems: comprehensive integrated approach, 176; land-use management approach, 177; regional economic approach, 176; urbanism approach, 177

spatial technologies: central role of, 441; evolution of, 440; use of, 441

spatial tools, availability of, 442

spatially enabled government (SEG), 441

state lands: alienation of, 349; protection of, 17

Statement on the Cadastre, 76

statistical data, management of, 18

STDM (Social Tenure Domain Model), 392

Steudler, D.: evaluation framework, 430–33; reengineering framework, 112

stovepipes. See silos

strata titles and units, 278–83

Structural Adjustment Programs (SAPs), 80

subdivision: allowing, 196; permits in development control, 195; processes, 107

subsidiarity, role in good governance, 31

survey marks, visible and hidden, 363

surveying. See also cadastral surveying tools; history of, 71; in TLTP (Thailand Land Titling Project), 100–101

surveyors: educational challenge for, 310–11; educational trends, 308–10; and lawyers in cadastral template project, 286–89; technical focus of, 72

sustainability: components of, 84; delivering land information for, 22–23; in institutional development, 304; role in good governance, 31; translating into LAS strategies, 86–87

"sustainability accounting," 87

sustainable development: achieving in integrated land-use management, 193; and administrative processes, 87; basis of, 31; and built environment, 87; delivering, 19; dimensions of, 19; facilitating, 233; implementation policies, 86–87; implementation through LAS, 84–86; influence of, 77; international efforts, 84–86; and natural environment, 87; overview, 8–10; reducing to achievable outcomes, 86; support for, 439–40; "triple bottom line," 19; and virtual environment, 87

Swaziland's Surveyor General's Department (SGD), 306–7

Sweden: Land Survey of seventeenth century, 48; planning system in, 177

SWOT matrix, use in evaluation matrix, 430–33

systems: design of, 121; informality of, 15

T

"takings" phrase in Fifth Amendment, 174

tax burdens, equitable, 167–68

taxation principles: capital gains tax, 167; development gains/betterment tax, 166; land tax, 166; property tax, 166; property transfer tax, 166; service tax, 166

taxation systems: efficiency in Europe, 48–49; state-based, 164

taxes, collecting in developing countries, 168

technical interoperability, 234–35, 254

technical journals, international articles in, 81

technology: demands of, 83; encouraging use of, 23–24; principle, 35

tenure improvement path (figure), 392

tenure processes, 98; determining boundary processes, 107, 109–10; engaging community in, 99–100; forming interests and properties, 107; subdivision processes, 107; systematic land titling, 99–100; transferring land by agreement, 100–104; transferring land through social events, 104–7

tenure systems: articulation of rights, 336–37; beneficiaries of, 153; identification of interests, 337; layering of tenures, 337; management of risks, 338; prioritization of interests, 338; responsibilities for land, 337; restrictions on land, 337

tenure tools. See also building tenures: common property, 334; customary, 333; formal, 333; group tenure, 334; illegal squatting, 334; indigenous, 333; informal, 333; leasehold, 334; license, 334; native, 333; occupation right, 334; overview, 333–334; possessory, 334; private ownership, 333; state ownership, 333; traditional, 333; trust ownership, 334

tenures. See also security of tenure: "bundle of sticks" analysis, 335; documenting in LAS design, 411; in land administration, 335–38; layering of, 337; theory, 335–36

Teresian Cadastre, 48

territorial waters, 208

territoriality, drivers for, 44

testamentary inheritance system, 105–6

Thailand Land Titling Project (TLTP), 146–47, 348

third parties, role in development process, 198

Three Pillars diagram: development of, 158; fourth pillar of (figure), 159

Ting, L., 112

title delivery in TLTP (Thailand Land Titling Project), 101

title insurance, precautions, 331

title registration, origin of, 60

title tools: 3D cadastre, 386; building tenures, 381–86

title: vs. deeds registration, 341–42; vs. deeds system, 59

titles: overriding interests, 346; in TLTP (Thailand Land Titling Project), 101

titling: and adjudication tools, 348–50: analysis of, 148; land held by poor, 148–49

titling programs: failure of, 147; perception of, 147

titling projects, LogFrame analysis for, 417–19

titling systems, potential of, 438–39

titling vs. land management tools, 395

TLTP (Thailand Land Titling Project), 146–47, 348

tool emergence: gender equity tools, 396–98; human-rights tools, 398–400; need for, 387; noncadastral approaches and tools, 394–96; pro-poor land management tools, 388–93

toolbox approach, 7–8. See also professional tools; business models, 330–32; capacity-building tools, 327–28; funding tools, 330–32; governance tools, 326; ICT tools, 327; institution-building tools, 327–28; land development tools, 327; land information tools, 327; land market tools, 326; land policy tools, 320–26; land use tools, 327; land valuation tools, 327; legal framework tools, 326; marine administration tools, 326; overview, 316–20; project management monitoring tools, 328–29; risk management tools, 330–32; SDI tools, 327

tools. See also cadastral surveying tools; professional tools: to administer continuum of tenures, 443; boundary, 358–65; business models, 330–32; capacity-building, 327–28; funding, 330–32; ICT (information and communications technology), 327; institution-building, 327–28; land development, 327; land information, 327; land market, 326; land policy, 320–26; land unit, 350, 355–58; land use, 327; land valuation, 327; to manage RRRs, 443–44; marine administration, 326; need for, 387; project management, 328–29; pro-poor vs. market, 388; risk management, 330–32; scaling up, 393; SDI (spatial data infrastructure), 327; stages of development, 391; titling and adjudication, 348–50

topographic datasets, 234

Torrens systems, 50–51, 66; benefits of, 103; function of land registry in, 124; popularity of, 60; publicly funded pools for, 331; use of, 123

Torrens title certificate (figure), 343

traditional ways of life, protection of, 8

transfer, evidence of, 102

transparency, role in good governance, 31

Treml, Eric, 210

U

UK model for organizational success, 305

UML (Unified Modeling Language), 242

UNCLOS (United Nations Convention on the Law of the Sea), 208–9

undeveloped countries. See developing countries

UNECE (UN Economic Commission for Europe), 78. See also Europe; hierarchy, 357; initiatives, 270–71; land administration principles, 30–31; view of land administration, 116

UNESCAP, implementation of readjustment, 191

UN–HABITAT, 73–74, 85–86; estimates of slum population, 183; predictions, 21; Web site, 21

Unified Modeling Language (UML), 242

United Kingdom, discretionary system in, 177

United Nations Convention on the Law of the Sea (UNCLOS), 208–9

United Nations Regional Cartographic Conference for Asia and the Pacific (UNRCC–AP), 221

United States: deeds registration in, 59–60; eminent domain legislation in, 175; formal surveying in, 62; LAS (land administration systems) in, 62–64; Measure 37 (2004 ballot initiative), 174; meridians and baselines in, 62; protection of private property in, 175; shift from rural to urban economy, 174

UNRCC (UN Regional Cartographic Conferences), 73

UNRCC–AP (United Nations Regional Cartographic Conference for Asia and the Pacific), 221

urban areas: land parcels in, 284–85; retrieval of space in, 44. See also cities; modern cities

urban conservation, 181–82

urban development, 200–201, 203; acceleration period of, 179; focus during 1960s and 1970s, 180; informal, 182–83; traditional example of, 202–3

urban environments, control of, 179–80

urban growth, addressing in next decade, 443

urban heritage and regeneration, 181

urban planning, as global problem, 183

urban planning controls, 179–80; building permit
control, 180–81; informal settlements, 183–84

"urban sprawl," 201

urban vs. rural land, 267

urban vs. rural populations, 277

U.S. law, landownership vs. government, 174

V

valuation principles, individual vs. mass, 165–66. See
also land valuation

value, 120

Victoria, Australia marine boundaries, 220

Vietnam: challenges related to, 8; mortgage process, 103

W

Wallace, J., 112

Wallace and Williamson 2005, 217

water, access to, 20. See also clean water

water table, decline beneath North China Plain, 20

wealth (personal), growth of, 19

Web access to data, enabling, 83

Web services tools, using in ICT for LAS, 244

Web sites: LandXML standard, 243; LinkPopularity, 251;
Millennium Development Goals (MDGs), 183;
OICRF (International Office of Cadastre and Land
Records), 78; PCGIA–FIG cadastral template project,
272; UN–HABITAT, 21; UN–HABITAT site for GLTN,
391; World Bank's Land Policy Network, 80

Western Europe: land records in, 61; registries in, 61

Western land administration trends (figure), 47

Western landowners, concepts built by, 160–61

Western LAS, development of, 47

Western property, characteristics of, 155–56

Western systems, 42

Williamson, I. P., 112

women, improving access to land, 20, 24. See also
gender equity

World Bank: "Comparative study of land administration
systems," 271; "Land policies for growth and poverty
reduction," 85–86; land policy approach, 73, 80;
lending and project policies, 423–25

world nations, number of, 138–39

world population estimate, 20

World War II: land administration after, 72–74;
reconstruction period, 51

Worldwide Cadastral Template Project. See
PCGIAP–FIG cadastral template project

WPLA (Working Party on Land Administration), 78, 270

Z

Zambia boundary process, 109, 111

zones, dividing countries into, 185